Statistical Analysis of Behavioural Data

Authors' address:

Institute of Theoretical Biology
P.O. Box 9516, 2300 RA Leiden
The Netherlands

Statistical Analysis of Behavioural Data

An approach based on time-structured models

Patsy Haccou and Evert Meelis

Institute of Theoretical Biology
Leiden University
The Netherlands

Oxford   New York   Tokyo
OXFORD UNIVERSITY PRESS

Oxford University Press, Walton Street, Oxford OX2 6DP

Oxford New York
Athens Auckland Bangkok Bombay
Calcutta Cape Town Dar es Salaam Delhi
Florence Hong Kong Istanbul Karachi
Kuala Lumpur Madras Madrid Melbourne
Mexico City Nairobi Paris Singapore
Taipei Tokyo Toronto
and associated companies in
Berlin Ibadan

Oxford is a trade mark of Oxford University Press

Published in the United States by
Oxford University Press Inc., New York

First published 1992
First published in paperback 1994 (with corrections)
Reprinted 1995

A catalogue record for this book is available from the British Library

Library of Congress Cataloging in Publication Data
(Data available on request)
ISBN 0 19 854850 8

Printed in Great Britain by
Bookcraft (Bath) Ltd
Midsomer Norton, Avon

PREFACE

Time is but an illusion

In the last few decades, behavioural science has changed from a mainly descriptive to a highly technological science. Pencil and paper have been replaced by advanced event-recorders. Visual scanning of scatter diagrams and histograms has given way to complicated data analyses. High-speed computers and software packages have considerably enlarged the possibilities for studying fascinating behavioural phenomena. Each of the developments in ethology, mathematical modelling, statistics, computer science and technical disciplines have to a certain extent occurred independently, without benefiting from the others. It is thus time for an integration, at least from the ethological standpoint (other disciplines may of course profit from this too).

Behavioural science is quite diverse in respect to subject, research questions, methods of experimentation, and observation and recording methods. This diversity is reflected in the ways data are analysed. Many analyses are of an explorative nature, based, for instance, on principal component analysis or factor analysis. Although the number of advanced data-analytical methods continues to increase, this process is not governed by a unifying viewpoint that suits the nature of the data and provides possibilities for a better understanding and interpretation. In this book we attempt to bridge the widening gap between such requirements and the possibilities offered by modern recording facilities and methodological tools. This book differs from other books on quantitative methods in behavioural science in that the methods are based on explicit models of the time structure of behaviour, i.e. stochastic processes which are characterized by parameters that can easily be interpreted as the tendency to switch from one act to another. Such models can, in principle, be used for any continuous time observation of behavioural processes, in causal as well as functional contexts (e.g. Haccou *et al.*, 1991). The initial development of this type of modelling in ethology was largely due to Hans Metz (Metz, 1974, 1981), inspired by Nelson's work (Nelson, 1965).

The models and methods described and discussed in this book enable a major advancement to a better understanding of animal and human behaviour as we observe it. The precise description of the recorded behaviour permit a refined analysis of for instance the effects of chemical, electrical or surgical treatments, gradual changes in the social or physical environment, ageing, etc. Interactions between two or more individuals of the same species, between a predator and its prey, or between a parasite and its host can be analysed and understood much better within the framework of a parsimonious and good fitting model which is well interpretable from an ethological point of view as well as in terms of causal mechanisms.

The book is intended for graduate students and researchers of behavioural science in a wide sense, including mathematicians who are interested in modelling behaviour, and statisticians involved in the analysis of behavioural records. The reader is assumed to have a basic knowledge of calculus, mathematical statistics and probability theory. Since the different topics are considered from the perspective of behaviour rather than mathematics, the level of mathematics varies throughout the text.

We discuss many examples, based on practical experience, to facilitate the application of the methods. At present most behavioural scientists do not have ready access to these procedures, since often these are insufficiently worked out in the literature or formulated in mathematical jargon. Most examples are simplified to emphasize their essence. Where simulated data are used, they are based on parameter values which have been determined empirically.

We begin with the consideration of a simple stochastic model, the continuous time Markov chain, and several generalizations of it which have been recently applied in ethological research. With this class of mathematical models it is possible to quantify alternations between behavioural categories as well as the distributions of durations of acts. The models, as well as the methods for data analysis derived from them, can be extended and adjusted when necessary. Although we cannot consider all possible situations, we expect that our selection of examples is such that the reader can derive the desired modifications by applying of the principles described.

Most of the results have been published elsewhere, sometimes in sources which are not widely available. For instance, many results are based on the Ph.D. theses of Haccou (1987) and Metz (1981). Where possible, we refer to sources which are readily accessible.

In our (long) experience with the application of mathematical models and statistical methods in the behavioural sciences, we have frequently encountered the question: what extra can be gained from all these complicated models and analyses? Our answer is clear: good methods of analysis cannot turn a bad scientist into a good one, except perhaps that he or she may be more successful in convincing a referee that a paper is worthy of publication. A good scientist, however, can make productive use of more powerful tools. Furthermore, without an explicit formulation of model assumptions, many kinds of problems may go unnoticed, but may lead more frequently to false negatives and false positives than is necessary. Moreover, most estimated parameter values will not be as accurate as is possible in principle. Of course, these statements need not to be true in any specific case: all we want to point out is that, on average, the available information is used suboptimally. By analysing data in the way described here, information contained in the time structure is used as much as possible. Furthermore, it provides formal testing and estimation procedures, with known properties (examples of applications are given in Chapters 7 and 8).

This book cannot contain the last word on this way of studying behaviour. We hope that it inspires other people to develop the necessary methods further. In particular, more advanced methods for special cases of semi-Markov chains, inhomogeneous (semi-)Markov chains, and functions of Markov chains are needed. Furthermore, more work has to be done on studying inter-rater reliability. In Chapter 2 we indicate a (in our opinion) fruitful way of thinking about this problem. We expect that such developments will have an important impact on a variety of applications.

Many people have contributed to this book. A large part of our results is based on and inspired by the work of and discussions with Hans Metz. We also greatly appreciate the cooperation and discussions with Menno Kruk and Mientje Bressers. Lolke Dijkstra supplied software. We thank Marianne van Dijken for carefully reading the manuscript, Janine Pijls, Annemoon van Erp and Danielle Willekens for providing practical examples, Ian Hardy for suggesting improvements to the English, Wout Meelis for technical assistance, Yvonne Zitman for typing and editing, Henny Regeer for typing, Peter Hock for drawing the figures, Henk Heijn and Peter Hock for designing the cover, and all our friends and relatives for being so patient with us. For financial support, we thank Prof Dr Jan van der Hoeven Foundation for Theoretical Biology. The research of the first author has been made possible by a fellowship of the Royal Netherlands Academy of Arts and Sciences.

Leiden, January 1992   P.H.
                                                                       E.M.

In the paperback edition a few errors have been corrected.

Leiden, August 1993   P.H.
                                                                       E.M.

OVERVIEW OF THE BOOK

For a quick overview of the main contents of the book, we recommend chapter 1 (except section 1.5), sections 3.1, 3.2, 4.1, 4.4 and 5.1, and the introductions of sections 5.2, 5.3 and 5.4. For the next stage, sections 7.1, 7.3, 7.4 and 7.5 are recommended. Readers who are anxious to know more about the practical use of sophisticated analysis are referred to chapter 8, in which examples of applications are given. The other sections can be consulted when needed. Chapters 2 and 6 can be read independently.

Chapters 3, 4, 5, 6 and 7 contain many different methods of analysis. These chapters are not intended to be read throughout, but rather to give a survey of the available methods. In the introductions to these chapters, we provide guidelines for choosing methods that can be of use in specific situations. We have provided worked examples of procedures in 'Boxes', which can be skipped (if desired) when reading the text.

CONTENTS

3   Analysis of time inhomogeneity

4   Tests for exponentiality

8   Examples of analyses based on continuous time Markov chain modelling

LIST OF SYMBOLS

In this section we give a list of frequently used notation. When possible notation is used consistently. However, the result is still inevitably a compromise between readability, logical flow and consistency with standard practice of the several disciplines brought together in this book. In such instances the notation is explained when introduced.

ANOVA	analysis of variance
c	observed censoring time
c_α	upper critical value of level α
C	censoring time
A, B, C,...	acts or states
CTMC	continuous time Markov chain
D_n	Kolmogorov-Smirnov test statistic
$E\,X$	mathematical expectation of the random variable X
f	transition frequency
$f(x)$	probability density function
$F(v_1, v_2)$	F-distributed variable with v_1 and v_2 degrees of freedom
$F(x)$	cumulative distribution function: $Pr\{X \le x\}$
$F_n(x)$	empirical distribution function
$\bar{F}(x)$	survivor function: $Pr\{X > x\}$
$\bar{F}_n(x)$	empirical survivor function
H_0	null hypothesis
H_1	alternative hypothesis
i	information matrix
K_m	Kruskal-Wallis test statistic
$l(x, \theta)$	likelihood function
$L(x, \theta)$	log-likelihood function
log	natural logarithm
MLE	maximum likelihood estimate/estimator
$N(\mu, \sigma^2)$	normal distribution with mean μ and variance σ^2
p	p-value
$Pr\{A\}$	probability of event A
r	estimated correlation coefficient
r_s	Spearman's rank correlation coefficient
R_j	autocorrelation coefficient with lag j

$r_1,..., r_n$	ranks in increasing order of n observations
s	standard deviation
S	sum of a series of observations
t	variable with a Student distribution
t	time from onset of observation
T	total observation time
V_n	'total time on test' statistic
$Var\ X$	variance of X
w_i	ith weight coefficient
W^2	Cramér-von Mises test statistic
$x_{(1)},..., x_{(n)}$	ordered set of observations: $x_{(1)} \le ... \le x_{(n)}$
$X, Y, Z,...$	random variables are denoted by capitals, the observed values by the corresponding lowercase characters $x, y, z, ...$
$\bar{x}$	average or mean value: $(x_1+...+x_n)/n$
z_α	critical value of the standard normal distribution
Z	standard normally distributed random variable
α_{AB}	transition rate from act A to act B: the transition probability per time unit
$\hat{\alpha}_{AB}$	estimated transition rate from act A to act B
α	level of significance
$B(a,b)$	beta distribution with parameters a and b
$\Gamma(k)$	gamma function with parameter value k; for integer values of k: $\Gamma(k) = (k - 1)!$
δ	small positive number
δ	indicator for censoring: when $\delta = 0$, the observation, e.g. a bout length, is censored; when $\delta = 1$ the observation is not censored
δ_{ij}	Dirac delta function: it is equal to zero when $i \ne j$ and equal to one otherwise
∂	partial derivative
ε	small positive number
ε	random error term
θ	parameter
θ	vector of parameters $(\theta_1,...,\theta_k)$
Θ	parameter space
λ	termination rate: termination probability per time unit
Λ	likelihood ratio test statistic
μ	expected value
ν	number of degrees of freedom

Π	product symbol
ρ	Weibull parameter
σ^2	variance
Σ	summation symbol
Σ	covariance matrix
τ	change point
Φ	cumulative standard normal distribution function
χ^2_k	chi-squared distributed variable with k degrees of freedom
ψ	contrast (in multiple comparison methods)
$\cap$	intersection
$\cup$	union
$\wedge$	conjunction of propositions: $p \wedge q$ means that the conjunction of the propositions p and q is true if and only if both p and q are true
$\vee$	disjunction of propositions: $p \vee q$ means that the disjunction of the propositions p and q is true if one/both of the propositions is/are true
10 (10) 50	10, 20, 30, 40, 50

1 INTRODUCTION

1.1 INTRODUCTORY REMARKS

Ethological research can have widely different goals, and accordingly, the collected data have different forms. In most instances, however, it involves detailed observation and recording of the behaviour of one or more individuals. In this book, we will consider the analysis of such detailed, continuous time records of behaviour. (For an example of an advanced statistical method for a different type of ethological data, see Hemelrijk and Ek, 1991.) The types of research questions may vary widely. One may, for instance, be interested in the way in which different individuals affect each others' behaviour, or the effects of seasonality, or experimental treatment, e.g. drug administration. In spite of these differences in research questions, there is a universal requirement of accurate observational data and methods of analysis for such data. Over the past few decades, these requirements have led to improved facilities for acquiring and processing data, turning ethology into an increasingly quantitative science. Most behavioural experiments involving direct observations of sequences of events and the time at which they occur require (computer) event recorders. See Appendix II for examples of hard- and software, available under the names Camera, Observer and Analyst, see also Fig. 1.1. The application of these recorders saves time, increases the accuracy and reduces the number of errors considerably. A huge amount of data is readily produced and to analyse these data, statistical procedures are needed. Accordingly, the analysis of behavioural data has received much attention in the literature (e.g. Hazlett, 1977, Colgan, 1978, van Hooff, 1982, Colgan and Zayan, 1986, Martin and Bateson, 1986, Siegel and Castellan, 1988).

Many statistical methods used in ethology, such as principal components analysis (PCA) or factor analysis, are not, however, primarily designed to analyse the time structure of behaviour. Whereas such methods can indicate whether the investigated factors have effects on behaviour, their results usually do not allow unambiguous conclusions about the exact effects on the time structure. For instance, a change in the mean bout length of an act may mean that the act indeed becomes shorter over the whole observation period. On the other hand, the bout lengths of that act may shift from short to long or vice versa during the observation period, and the place of this shift may be altered, causing a change in the overall average. In most cases, it is important to be able to discriminate between the two types of effect, since they obviously imply different types of behavioural mechanisms. Therefore, methods like PCA are more appropriate for the initial stages of analysis. For more detailed analyses of the time structure of behaviour we need methods that take the special nature of ethological data into account.

The methods given in this book are designed for the analysis of continuous time records of behaviour. Such records consist of the beginning and ending times of several acts that may occur simultaneously (examples of behavioural records are given in Boxes 1.1, 1.4 and 1.5). The methods are based on a class of mathematical models (namely, continuous time Markov chain models and generalizations thereof) suited for taking account of both the durations and the sequence of acts.

Figure 1.1 Event recorder designed by the Ethopharmacology Group, Leiden University: Top Left: Video Screen; Top Middle: Video Time Code Generator (VTG); Top Right: PC Monitor; Bottom Left: Encoding Keyboard, Remote Control and Video recorder (partially hidden); Bottom Right: PC Keyboard.

If a model gives an accurate description of the behaviour, the effects of various factors can be studied in terms of its parameters, which represent behavioural tendencies to switch between acts. The advantages of this approach over methods not designed for ethological data are that it provides better insight into the behavioural process itself, and that it leads to more powerful methods for testing (experimental) effects, since the methods are based on previously validated models. Moreover, the results are readily interpretable in terms of behavioural mechanisms. This allows us to make more explicit conclusions about effects on the time structure. For instance, whether behaviour is homogeneous during observation periods and, if not, when and how it changes. We can distinguish whether e.g. a drug affects shifts in bout lengths during the observation, affects the mean bout length, or an animal's ability to switch between acts (section 8.3). When investigating social interactions, we can find out e.g. what is the quantitative effect of an act of one individual on the other's behaviour, and how long the effect lasts. It is even possible to distinguish the effect of experimental treatment from that of another individual's acts, and to study interactive effects of these two factors (section 8.2).

The modelling approach followed in this book is based on one of the most essential features of behaviour, namely that it can be studied on several time scales, for instance at the level of gross periods (e.g. sleep-wake cycles) or more subtle changes (e.g. variations in arousal) or at the level of acts or parts of acts. This property implies a certain arbitrariness in the definition of behavioural categories. For instance, '*Exploring*' can be considered as an act, but we may equally well distinguish the components of exploring, such as '*Sniffing*', '*Walking*', '*Scanning*', etc. This has led to many discussions about the choice of behavioural categories (see e.g. Huntingford, 1984, Chapter 2). It also has consequences for the formulation of descriptive models. The class of models which applies depends on the time scale at which behaviour is considered. Therefore it is crucial that analysis proceeds within a well-defined, ethologically interpretable framework of models. Within this framework, the behavioural categories can be chosen according to a general methodological rule: the principle of parsimony. Following this, categories are chosen in such a way that the simplest possible model suffices to describe the resulting observed process (with the understanding that this model should be ethologically sound and in accordance with the research interest). When an accurately fitting model has been formulated for the behaviour on a certain time scale, other time scales can be studied using this model as a basis. For instance, global changes in behaviour in the course of the day can be described by means of changes in the parameters of a model that describes the behaviour accurately for a period of, say, 1 hour.

The class of models considered in this book is suitable for describing behaviour on time scales where, for a substantial proportion of the acts, neither the durations nor the sequence of acts appear predetermined. These are often short time scales that usually coincide with the level at which behaviour is studied by ethologists. The simplest model of the class, used as the basic model in the analysis, is the continuous time Markov chain (CTMC).

Continuous time Markov chain models have been applied successfully in the analysis of behaviour of widely different species, e.g. rhesus monkeys (Dienske and Metz, 1977), juncos (Wiley and Hartnett, 1980), mosquitoes (Peterson, 1980) and barbs (Putters *et al.*, 1984). From an ethological point of view they are attractive models since their parameters subsume the information of both durations and alternations of acts. Thus they provide a complete, concise and ethologically well-interpretable summary of the observations. An additional advantage is that when interactions between one or more individuals are adequately described by these models it is possible to study contributions of the participants in the interaction separately and to analyse how they affect each other's behaviour.

Continuous time Markov chains are part of a large class of stochastic models. When the basic model does not fit, other models within this class may be used to describe a behavioural process. In section 1.3 we treat a number of such generalizations that are relevant for ethological data.

A survey of the principles of modelling and the procedure of data analysis is given in section 1.4. Methods for preliminary data inspection and correction of the raw records are described in Chapter 2. In Chapter 3 we give several visual scanning and formal methods for detecting time inhomogeneity. Time inhomogeneity is interesting in itself from an ethological point of view, since it can indicate major motivational shifts. Whether or not records are homogeneous is also of great importance from the point of view of modelling, since homogeneity reduces the complexity of the analysis considerably. In Chapters 4 and 5 tests for the main model assumptions are treated and illustrated by worked examples. In Chapter 6 procedures for combining dependent as well as independent test results from different data sets and multiple comparison methods are treated. In Chapter 7 we consider how to analyse the effects of one or more factors on behaviour. When available, asymptotic results for large observation times are given (see also section 1.5). These results can be used for tests when sample sizes are large.

Note that throughout this book each behavioural record is treated as a separate data set. An entirely different approach is to treat different records as realizations of the same stochastic process and derive asymptotic results for large numbers of observations. This line is usually followed if there are many short records; for example, Ridder (1987) applies it in an econometric study. To a certain extent, the derivation of asymptotic results is simpler with this approach, since the different records are independent, whereas in a single process there are dependencies (see e.g. Basawa and Prakasa Rao, 1980). However, in ethological studies there is usually much inevitable variation between different individuals and trials, which cannot be removed by adjusting the experimental set-up. Therefore, it is often not realistic to assume that the records are realizations of one stochastic process, and it is preferable to take the approach outlined in this book.

In Chapter 8 we discuss a few applications of an analysis based on (semi-)Markov chain models, as an illustration of the powerful tools which are now available if one has modern event recorders and computer facilities with an adequate software package.

Some terminology used throughout the book is explained in Box 1.1.

Box 1.1   Example: a continuous time record of solitary rat behaviour

Table 1 contains a simplified example of a continuous time record of rat behaviour. We will use these data to illustrate some of the terminology used throughout the book.

Table 1   Continuous time record of solitary rat behaviour

Act	Begin	Duration	Act	Begin	Duration
Shake	0.00	0.45	*Sit*	924.41	2.90
Walk	0.45	0.59	*Shake*	927.31	0.45
Sit	1.04	10.51	*Sit*	927.76	18.14
Walk	11.55	0.89	*Walk*	945.90	0.72
Sit	12.44	20.25	*Care*	946.62	10.47
Walk	32.69	6.19	*Shake*	957.09	0.45
Care	38.88	40.29	*Walk*	957.54	0.27
Sit	79.17	36.20	*Sit*	957.81	17.58
Walk	115.37	6.25	*Walk*	975.39	0.33
Sit	121.62	10.79	*Sit*	975.72	15.06
Walk	132.41	0.30	*Care*	990.78	75.45
Sit	132.71	3.18	*Sit*	1066.23	90.28
Walk	135.89	0.95	*Care*	1156.51	14.62
Sit	136.84	2.67	*Sit*	1171.13	11.18
Rear	139.51	1.41	*Walk*	1182.31	0.50

Table 1 Continued

Act	Begin	Duration	Act	Begin	Duration
Sit	140.92	1.74	*Sit*	1182.81	13.08
Walk	142.66	5.56	*Rear*	1195.89	0.17
Sit	148.22	28.09	*Sit*	1196.06	5.64
Walk	176.31	1.02	*Shake*	1201.70	0.45
Sit	177.33	11.76	*Sit*	1202.15	15.44
Rear	189.09	6.67	*Care*	1217.59	68.18
Sit	195.76	22.97	*Sit*	1285.77	53.81
Rear	218.73	11.22	*Walk*	1339.58	6.09
Sit	229.95	53.13	*Sit*	1345.67	4.41
Walk	283.08	3.22	*Shake*	1350.08	0.45
Sit	286.30	1.90	*Sit*	1350.53	5.46
Shake	288.20	0.45	*Walk*	1355.99	0.78
Walk	288.65	2.19	*Rear*	1356.77	5.87
Sit	290.84	3.47	*Sit*	1362.64	51.27
Care	294.31	15.80	*Walk*	1413.91	2.78
Sit	310.11	22.10	*Sit*	1416.69	17.71
Walk	332.21	8.03	*Shake*	1434.40	0.45
Rear	340.24	3.67	*Care*	1434.85	28.71
Sit	343.91	55.11	*Sit*	1463.56	12.48
Walk	399.02	0.15	*Walk*	1476.04	2.20
Rear	399.17	0.96	*Sit*	1478.24	4.31
Sit	400.13	12.71	*Shake*	1482.55	0.45
Care	412.84	33.15	*Sit*	1483.00	4.64
Sit	445.99	67.96	*Rear*	1487.64	1.35
Care	513.95	173.53	*Sit*	1488.99	19.51
Sit	687.48	64.47	*Care*	1508.50	8.30
Care	751.95	105.83	*Shake*	1516.80	0.45
Shake	857.78	0.45	*Sit*	1517.25	14.20
Walk	858.23	1.42	*Walk*	1531.45	2.51
Sit	859.65	17.05	*Sit*	1533.96	12.06
Care	876.70	31.88	*Rear*	1546.02	6.97
Sit	908.58	3.31	*Sit*	1552.99	32.01
Walk	911.89	1.18	*Care*	1585.00	2.51
Sit	913.07	9.93	*Shake*	1587.51	0.45
Rear	923.00	1.41	*Walk*	1587.96	4.74

An ethogram is the set of behavioural categories that is considered. In the example, the ethogram consists of *Care*, *Walk*, *Sit*, *Rear* and *Shake*.

A bout is a time interval during which a certain act is performed. A bout length is the duration of such a time interval. For instance, the first *Sit* bout starts at observation time 1.04, stops at 11.55, and the bout length is 10.51 seconds.

A gap is a time interval between two successive occurrences of an act and the gap length is the duration of such an interval. For example, the first gap between *Sit* bouts occurs from 11.55 to 12.44 and has length 0.89 seconds.

Sometimes, we use the term bout in a wider sense to denote time intervals of various types, such as gaps between successive occurrences of an act, or an interval during which a sequence of acts belonging to a certain group are performed. For example, bouts of the group '*Shake*, *Walk*, *Sit*' occur in Table 1 from $t = 0.00$ to $t = 38.88$, from $t = 79.17$ to $t = 139.51$, etc.

We will often consider the sequence of acts in the behavioural record. This is the observed sequence of acts, disregarding the durations. In Table 1, the sequence starts with '*Shake*, *Walk*, *Sit*, *Walk*'. From this sequence, the transition matrix between acts can be calculated. This is a matrix of observed frequencies of transitions from each act to each other act, divided by the total number of occurrences of the first act. These quantities are the estimated transition probabilities between acts (see eqn (1.2)). The transition matrix of the record in Table 1 is given in Table 2. Since transitions to the same act are impossible, the diagonal elements are all zero. The sums of the row elements of the matrix are all equal to one.

Table 2 Transition matrix of the data in Table 1

To: From:	*Care*	*Walk*	*Sit*	*Rear*	*Shake*
Care	0	0	0.69	0	0.31
Walk	0.08	0	0.75	0.13	0
Sit	0.24	0.45	0	0.17	0.14
Rear	0	0	1	0	0
Shake	0.09	0.45	0.45	0	0

An often-used graphical representation of a behavioural record is the so-called bar plot, where all the acts in the ethogram are arranged along the

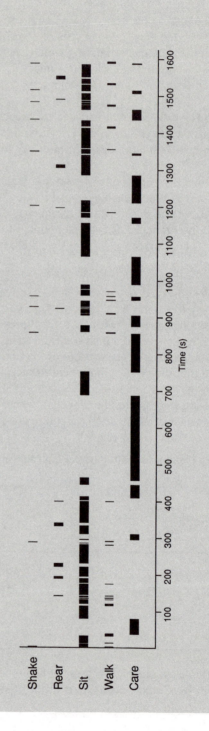

Figure 1 Bar plot of the behavioural record of Table 1.

vertical axis, and the observation time is given on the horizontal axis. All occurrences of each act are represented by means of bars. The bar plot corresponding to our example is given in Fig. 1.

1.2 THE CONTINUOUS TIME MARKOV CHAIN (CTMC)

When variability in the durations and alternation of acts makes a deterministic causal model unfeasible, a CTMC is the simplest possible generalization. Therefore, this type of model is a good starting point for analysis in such cases. A CTMC a process in which several distinct states are visited successively. Figure 1.2 gives an example of how a bar plot of a short observation of such a process might look. The main property of the process, the so-called 'Markov property', is that the probability of any future behaviour of the process, when its present state is known exactly, is not altered by additional knowledge concerning its past. The process is completely characterized by a set of parameters called 'transition rates', which are the probabilities per unit of time of switching from one state to another. Thus, the process can be represented by a set of states and the transition rates between the states (the so-called 'state-space representation', see Fig. 1.2).

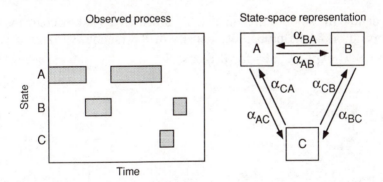

Figure 1.2 Example of a barplot of a short record of a continuous time Markov chain with three states, and a state-space representation.

When the model is applied to behaviour, the states represent behavioural categories. The Markov property implies that the model describes behavioural sequences in which transitions between acts are independent of both the time a bout has already endured and previous behaviour. This assumption must, however, be justified before the model can be applied to a particular behavioural process. Chapters 3-6 are concerned with various methods for testing this assumption. When the model does fit, the behavioural record can be represented by means of the transition rates between behavioural categories, which subsume the information of alternation and duration of acts. These parameters can be interpreted as behavioural tendencies to switch from one act to another.

Throughout this book we will only consider processes of the ergodic type with finite state space. This means that there is a finite number of states and all states can eventually be reached (see Feller, 1968, Chapters 15 and 17). The formal definition of the *transition rate* from state A to state B, α_{AB}, is

$$\alpha_{AB} = \lim_{\Delta t \downarrow 0} \frac{Pr\{\text{trans. A to B in } (t, t+\Delta t) \,|\, \text{proc. is in A at } t\}}{\Delta t} , \quad (1.1)$$

where $Pr\{\text{event 1} \,|\, \text{event 2}\}$ denotes the chance on event 1 given that event 2 occurs. Note that transition rates can have any value larger than or equal to zero (thus even larger than one), since they are chances per unit of time. The *termination rate* of state A, λ_A, is the chance per unit of time that state A is left and hence that the process goes to any of the other states, i.e.:

$$\lambda_A = \lim_{\Delta t \downarrow 0} \frac{Pr\{\text{A is left in } (t, t+\Delta t) \,|\, \text{process is in A at } t\}}{\Delta t} . \quad (1.2)$$

Let X denote the sojourn time in state A. Since, in a Markov chain, λ_A does not depend on X, the probability density of the random variable X is

$$f(x) = \lambda_A \exp[-\lambda_A x] , \text{ for } 0 \leq x < \infty \quad (1.3)$$

with

$$\lambda_A = \sum_B \alpha_{AB} . \quad (1.4)$$

Box 1.2   Exponential distribution

1. Properties of the exponential distribution

The probability density function $f(x)$, $0 \leq x < \infty$, of an exponentially distributed random variable X with parameter λ is defined by

$$f(x) = \lambda \exp[-\lambda x]. \quad (1)$$

See Fig. 1.

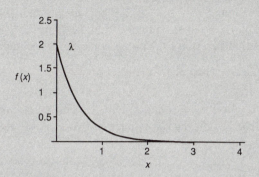

Figure 1 The exponential distribution with parameter $\lambda = 2$.

The cumulative distribution function $F(x)$ is defined by

$$F(x) = Pr\{X \le x\}. \tag{2}$$

In the case of an exponential distribution it follows that

$$F(x) = \int_0^x \lambda \exp[-\lambda y] dy = 1 - \exp[-\lambda x]. \tag{3}$$

The survivor function of X, denoted by $\bar{F}(x)$, is defined by $\bar{F}(x) = Pr\{X > x\}$, which in this case is equal to $\exp[-\lambda x]$. By taking logarithms it follows that

$$\log \bar{F}(x) = -\lambda x, \tag{4}$$

and hence the log-survivor function is linear in x. See Fig. 2.

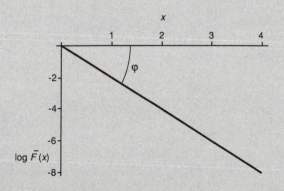

Figure 2 The log-survivor function. Note that tg $\varphi = \lambda = 2$.

This well-known property can be used as a visual test of the exponentiality of the data. Suppose that we have n bout lengths $x_1,..., x_n$. Denote the ordered observations by $x_{(1)} \leq x_{(2)} \leq ... \leq x_{(n)}$ (see the first column of Table 1 for an example). The total number of observations, n, is equal to 20.

The distribution function $F(x)$ in a point x can be estimated by

$$F_n(x) = \frac{i-1}{n} \quad \text{when } x_{(i-1)} \leq x < x_{(i)}, \tag{5}$$

for $i = 1,..., n$, where $x_{(0)} = 0$ and $F_n(x) = 1$ when $x \geq x_{(n)}$ (see the second column in Table 1). The survivor function (given in the third column) is estimated by

$$\bar{F}_n(x) = 1 - F_n(x). \tag{6}$$

Table 1 Calculation of the log-survivor function based on a sample of 20 observed bout lengths.

x	$F_n(x)$	$\bar{F}_n(x)$	$\log \bar{F}_n(x)$
0.0302	0.05	0.95	−0.051
0.0401	0.10	0.90	−0.105
0.0604	0.15	0.85	−0.163
0.0863	0.20	0.80	−0.223
0.160	0.25	0.75	−0.288
0.252	0.30	0.70	−0.357
0.387	0.35	0.65	−0.431
0.876	0.40	0.60	−0.511
0.895	0.45	0.55	−0.598
0.935	0.50	0.50	−0.693
1.08	0.55	0.45	−0.799
1.20	0.60	0.40	−0.916
1.34	0.65	0.35	−1.050
1.50	0.70	0.30	−1.204
1.57	0.75	0.25	−1.386
1.60	0.80	0.20	−1.609
1.69	0.85	0.15	−1.897
1.98	0.90	0.10	−2.303
2.16	0.95	0.05	−2.996
2.69	1.00	0.00	-

A method for calculating confidence intervals for the estimated survivor function is given in subsection 4.5.1 (note the distinction in notation between $F(x)$, the theoretical cumulative distribution function, and $F_n(x)$, its empirical counterpart, and similarly between $\bar{F}(x)$ and $\bar{F}_n(x)$). The plot (see Fig. 3) of x_i against the log $\bar{F}_n(x_i)$ (the fourth column) should approximate a straight line through the origin if the x_i are a sample from an exponential distribution. In the following chapters, especially Chapter 4, we discuss the ethological significance of the most important types of departures from a

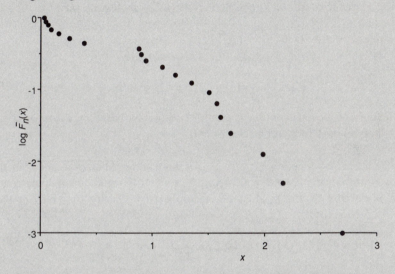

Figure 3 Empirical log-survivor plot based on the data from Table 1.

straight line. When an act has an exponentially distributed bout length, its future duration, given that it has already lasted x_0 time units, does not depend on x_0. More formally:

$$Pr\{X \le x_0 + x | X > x_0\} = Pr\{X \le x\}, \qquad (7)$$

for all $x_0 > 0$ and $x > 0$, as can easily be shown. The conditional cumulative distribution function of X, given that the bout length is at least x_0 is

$$Pr\{X \le x_0 + x | X > x_0\} = \frac{Pr\{X \le x_0 + x \cap X > x_0\}}{Pr\{X > x_0\}}$$

$$= \frac{\displaystyle\int_{x_0}^{x_0 + x} \lambda \exp[-\lambda s] ds}{\exp[-\lambda x_0]} = \frac{\exp[-\lambda x_0] - \exp[-\lambda(x_0 + x)]}{\exp[-\lambda x_0]}$$

$$= 1 - \exp[-\lambda x]. \qquad (8)$$

Hence, it follows by differentiation that the conditional probability density is equal to $\lambda\exp[-\lambda x]$, the unconditional density.

Conversely, if a random variable has this so-called lack of memory property, it must have an exponential distribution. This is proved in Box 1.3. The expectation of X is equal to

$$EX = \int_0^\infty x\lambda\exp[-\lambda x]\,\mathrm{d}x = \frac{1}{\lambda}\int_0^\infty y\,\exp[-y]\,\mathrm{d}y = \frac{1}{\lambda}. \tag{9}$$

The variance of X, $VarX$, can be derived in an analogous way:

$$VarX = EX^2 - (EX)^2 = \int_0^\infty x^2\lambda\exp[-\lambda x]\,\mathrm{d}x - \frac{1}{\lambda^2}$$

$$= \frac{1}{\lambda^2}\int_0^\infty y^2\exp[-y]\,\mathrm{d}y - \frac{1}{\lambda^2} = \frac{2}{\lambda^2} - \frac{1}{\lambda^2} = \frac{1}{\lambda^2}. \tag{10}$$

2. Maximum likelihood estimator (MLE) of λ

The likelihood function $l(x_1,\dots, x_n;\lambda)$, or more simply $l(\lambda)$, of n independent exponentially distributed random variables $X_1,\dots, X_n$ with common parameter λ is defined by the product of the probability density functions at the observed values $x_1,\dots, x_n$:

$$l(\lambda) = f(x_1)\dots f(x_n) = \lambda^n\exp[-\lambda\sum_{i=1}^n x_i]. \tag{11}$$

The maximum likelihood estimator (MLE) for λ, denoted by $\hat{\lambda}$, is the most likely value given the observations $x_1,\dots, x_n$, i.e. $l(\lambda)$ attains its maximum for $\lambda = \hat{\lambda}$. The MLE can be found by solving the equation

$$\frac{\mathrm{d}l}{\mathrm{d}\lambda} = 0 \tag{12}$$

or

$$\frac{\mathrm{d}\log l}{\mathrm{d}\lambda} = 0, \tag{13}$$

since the logarithm gives a monotonic transformation.

Applying this to eqn (11) we get:

$$\frac{\mathrm{d}}{\mathrm{d}\lambda}\left(n\log\lambda - \lambda\sum_{i=1}^n x_i\right) = \frac{n}{\lambda} - \sum_{i=1}^n x_i. \tag{14}$$

Equating (14) to zero we find

$$\hat{\lambda} = \frac{1}{\bar{x}}, \tag{15}$$

where $\bar{x}$ is the mean bout length. (Note that $(d^2\log l)/d\lambda^2 = -n/\lambda^2$, thus it is indeed a maximum.) Hence the MLE of λ is equal to the reciprocal of the mean bout length.

3. Confidence interval for λ

A $(1 - \alpha)$ confidence interval for λ is given by

$$\frac{\chi^2_{2n}(1-\tfrac{1}{2}\alpha)}{2n\bar{x}} \leq \lambda \leq \frac{\chi^2_{2n}(\tfrac{1}{2}\alpha)}{2n\bar{x}} \tag{16}$$

(Patel *et al.*, 1976). This is based on the fact that the sum of n independent exponentially distributed random variables with common parameter $\lambda = 1$ follows a chi-squared distribution with $2n$ degrees of freedom. $\chi^2_{2n}(\tfrac{1}{2}\alpha)$ and $\chi^2_{2n}(1 - \tfrac{1}{2}\alpha)$ denote, respectively, the $\tfrac{1}{2}\alpha$ and the $1 - \tfrac{1}{2}\alpha$ points of a chi-squared distribution with $2n$ degrees of freedom. For instance, from Table 1 in Box 1.1 it can be calculated that the mean bout length of *Care* in the example is equal to 46.82 seconds and its estimated termination rate is 0.021 s^{-1}. The number of *Care* bouts is 13. To calculate the 95% confidence interval of the termination rate, look up the values of $\chi^2_{2n}(0.025)$ and $\chi^2_{2n}(0.975)$ in Table A2. Since $2n = 26$, it follows that these are equal to 41.9 and 13.8, respectively. Hence, with a probability of 95%, the interval from $13.8/(26 \times 46.82) = 0.0113$ s^{-1} to $41.9/(26 \times 46.82) = 0.0344$ s^{-1} covers the real value of λ.

This is proved in Box 1.2. The density function (1.3) is called the 'one-parameter exponential distribution' or simply the 'exponential distribution'. See Box 1.2 for a description of the most important properties of the exponential distribution, a derivation of the maximum likelihood estimator of the parameter λ_A, and its confidence interval. The relation between the Markov property and the exponential distribution is elucidated in Box 1.3. See e.g. Feller (1968) for further details.

Box 1.3   Relation between Markov property and the exponential distribution

Let $\lambda(x)$ denote the termination rate of state A, i.e. the probability per time unit that the state is left at x (e.g. an act is terminated) given that it is not left before:

$$\lambda(x) = \lim_{\Delta x \downarrow 0} \frac{Pr\{A \text{ is left in } (x, x+\Delta x) \mid \text{process is in A at } x\}}{\Delta x}$$

$$= \lim_{\Delta x \downarrow 0} \frac{Pr\{A \text{ is left in } (x, x+\Delta x) \wedge \text{process is in A at } x\}}{\Delta x \, Pr\{\text{process is in A at } x\}}$$

$$= \frac{f(x)}{\bar{F}(x)}. \tag{1}$$

The termination rate specifies the distribution of X, the time that the process is in state A, since

$$\lambda(x) = -\frac{d \log \bar{F}(x)}{dx}. \tag{2}$$

By integrating $\lambda(x)$ and using the fact that $\bar{F}(0) = 1$ we obtain

$$\bar{F}(x) = \exp[-\int_0^x \lambda(s) ds]. \tag{3}$$

By differentiation it follows that

$$f(x) = \lambda(x) \exp[-\int_0^x \lambda(s) \, ds]. \tag{4}$$

If the termination rate $\lambda(x)$ is constant, i.e. equals λ, it follows that $f(x) = \lambda \exp[-\lambda t]$; hence the distribution of the bouts is exponential. Conversely, if the distribution is exponential it follows by substitution in (1) that $\lambda(x)$ is constant:

$$\lambda(x) = \frac{\lambda \exp[-\lambda x]}{\exp[-\lambda x]} = \lambda. \tag{5}$$

Hence, the termination rate of an act is constant (i.e. independent of the past) if and only if the duration of the bouts follows an exponential distribution.

Given the sequence of states, the residence times in the states are independent. Furthermore, it can be proved that the sequence of states (for instance: A, B, A, C, B as in Fig. 1.2) is a first-order discrete Markov chain. This process is completely characterized by the transition probabilities

$$p_{AB} = Pr\{\text{the next state is B} \mid \text{process is in A}\}$$

$$= \frac{\alpha_{AB}}{\lambda_A}. \tag{1.5}$$

(See section 5.1 and Box 5.1 for a more extensive treatment of discrete Markov chains.) The transition probabilities in a first-order discrete Markov chain are independent of the previous sequence of acts and constant in time.

From the property of a CTMC that the sojourn times are independent and that the sequence of states is a discrete first-order Markov chain it follows that the transition rates contain all the information concerning distributions of the sojourn times and the probabilities of alternations between the states. Thus, conditionally on the initial state, the whole process can be represented by these parameters (as in Fig. 1.2).

In ethological applications the states of the model are assumed to correspond to behavioural categories. The transition rates represent the tendencies to switch between different acts. The model can be applied to solitary behaviour as well as to interactions between one or more individuals. In the latter case the behavioural categories represent combinations of acts of different individuals (this is illustrated in Box 1.4). Thus, the methods for analysing social interactions are not essentially different from those for analysing solitary behaviour.

Box 1.4   Example: interactions of male rats

Table 1 shows a record of a dyadic interaction between two male rats. Per individual, recorded behavioural categories are exclusive. Between individuals, behavioural categories may overlap. The original record is transformed to one with exclusive behavioural categories which consist of combinations of acts of both individuals. The corresponding bar plots are given in Fig. 1. If the record is described by means of a CTMC, the states represent these combinations of acts. The transition rates still represent tendencies of the individuals to switch between acts.

For instance, $\alpha_{(Crawl\ over\ +\ Immobile)(Rear\ +\ Immobile)}$ represents the tendency of individual 1 to switch from *Crawl over* to *Rear* while individual 2 is *Immobile*. The estimation of transition rates, termination rates and transition probabilities proceeds as in the case of solitary behaviour.

Table 1 Dyadic interaction between two male rats. (Numbers in the first column refer to the two individuals)

Original record		Transformed record		
Event	Observation time	Behavioural category	Begin	Duration
1: *Crawl over*	0			
2: *Immobile*	0	*Crawl over*	0	10
1: end *Crawl over*	10	+ *Immobile*		
1: *Rear*	10	*Rear*	10	5
1: end *Rear*	15	+ *Immobile*		
1: *Walk*	15	*Walk*	15	5
2: end *Immobile*	20	+ *Immobile*		
2: *Sniff*	20	*Walk*	20	2
1: end *Walk*	22	+ *Sniff*		
1: *Sniff*	22	*Sniff*	22	11
2: end *Sniff*	33	+ *Sniff*		
2: *Walk*	33	*Sniff*	33	2
1: end *Sniff*	35	+ *Walk*		
1: *Rear*	35	*Rear*	35	4
1: end *Rear*	39	+ *Walk*		
1: *Walk*	39	*Walk*	39	5
1: end *Walk*	44	+ *Walk*		
1: *Groom Opponent*	44	*Groom Opp*	44	1
2: end *Walk*	45	+ *Walk*		
2: *Immobile*	45	*Groom Opp*	45	16
1: end *Groom Opp*	61	+ *Immobile*		
1: *Walk*	61	*Walk*	61	2
2: end *Immobile*	63	+ *Immobile*		
2: *Walk*	63	*Walk*	63	2
1: end *Walk*	65	+ *Walk*		

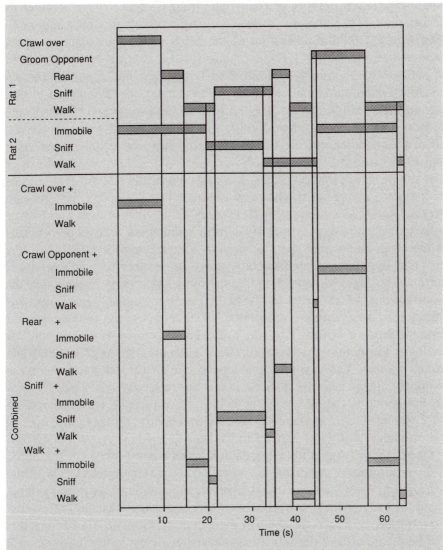

Figure 1 Bar plots of the record of Table 1. Top: separate behavioural categories for each individual (original record). Bottom: combined behavioural categories (transformed record).

If the model gives a good description, the behavioural records can be represented by estimates of the transition rates. α_{CD} can be estimated by the maximum likelihood estimator (MLE):

$$\hat{\alpha}_{CD} = \hat{p}_{CD}\, \hat{\lambda}_C = \frac{N_{CD}}{N_C} \frac{1}{\bar{x}_C}, \tag{1.6}$$

where N_{CD} is the number of transitions from C to D, N_C the number of occurrences of C and $\bar{x}_C$ the mean bout length of behavioural category C. (See Box 1.2 for the derivation of the MLE of the parameter λ_C of an exponential distribution.) The estimator for α_{CD} is equal to the estimator of the transition probability from C to D (N_{CD}/N_C) multiplied by the estimator of the termination rate of act C. For instance, in the example in Box 1.1, the estimated transition probability from *Care* to *Sit* is 0.69 (see Table 2, Box 1.1). Since the mean duration of *Care* is 46.82 seconds, the estimated termination rate is 0.021 s^{-1}. Consequently, the estimated transition rate from *Care* to *Sit* is 0.015 s^{-1}.

It can be shown that $\hat{\alpha}_{CD}$ and $\hat{\alpha}_{EF}$ are asymptotically (i.e. for large observation time) independent when C $\neq$ E and/or D $\neq$ F (see e.g. Basawa and Prakasa Rao, 1980, Chapter 5). This is a great advantage over many other behavioural parameters frequently used by ethologists, such as percentages of the observation time spent on certain acts. It makes it possible to test different hypotheses independently when the number of observations is sufficiently large. Moreover, in interactions this property implies that the contributions of different individuals can be studied separately. For instance, in the example given in Box 1.4 the effect of the behaviour of rat 2 on the tendency of rat 1 to stop *Sniff* can be studied by comparing, for example, the termination rates of (*Sniff* + *Immobile*), (*Sniff* + *Walk*) and (*Sniff* + *Sniff*). The asymptotic independence of the estimators of these termination rates makes it possible to test this relatively easily (see Chapter 7). Furthermore, the effects of the behaviour of rat 1 on the behaviour of rat 2 can be studied analogously and independently. Further properties of the estimators will be discussed in Chapter 7.

One aspect of social interactions that needs some further consideration is the occurrence of simultaneous transitions by different individuals. Some behavioural categories can only be performed simultaneously, an example being the body contact between mother and infant rhesus monkeys (see Box 1.5). If it can be observed which of the individuals initiates transitions, this can be recorded, and the transition tendencies of the individuals can still be studied separately. For instance, in the example of Box 1.5, the tendency of the infant to stop body contact with its mother is estimated by

$$\hat{\alpha}_{CIN} = \frac{N_{CIN}}{N_C \, \bar{x}_C} \, , \qquad (1.7)$$

where N_{CIN} is the total number of transitions from 'Contact' to 'No Contact' that are initiated by the infant. (See e.g. Dienske and Metz, 1977.) In the example, the estimated tendency is 4/(5×74), since the mean duration of body contact bouts is 74 and the infant initiated 4 of the 5 observed transitions. It sometimes happens that a combination of acts can only be performed simultaneously, but it cannot be observed which individual initiates and/or terminates the transitions. In that case there is nothing left to do but to accept that it is impossible even to define separate transition tendencies of the different individuals.

Box 1.5 Example: mother-infant interaction of rhesus monkeys

Table 1 gives an example of a record of the alternations between body-contact states in the mother-infant interaction of rhesus monkeys (see also Dienske and Metz, 1977). In first instance, two states are distinguished, namely *No contact* and *Contact*. Mother as well as infant can initiate switches from one state to the other. At each alternation it was recorded which animal took the initiative.

Table 1 Alternation of body-contact states in mother-infant interaction of rhesus monkeys

Body-contact state	Started by	Begin	Duration
No contact	Infant	0	30
Contact	Infant	30	64
No contact	Infant	90	121
Contact	Infant	215	11
No contact	Mother	226	20
Contact	Infant	246	221
No contact	Infant	467	35
Contact	Mother	502	23
No contact	Mother	525	101
Contact	Infant	626	51
No contact	Infant	677	10

Simultaneous transitions in the record can also be artefacts due to limited recording accuracy. When acts by different individuals do not have to be performed simultaneously, the chance of an exactly simultaneous transition is zero. Usually, transitions between such acts are not recorded as simultaneous. However, when transitions occur very close to one another, the recorded times of transitions can sometimes be equal due to the inevitable discrete nature of the observation times. One possible way to deal with this is to divide the initiatives of simultaneous transitions over the different individuals involved according to the proportions of observed initiatives by these individuals. In Chapter 2, problems connected with a limited recording accuracy are treated in more detail.

1.3   GENERALIZATIONS

Continuous time Markov chains are a good starting point for analysis when deterministic models are not feasible. It is often necessary, however, to modify the model in order to accommodate the special features of a specific (set of) observation(s). For example, one might want to include some deterministic elements. The class of CTMC models can be considered as a subset of a much larger class of stochastic models. In this section we discuss the main types of possible generalizations of practical importance. In subsection 1.3.1 we drop the assumption of constant termination rates and list the most important alternative bout length distributions. In subsection 1.3.2 we abandon the discrete time Markov property of first-order dependence in the sequence of acts. In the next subsection we treat time inhomogeneity and dependence of one or more internal or external covariables. Finally, almost all types of generalizations can be combined in one way or other. A few remarks on this topic are made in subsection 1.3.4.

1.3.1   Semi-Markov chains

In this subsection we generalize the class of CTMCs with respect to the bout length distribution of one or more acts. We maintain the assumption of independence of the past as far as it concerns the order and durations of preceding acts. This implies that the endpoints of an act are so-called renewal points. Such a process is called a semi-Markov chain. As in a CTMC, the sequence of states in a semi-Markov chain is first-order dependent and, given the sequence, the residence times in the states are independent. The only difference is that the distributions of the residence times are not necessarily exponential.

Constant bout lengths

In a certain sense, the most extreme alternative to exponentially distributed bout lengths is that the duration of an act is (approximately) constant. An example is the act *Shake* in the example in Box 1.1. Since *Shake* always lasts exactly 0.45 seconds, the termination rate of *Shake* is zero for $t < 0.45$, infinite at $t = 0.45$ and otherwise undefined. The transition probability from *Shake* to any other act can still be defined as in eqn (1.5) and can be estimated in the normal way. For instance, the estimated transition probability from Walk to Sit is

$$\hat{p}_{Walk,Sit} = \frac{N_{Walk,Sit}}{N_{Walk}} = \frac{18}{24} = 0.75. \tag{1.8}$$

(See Table 2, Box 1.1.) Note that transition rates from an 'ordinary' Markov state towards a state with a constant duration are well defined too and can be estimated as in eqn (1.6). If the other CTMC assumptions, mentioned in section 1.2, are met, straightforward generalizations of the methods of analysis for ordinary CTMC can be used to deal with states with a constant bout length. Therefore, such states are not considered separately. It may happen, for instance, that experimental conditions in-fluence the otherwise fixed duration of an act. In that case the value of this duration as well as the other parameters have to be analysed. For instance, one might study the effects of experimental conditions on the means and variances of bout durations of such an act.

When each act of the ethogram has a (nearly) constant duration it is preferable to describe the process by a discrete Markov chain and base the analysis on that model. In that case the experimental data can be represen-ted by the matrix of transition probabilities and the (mean) values of the durations of each act.

Point events

A special case of constant duration is a duration of (nearly) zero time units. Examples of acts with such short durations are vocalizations or bites. Such acts can be represented by instantaneous states. For instance, Dienske *et al.* (1980) described conflicts between mother and infant rhesus monkeys by means of instantaneous states. In general, an instantaneous state can be treated as the limiting case of a state with small, but finite, sojourn time (see Metz, 1981, subsection 7.2.1). As is the case with states of constant duration in general, transition rates towards instantaneous states are still well defined, as are transition probabilities from such states.

Time-lags

Distributions of durations of acts often deviate systematically from ex-ponential distributions in the sense that each bout has a minimum duration. This type of departure is especially relevant in later stages of the analysis and arises when there are constraints on how quickly a switch between behavioural categories can occur. Such constraints can be due to processes in the central nervous system of an animal or they can be purely mechanical, e.g. due to the time required to make certain motions. To a small degree the reaction time of an observer may also contribute to this effect. Although most experienced observers often correct automatically for a delay in their reaction, it may require some time before certain acts can be recognized. For instance, it takes a second or so to determine whether a rat is moving its head slowly to and fro or whether it is holding its head still. In the CTMC model switches between states can occur arbitrarily quickly. Speed constraints of animals generally cause a shortage of short bouts. This does not create any problems as long as the mean duration of the time during which no switches occur is small compared with the mean bout length. If this is not the case an adjustment is needed. A further reason for adjusting the model in this way can be that one wants to study the minimum durations themselves, e.g. when motoric abilities are studied.

The length of minimum durations, or time-lags, can be fixed or stochastic. If the minimum duration is fixed then the bout lengths have a 'displaced' exponential distribution, also called (in the statistical literature) a two-parameter exponential distribution:

$$f(x) = 0 \qquad \text{for } 0 \leq x < m$$

$$= \lambda \exp[-\lambda(x-m)] \quad \text{for } x \geq m. \tag{1.9}$$

This is a model for a constant time-lag m. It can be applied when there is a strict constraint on the speed of switching between acts. If m is known then it can be subtracted from each bout of the act under consideration. If m is not known it can be estimated by the MLE x_{min}, the minimum of the bout lengths $x_1,..., x_N$ (Johnson and Kotz, 1970). Next, the data can be transformed by subtracting x_{min} from the x_i's:

$$x_i^* = x_i - x_{\min} , \quad i = 1,..., N. \tag{1.10}$$

After removing bouts x_i^* of length zero, the resulting positive durations have an exponential distribution with parameter λ, since the distribution of an exponentially distributed variable X with parameter λ, on the condition that $X > c$, where c is a positive constant, equals the unconditional distribution (see Box 1.2). The MLE of λ is equal to $1/(\bar{x} - x_{min})$. The estimates of the minimum duration and the adjusted termination rate can serve as ingredients for further analysis.

To account for stochastic time-lags we let the transition rates depend on the time since entry, x. We maintain the assumptions that the sequence of states is first-order dependent and that, given the sequence of states, the residence times in the states, the bout durations, are independent. The only difference is that the distributions of the bout lengths, X, are not necessarily exponential. We assume that the time-lags only depend on the current state.

$\lambda_A(x)$ denotes the chance per unit of time to leave state A, the termination rate of A (see Box 1.3). There is a one-to-one relation between the probability density of the bout lengths, $f_A(x)$, and the termination rate:

$$f_A(x) = \lambda_A(x)\exp[-\int_0^x \lambda_A(s)ds] \qquad (1.11)$$

(see Box 1.3). Therefore, models for time-lags are easily formulated in terms of the bout length distributions. In a CTMC the termination rate is constant: $\lambda_A(x) = \lambda_A$. When this model is generalized to incorporate time-lags, it is reasonable to assume that

$$\lambda_A(0) = 0 \quad \text{and} \quad \lim_{x \to \infty} \lambda_A(x) = \lambda_A. \qquad (1.12)$$

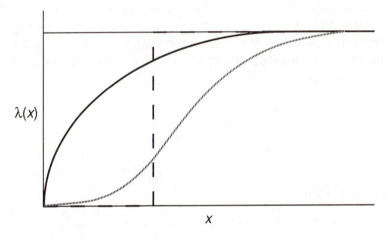

Figure 1.3 Three models with monotonically increasing termination rate:
solid line: the termination rate is a convex function of x
dotted line: the termination rate is a sigmoid function of x
broken line: termination rate with a fixed minimum duration.

Furthermore, we assume that the termination rate increases monotonically with x. In the following we will omit the subscript for convenience.

There are several types of models that have the required properties. We distinguish two main types which are as different as possible within the class of models with monotonically increasing termination rates (see Fig. 1.3). The first type of termination rate that we consider is convex, and hence it has a negative second derivative. A termination rate of such a form is generated by, for example, exponentially distributed time-lags, where the parameter of the time-lag distribution is much larger than the asymptotic termination rate:

$$\lambda(x) = \frac{\lambda(1 - \exp[-(\eta - \lambda)x])}{1 - \frac{\lambda}{\eta}\exp[-(\eta - \lambda)x]} \ , \ \eta \gg \lambda. \tag{1.13}$$

This model would imply that the process is in fact a CTMC in which the transitions between a 'time-lag state' and the 'real' bout are not observed. It is a plausible model if there is some kind of behaviour that can be associated with the time-lag and can be considered as a Markovian state. For instance, if *Grooming* is always preceded by a bout of *Nose scratching*, and both have exponential durations whereas the mean duration of *Nose scratching* is very short, the termination rate of the combined act *Nose scratching followed by grooming* may be modelled as in eqn (1.13).

The second type of model concerns termination rates with one inflection point where the second derivative changes from positive to negative. Termination rates with such a sigmoid form can occur, e.g. when the time-lags have a normal distribution (conditioned on positiveness). In these cases the termination rate is

$$\lambda(x) = \frac{\lambda g(\lambda, \sigma^2, \eta, x)}{1 - \Phi(\frac{x - \eta}{\sigma}) + g(\lambda, \sigma^2, \eta, x)} \tag{1.14}$$

with

$$g(\lambda, \sigma^2, \eta, x) =$$
$$\exp[-\lambda x + \tfrac{1}{2}\lambda^2 \sigma^2 + \lambda \eta \sigma]\{\Phi(\frac{x - \eta + \lambda \sigma^2}{\sigma}) - \Phi(\frac{-\eta + \lambda \sigma^2}{\sigma})\}, \tag{1.15}$$

where $\Phi(x)$ is the standard normal distribution function, η is the expected lag duration and σ^2 is the variance of the time-lags. The location of the inflection point in $\lambda(x)$ is determined by η whereas the rate at which λ is approached depends on σ^2. (An example of such a termination rate is given in Fig. 1.4). The assumption of normally distributed time-lags is intuitively appealing. However, there is no simple explicit expression for the probability density of X. Consequently, estimation and test procedures based on models of this form are rather complicated. It is convenient to use a more tractable model for termination rates of this form as a first approximation. One such model that closely resembles the former is

$$\lambda(x) = \frac{\lambda(1 - \exp[-x/\beta])}{1 + \exp[1 - x/\beta]}, \tag{1.16}$$

where β is the point of inflection. Since $\lambda(\beta) = (1 - e^{-1})/2$, the relative magnitude of β does not depend on the scale of measurement. It can be

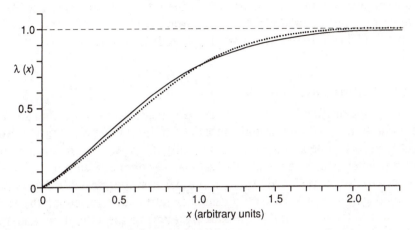

Figure 1.4 Termination rates according to eqn (1.16) with $\beta = 0.4$ (solid line) and eqn (1.14) with $\eta = 0.4$, $\sigma^2 = 0.24$ (dotted line). In both cases the asymptotic termination rate is 1.

shown numerically that such a logistic termination rate is nearly indistinguishable from that of eqn (1.14) with $\eta \approx \beta$ and $\sigma^2 \approx 1.5\beta^2$ (see Fig. 1.4). With this model the probability density of X can be formulated explicitly. In the light of the previous discussion, β can be interpreted as a measure of the average duration of the time-lag. Since the two models cannot usually be distinguished on the basis of the data unless there are many observations at small x, the latter model is an attractive alternative: it is a more tractable and sufficiently accurate approximation of the more appealing model based on the assumption that the time-lag has a normal distribution. In section 8.3 this model is used to analyse low-dose drug effects on motoric abilities (indicated by effects on the time-lags) as well as the tendencies to stop certain acts.

Gamma distributed bout lengths

The occurrence of gamma distributions when analysing behavioural data can be understood from the fact that the sum of k ($k \geq 2$) independent and identical exponentially distributed random variables with parameter λ follows a gamma distribution with probability density function

$$f(x) = \frac{\lambda(\lambda x)^{k-1} \exp[-\lambda x]}{(k-1)!}, \tag{1.17}$$

(Feller, 1971). Hence, if acts A and B have approximately equal mean durations and A is almost always succeeded by B and, furthermore, if the combination (AB) is recorded as one act, denoted by C, the durations of

C are often approximately gamma distributed. By distinguishing acts A and B the semi-Markov model reduces to a Markov model. If it is not possible to distinguish the two acts, methods of analysis for semi-Markov chains must be used (see Chapter 7).

Normally distributed bout lengths

It often happens that bout lengths can be considered as fixed plus an error term due to internal or external causes and to inaccurate registration. Such bout lengths can be assumed to be normally distributed with parameters μ and σ^2, where σ is in general relatively small compared with μ (i.e. $E X = \mu + \varepsilon, \varepsilon \approx N(0,\sigma^2)$; that is, ε has a normal distribution with expectation zero and variance σ^2). In such a case it is impossible to reduce the semi-Markov model to a Markov model by altering the ethogram and thus statistical methods have to be adjusted as indicated in Chapter 7. In some cases, a log-normal distribution may be more appropriate, i.e. it is assumed that the log-transformed bout lengths rather than the bout lengths themselves have a normal distribution, since the bout lengths cannot be negative.

1.3.2 Functions of Markov chains

When several states of a Markov chain are combined into one state, the resulting process is a function of a Markov chain (see Metz, 1981). This type of model can be applied when there is a group of behavioural categories, say A and B, which can be considered as states in a Markov chain, but which are recorded as one act, say C. When A and B always occur together in a specific order, e.g. a recording of C always means that (AB) has occurred, the process is a semi-Markov chain as well as a function of a Markov chain and the methods given in Chapter 7 for analysing semi-Markov chains can be used. However, it may also occur that, for instance, acts A and B are recorded as one act, C, whereas they do not necessarily follow each other. In this case a C bout corresponds to a bout of A or B or (AB) or (BA). Such functions of Markov chains are difficult to handle. Therefore, it is usually advisable to try and distinguish the subacts (i.e. A and B in this case) on observable grounds and to adjust the behavioural record before further analysis. This may not always be possible, however. Since nothing is often known about the transitions between the states within a group of lumped states, there are usually several models that can equally well account for the observed process. Metz (1981, Chapter 7) discusses this 'identifiability problem' extensively. Sometimes, it can be solved by biological interpretation and/or experimental manipulation. An example is the work of Putters *et al.* (1984) on the mating behaviour of barbs, which is discussed in section 8.4. Putters and his coworkers were able to associate the lumped states with the female's state of willingness to mate.

1.3.3 Inhomogeneous Markov chains

Global behavioural changes often occur during observation periods. When the behaviour in homogeneous periods is adequately described by a CTMC, changes over longer periods can be described by means of parameter changes in the model. The resulting model is an inhomogeneous Markov chain. For instance, Dienske *et al.* (1980) and Haccou *et al.* (1983) studied the alternation of fully awake and drowsy phases of infant rhesus monkeys by means of an inhomogeneous Markov model with abrupt changes. In Chapter 3 methods are given for testing for homogeneity, estimating the locations of abrupt changes, and determining the number of changes in a record.

Parameter changes may also be gradual rather than abrupt. For instance, drug effects may change gradually in time. Some inhomogeneous models with gradual changes are considered in section 3.4.

1.3.4 Concluding remarks

Deterministic components

Some scientists object to stochastic models of behaviour, since they feel that behaviour has many deterministic components. However, deterministic descriptive models are only special cases of Markov chains, where transition rates have a specific form and/or value. For instance, we have shown in subsection 1.3.1 how to deal with acts that have fixed durations. Behavioural processes may also contain deterministic sequences of acts, for instance B may always follow A (some special cases have already been treated in the preceding subsections). In such cases, there are two possible options. The complex (AB) may be modelled as one behavioural state. If the other behavioural categories can be regarded as states in a Markov chain, the resulting process is a function of a Markov chain as well as a semi-Markov chain, and the methods of Chapter 7 can be applied. Alternatively, acts A and B may be distinguished and considered as separate states in a (semi-)Markov chain. In that case the transition probability from A to B is equal to one. Consequently, the transition rate from A to B is equal to the termination rate of A.

Combinations of models

As has already been mentioned, different types of generalizations may be combined. Thus, some behavioural categories can be modelled as proper

Markov states and others as semi-Markov states, or lumped states of a Markov chain. Furthermore, we may have to use inhomogeneous semi-Markov chains, etc. It is advisable, however, to keep the model as simple as possible and to analyse the observations according to the parsimony principle. Thus, instead of starting with the most general model, it is better first to look for global inhomogeneities, then to check for more subtle changes and consider bout length distributions of coarse behavioural categories. This research strategy is described in detail in section 1.4.

1.4   OUTLINE OF THE ANALYSIS

The analysis of behavioural records as described in this book proceeds within the framework of CTMC models and their generalizations thereof described in section 1.3. Here, we give a brief outline of the procedure, referring to later sections where necessary. A scheme is given in Fig. 1.5. We assume that the data have already been thoroughly screened (as described in Chapter 2) for peculiarities such as obvious recording errors and extreme outliers. We also assume that inadvertent gaps in the record have been removed and that the behavioural categories have been defined in such a way that they are exclusive and occupy 100% of the observation time. Note that this imposes no restriction: if certain acts can occur simultaneously, exclusive behavioural categories of combinations of these acts can be defined, analogous to what is done in the case of social interactions (see section 1.2, Box 1.4). This subject is treated in more detail in section 2.2, where we also describe how to deal with records that contain a few structural gaps. The idea is to inspect the different time scales and search carefully for behavioural categories that can be considered as Markov states. The initial choice of candidate categories depends largely on the research subject. However, during the procedure we may find that certain categories must be split up or lumped together.

The first step (indicated by (1) in the figure) is to search for homogeneous periods, since it seldom occurs that the time structure of behaviour remains constant during the whole observation period. This should be taken into account, since an inadvertent mixture of different periods usually leads to extra stochasticity, i.e. larger variances. This enhances the chance of drawing wrong conclusions at later stages of the analysis (e.g. concerning experimental effects). Gross periodicity is usually evident, but if not it can be detected by means of bar plots of behavioural records (see Box 1.1 and subsection 3.2.1). More subtle behavioural changes (step (2)) can be detected by various visual scanning methods (see subsections 3.2.2-3.2.4). These methods indicate whether changes are more or less abrupt or gradual.

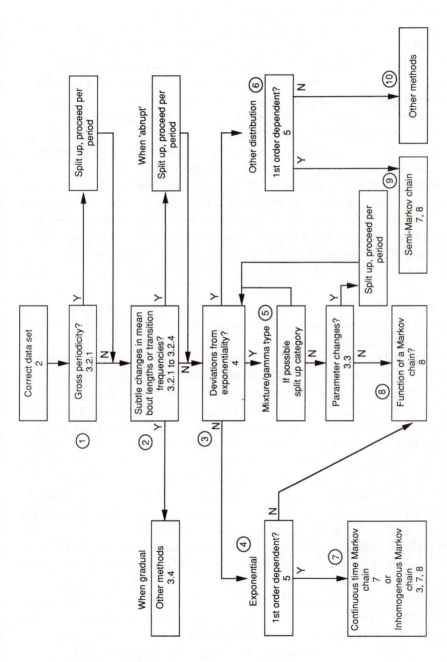

Figure 1.5 Scheme of the analysis with (sub)section numbers.

General methods for the analysis in the presence of gradual changes over relatively long periods are not available, since the class of possible models is too large and analytically intractable. In section 3.4 we treat a few examples for the analysis of ethological data with gradual changes. Other examples of adjusted methods are those used in the study of internal 'clocks' (e.g. see Winfree, 1980, Strogatz, 1986, Carpenter, 1987).

If changes can be considered abrupt, the periods can be divided into homogeneous subphases by applying non-parametric methods for the changes in bout lengths given in subsection 3.3.1 for each act. The results based on the bout lengths of the separate acts have to be combined in order to divide the behavioural record into homogeneous periods. Furthermore, this intermediate result must be in agreement with the results of the methods given in subsection 3.3.2 for detecting abrupt changes in the transition matrix. In the case of contradictory results a combined method for detecting abrupt changes in bout lengths and the transition matrix (see subsection 3.3.3) might be applied. However, this type of method is only applicable if, per period, all Markov assumptions are met, i.e. if within each period all bout length distributions are exponential and the sequence of transitions can be described by a first-order discrete Markov chain. Hence it is strongly advised not to apply these methods at this stage, but instead to proceed first with step (3): the analysis per (perhaps provisionally defined) period.

In each period the properties of behavioural categories are compared with those of the states of the CTMC model (step (3)). As stated before, we assume that the behavioural categories are exclusive and that together they occupy 100% of the observation time. From section 1.2 (Box 1.3) it follows that if the model fits, the bout length distribution of a behavioural category should be exponential. It is best to test this property first instead of sequential dependency properties, since deviations from exponentiality usually indicate more clearly which behavioural categories cause trouble and, more importantly, what may be the source of such trouble. Tests are given in Chapter 4. Tests of the sequential dependency properties (steps (4) and (6)) are described in Chapter 5.

The characteristics of the CTMCs mentioned above can be tested separately, but this usually leads to a large number of tests. Even when the test results are independent, which they need not be, it is to be expected that there is a certain proportion of significant results when the null hypothesis is true. For instance, when m different samples are tested, the expected number of significant results under the null hypothesis is equal to αm (where α is the level of significance of each test). Moreover, when m tests are performed on the same data set, the chance of at least one significant result under the null hypothesis is much larger than α. In Chapter 6 we discuss how individual test results can be combined and how the results of overall tests of Markov properties can be further analysed by means of multiple comparison methods. If the CTMC model gives a good description, effects of, for example, experimental factors on behaviour can be assessed by means of the methods described in Chapter 7 (step (7)). If there are parameter changes over longer periods, the models and methods of Chapter 3 apply.

Systematic deviations from exponentiality may be in the direction of a mixture of exponentials or a gamma-type distribution (step (5)). This type of departure indicates that a behavioural category can be considered as a combination of two or more Markov states. Usually the category can be split up according to observable differences (this approach was first followed by Nelson, 1965; see also Dienske and Metz, 1977). Subsequently the bout lengths of the subcategories are tested for exponentiality.

Deviations in the direction of mixtures can also be caused by parameter changes that have been previously overlooked. If it can be assumed that the distributions in the sub-periods are exponential, normal or displaced exponential, the more powerful parametric detection and test procedures for change points, described in subsection 3.3.1, can be used. In the exponential case a method for combining two or more acts is available. If significant changes are found, the period must be split up and further analysis must proceed for each subperiod separately. If there are no apparent time inhomogeneities and a behavioural category cannot be split up, the process can sometimes still be described by a function of a Markov chain (step (8)), see Metz (1981). The same applies when bout lengths are exponential but there are deviations from first-order sequential dependency. However, such deviations sometimes disappear when certain behavioural categories are combined (see Chapter 5 and Haccou *et al.*, 1985).

If the bout length distributions are not (all) exponential and the deviations are not in the direction of a mixture of exponentials or a gamma distribution we proceed with step (6). When there are no deviations from the sequential dependency properties, the process can be described by a semi-Markov chain (step (9)) and the methods described in Chapter 7 can be applied. If, moreover, the bout length distributions are mixtures of exponentials the process can also be described by a function of a Markov chain. Such special functions of Markov chains are easier to handle than general ones, since there is still first-order sequential dependency. If the bout length distributions are not exponential, mixtures or gamma type, and if there are deviations from first-order sequential dependency, one needs methods beyond the scope of this book. Such processes are difficult to handle on a general basis. Usually, all one can do is use *ad hoc* methods (step (10)).

The procedure described in this section can be viewed as a trial and error method (by dividing the record in homogeneous parts or by redefining the ethogram) for finding the most parsimonious description of the observations that is in agreement with existing knowledge and well interpretable ethologically.

1.5 TECHNICAL PRELIMINARIES: OBSERVING FOR A FIXED PERIOD

1.5.1 Introductory remarks

Usually, behaviour is observed for a fixed period T. This has certain implications for some of the test procedures considered in the next chapters. One consequence is that the beginning of the first bout and the end moment of the last observed bout is usually not recorded (this is called censoring). It follows from the results of Gill (1980) that these bout lengths can be treated as if censoring were random (see also section 4.4). However, if the observation time is sufficiently large, these bout lengths can also be neglected without a great loss of information. In the following we shall only consider the set of completely observed bouts, since most of the time we shall be dealing with asymptotic results (for large T). Only in Chapter 4 shall we return to the issue of random censoring (albeit in a different context).

The problem of a censored first and/or last bout length can be avoided by starting the observation period when a transition occurs and/or waiting until the behavioural category that is performed at time T is terminated, instead of finishing observation at the preset time. This is an example of a so-called 'stopping time' (see e.g. Karlin and Taylor, 1975, Chapter 6). In the following we shall refer to this time as T'.

Another consequence of observing until time T, or T', is that the number(s) of completely observed bouts of each act as well as the number(s) of transitions between acts are random variables, whereas most test procedures found in the literature are derived for the case of a fixed sample size. Likelihood-based test procedures against deviations in the direction of semi-Markov chains (Chapter 4) or inhomogeneous Markov chains (Chapter 3) are the same when the process has been observed for a fixed time T, until a fixed number of bouts has been observed, or until time T' or any other 'stopping time', except possibly for the inclusion of the last bout length as a censored observation. Other tests are performed conditionally on the observed sample sizes (more details are given in section 3.1). However, the critical values of the test statistics reported in the literature are usually calculated from the asymptotic distributions derived for the case when the sample size goes to infinity in a given manner. Yet, in the case of an ergodic (semi-)Markov chain, where all states can in principle be reached (see section 1.2), the numbers of visits to each state as well as the numbers of transitions between states go to infinity when T does. Therefore, the usual limit results continue to hold and the asymptotic distributions can often still be used. This idea is elaborated in a more technical manner in the next subsection, which, if the reader wishes, can be skipped without further consequences.

1.5.2 Technical remarks

Let $N(T)$ denote the random 'sample size'. This can be either the number of bouts of a certain act or the number of transitions between certain acts, depending on which test we are considering (see the following chapters). In a (semi-)Markov chain, $N(T)/T$ almost surely goes to a positive constant when T tends to infinity. This implies that when a sequence of test statistics denoted by $Z_1, ..., Z_n, ...$ (Z_n being based on the first n bout lengths under consideration) converges in distribution to Z as n tends to infinity, and furthermore for each $\varepsilon > 0$ there is a $\delta > 0$ such that

$$\limsup_{n \to \infty} Pr\{ \max_{|i-n| \le \delta n} d(Z_i, Z_n) \ge \varepsilon \} \le \varepsilon, \tag{1.18}$$

(where $d(Z_i, Z_n)$ is the distance between Z_i and Z_n), then the sequence $\{Z_{N(T)}\}$ converges in distribution to Z as well when T goes to infinity (Aldous, 1978, Proposition 1). In principle the validity of condition (1.18) should be verified for each sequence $\{Z_n\}$ under consideration. However, it holds in most practical cases. For instance, it holds when Z_i is the normed sum of independent random variables with finite expectation and variance, and for dependent Z_i when $Z_1, ..., Z_n, ...$ is a mixing sequence (Guiasu, 1971). In the following chapters the observed value of $N(T)$ will be denoted by n or N, unless the dependence of T is stressed, in which case it will be denoted by $n(T)$ or $N(T)$.

2 PRELIMINARY INSPECTION OF OBSERVATIONS

2.1 INTRODUCTORY REMARKS

The models described in sections 1.2 and 1.3 are idealized representations of behavioural records. Before tests can be applied on whether any of these models fits the data, records should meet certain conditions. One of these conditions, namely that behavioural categories should be exclusive, has already been mentioned in Chapter 1. We return briefly to this issue in subsection 2.2.1. Until now it has also been assumed that behaviour was observed continuously throughout the observation period, but sometimes it may be impossible to do this. For instance, animals may be out of sight for certain periods. This is discussed in subsection 2.2.2. (One of the consequences is the occurrence of censors, see section 4.4.) Yet, even if there are no such gaps and behavioural categories are exclusive, the methods described in the succeeding chapters can usually not be applied directly to the raw data, since these often contain recording errors. Furthermore, rare events may sometimes cause disturbances during observation periods (for instance, someone walks into the room during an experiment, or an animal gets hiccups). Therefore, in most cases it will be necessary to inspect and correct behavioural records. In section 2.3 we discuss the most common types of errors (and outliers) and ways in which they can be corrected (or removed). Methods for examining inter-rater reliability are given in section 2.4.

2.2 STRUCTURAL OVERLAPS AND GAPS

Recording errors may cause inadvertent overlaps and/or gaps. Such errors will be considered in detail in section 2.3. Here, we will discuss how to handle overlaps and gaps that are not due to errors, but are inherent to the experimental situation.

2.2.1 Definition of exclusive acts

The original ethogram may contain acts that can occur simultaneously, so that initially there are overlaps between acts in a record. For instance, categories belonging to totally different aspects of behaviour, such as the movements of an animal and its position in the course of time, can of course overlap. Another important example is interactive behaviour, where each of the participating animals has its own set of behavioural categories that can overlap with acts of the other animals. A simple way to transform a behavioural record so that bouts are exclusive is to redefine behavioural categories. The new ethogram consists of combinations of overlapping acts (see Box 1.4). The transformed data set can be analysed along the lines indicated in Chapters 3-7. In some cases, it is also possible to focus on one group of exclusive acts and leave overlapping acts out of consideration. For example, certain vocalizations that can occur simultaneously with other acts, like *Walking*, *Grooming*, etc., may be left out in the first instance. Such factors can be taken into account at a later stage of the analysis. If deviations from Markov assumptions are found, such acts can be used to simplify the model description. For instance, if *Walking* appears not to be exponentially distributed, it may be split up into *Walking with vocalization* and *Walking without vocalization*, and subsequently exponentiality of these categories can be tested (see Chapters 1, 4 and 5).

We may also (initially) take some overlaps into account and leave others out of consideration. For instance, an animal's spatial position may affect its *Groom* bouts, so we would want to distinguish *Groom in left side of observation area* from *Groom in right side of observation area*. If, on the other hand, spatial position is not expected to affect *Walk* bouts, it is not necessary to split *Walk* up into two different types. This approach is especially useful when many acts can overlap at the same time. In such cases it is not very practical to distinguish all possible combinations of acts (as was done in Box 1.4), since that would result in too many different behavioural categories. As a consequence, for most behavioural categories there would be too few bouts left to analyse. An example is provided in Box 2.1.

Box 2.1 Examining records with overlapping behavioural categories

Figure 1 shows a (fictitious) bar plot of squirrel behaviour. There are three sets of behavioural categories:
(1) *Sit, Move*
(2) *Vocalize, Non-vocalize*
(3) *On ground, In tree*.
Per set, the behavioural categories are exclusive, but between sets they can overlap. If all possible combinations of categories were considered, there

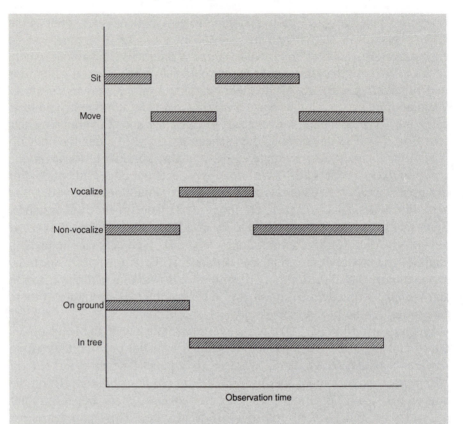

Figure 1 Bar plot of squirrel behaviour, with three sets of behavioural categories that can occur simultaneously.

would be eight behavioural categories, consisting of combinations of acts. For instance, *Sit+On ground+Vocalize*, *Sit+In tree+Vocalize*, etc. Of each combination, there would only be a few (or zero) bouts. Alternatively, we may choose to analyse only combinations of two sets of categories and leave the other set out. Figure 2 shows the resulting bar plot when *Vocalize* and *Non-vocalize* are ignored.

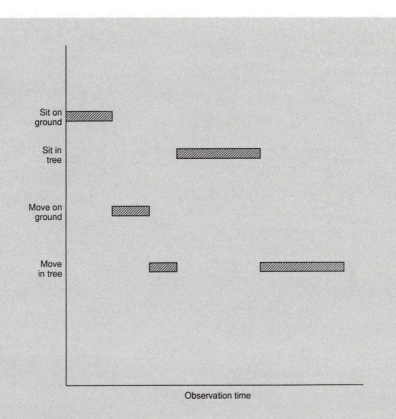

Figure 2 Resulting bar plot when *Vocalize* and *Non-vocalize* are ignored and the other two sets are combined.

Another possibility is to combine only some of the behavioural categories. For instance, if it is expected that vocalizations affect the temporal structure of *Sit* bouts, but not those of *Move* bouts, whereas the location of the squirrel affects *Move* but not *Sit*, we might consider the following set of exclusive behavioural categories:

Sit+Vocalize, Sit+Non-vocalize, Move+On ground, Move+In tree

The resulting bar plot is given in Fig. 3.

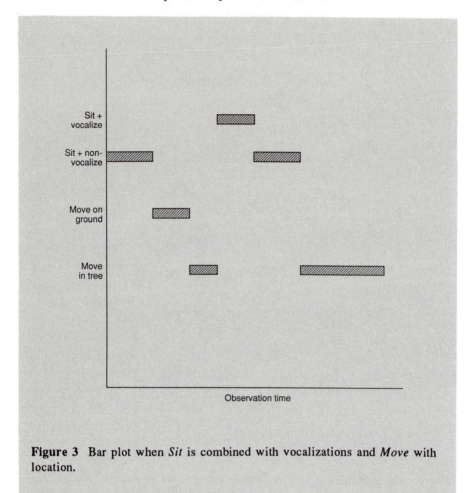

Figure 3 Bar plot when *Sit* is combined with vocalizations and *Move* with location.

2.2.2 How to handle records with gaps

Sometimes it is impossible to observe behaviour continuously. For instance, in field observations animals may be out of sight for a while. In such cases, records do not cover 100% of the observation time, but contain gaps (see Fig. 2.1).

When the duration of such gaps is relatively small, the parts of the record before and after a gap can presumably be considered similar. In such cases, bout lengths can still be studied with the methods given in Chapters 3 and 4, where bouts preceding or succeeding a gap may be censored (see Fig. 2.1 and section 4.4). When sequences are studied, some adjustments are

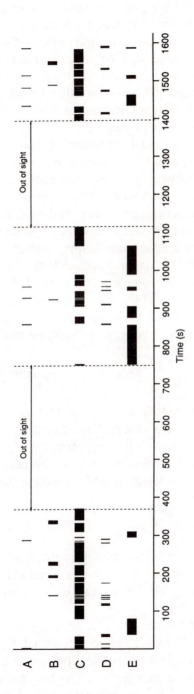

Figure 2.1 Bar plot of a record which contains gaps when the animal is out of sight.

needed, since it is not known what happens during a gap. The simplest way
to handle this is to consider a gap as a separate (behavioural) category.
However, since different acts will usually have been performed during
different gaps, it cannot be expected that the category *Gap* can be modelled
as a Markov state. Thus, some tests on sequential dependency properties
(e.g. the tests against second-order dependency given in subsections 5.2.1
and 5.2.2) should not be applied, since significant results of such tests can
be due to the occurrence of gaps instead of deviations from first-order
dependency in the behavioural process.

Especially when gaps are relatively long, it may not be justified to
assume beforehand that parts of the record before and after a gap are
similar. When the parts between gaps are long enough, they can be analysed independently, i.e. they are treated as separate records. In other cases
it is advisable to test whether certain aspects of the behavioural process are
similar in the different parts; for instance, whether bout length distributions
of a behavioural category are the same before or after a gap. If such tests
do not give significant results, the approach outlined in the previous para-
graph can be used. However, some caution is needed, since tests do not
have much power when the numbers of bouts in the intervals between gaps
are small. Therefore, it is advisable to do the tests on several records and
combine the test results (see Chapter 6) to increase the power.

2.3   DATA CORRECTION

Behavioural records may contain errors as a result of misclassification of
one or more acts or inaccurate observation of begin or end times of acts.
Furthermore, there may be outliers due to (internal or external) disturbances
during observation. In this section we discuss the consequences of such
errors and possible ways of detecting and correcting them.

2.3.1 Misclassification of acts

The extent to which the misclassification of acts affects the estimation of
behavioural parameters depends on the number of misclassifications relative
to the total number of bouts. We demonstrate by an example that the effects
may be severe.

Figure 2.2a gives a fictitious example of a bar plot with three acts,
denoted by A, B and C. In Fig. 2.2b the second act, B, is replaced by C
to show the effects of a single misclassification (for an inexperienced data
analyst these effects may be unexpected). The four transition frequencies
(see section 1.2 and Box 1.1) to and from act A will be affected as follows:

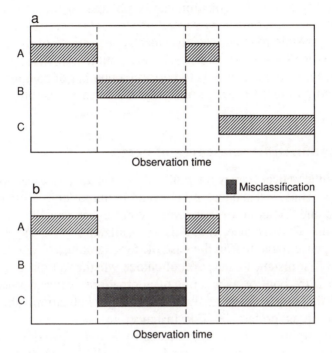

Figure 2.2 Bar plot of a behavioural record:
a. the correct plot
b. the plot with one misclassified act B.

$$\frac{n_{AB}}{n_A} \text{ is replaced by } \frac{n_{AB} - 1}{n_A}, \quad \frac{n_{BA}}{n_B} \text{ by } \frac{n_{BA} - 1}{n_B - 1},$$

$$\frac{n_{AC}}{n_A} \text{ by } \frac{n_{AC} + 1}{n_A} \text{ and, finally, } \frac{n_{CA}}{n_C} \text{ by } \frac{n_{CA} + 1}{n_C + 1}.$$

Moreover, the number of bouts of act B is reduced by one. Omission of this bout, especially if it is either very large or small, can have a large effect on the mean bout length of B. If the chance of misclassification and the duration of a bout of type B are correlated, the estimate of the average bout length of act B is biased. For act C the reverse holds: one bout length is added, which might also result in a biased estimation of the average bout length of act C. Accordingly, in short records even one misclassification may have a relatively large influence on estimates of parameters.

Except when it leads to inadmissible sequences (see subsection 2.3.2), misclassification is very difficult to detect. When possible, the inspection of videotapes is the most thorough and reliable method. Indications about the probability of misclassification can be obtained by comparing records of the same behaviour made by several observers (see section 2.4). In principle, it is possible to examine subsequently the effects of different types of misclassifications on parameter estimates, but it is difficult to do this in a formal way. One possibility is to perform a number of computer simulations, based on the estimated probability of misclassification.

2.3.2 Inadmissible sequences of acts

Within the theoretical framework outlined in Chapter 1, acts cannot succeed themselves. Thus, a sequence 'A,A' would be impossible. Either it would have to be one 'A' bout, or there was a point event, say 'P', between the two A bouts. For instance, if 'A' is a vocalization, 'P' may be a short breathing pause. Note that in the case of, for example, vocalization trains, it is in the first instance a question of choice whether a train is considered as one *Vocalize* bout or as a series of alternations between *Vocalize* and *Pause*. In the course of an analysis we can get indications about which representation is preferable. For instance, in some cases, considering vocalization trains as a behavioural category may remove deviations from first-order dependency (see Chapter 5).

Thus, a sequence like 'A, C, B, A, A, B, ...' is impossible. Either the observer has failed to observe a point event, say P, between the fourth and the fifth act so that the correct sequence would have been A, C, B, A, P, A, B, ..., or there is a recording error: one or both acts were misclassified or the two bouts are in fact one and the same act and the bout length should be the sum of the two separate bout lengths. When correction cannot be based on videotapes of the record, either the first or the second 'A' bout might be replaced by the most likely act. For instance, if the second 'A' bout is replaced, we substitute the act (or one of the acts) X for which $\hat{p}_{AZB}$ is maximal:

$$\hat{p}_{AXB} = \max_{Z \neq A; Z \neq B} \frac{n_{AZB}}{n_A}, \tag{2.1}$$

where n_{AZB} denotes the number of sequences A, Z, B in the total record and $\hat{p}_{AZB}$ the corresponding transition frequency. When there are more errors of the same type, A can be replaced by either of the other acts, proportional to their observed frequency.

A different type of inadmissibility of a sequence of acts is that act A cannot be succeeded by act B, due to mechanical or spatial constraints, for instance. In such cases either A or B may be replaced by the most likely act.

The complementary situation is that act A always has to be succeeded by B, but that it was not recorded in such a way. The correction is easily made.

2.3.3 Checking recorded begin and end times of acts

Inadvertent overlaps and gaps between subsequent acts

In this subsection we assume that there are no inadmissible sequences of acts in the record and that in theory the considered acts are exclusive and together occupy exactly 100% of the observation time. Thus, there are no structural overlaps and/or gaps, such as discussed in section 2.2. In practice most records are as illustrated in Fig. 2.3, with erroneous gaps between succeeding acts and with overlapping acts. These errors can be caused by the restricted reaction time of an observer or by the fact that for many acts it is not possible to determine an unambiguous begin or end time during the observation, because an act has to last for a little while before it can be recognized as such. Examples are acts like *Shaking* or *Walking*.

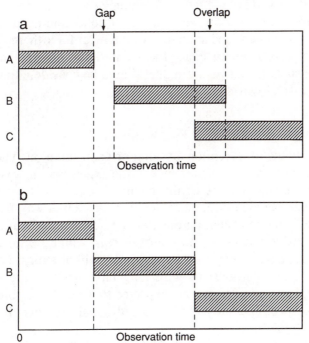

Figure 2.3 Synchronization of begin and end times of succeeding acts:
a. bar plot of the original data, with gaps and overlapping bouts
b. bar plot of the corrected record.

There are at least three different possibilities for correction of the time of transition between a pair of acts. It can become:

1. the original observed end time of the first act;
2. the original observed begin time of the second act;
3. the (weighted) average of these two observed times.

The preferred choice (and weights) depends on the relative accuracy that can be obtained in determining the begin and end times of a sequence of two acts. Hence, in one record different decision rules for correction may be used, depending on the combination as well as the order of acts concerned.

Whether or not (small) timing errors such as overlaps and/or gaps should be corrected in advance also depends on the computer software that is used in further analysis. In general, overlaps (such as between acts B and C in Fig. 2.3a) cause the most problems, since most computer programs do not detect the transition from B to C in such cases. Gaps (such as between A and C in Fig. 2.3a) can usually be ignored, provided that they are not too large.

When it is known that certain acts have a fixed duration (or are point events) the observed bout lengths can be replaced, when desired, by these fixed durations. However, in general we do not recommend this, since it affects begin and end times in the rest of the record. Furthermore, it rarely happens that acts have an exact fixed duration, although they may be modelled as such. (As always, a model should be considered as a simplification of the 'real world'.)

Outlier detection

Outliers are bouts of exceptionally long or short duration. These may be caused by, for example, a disturbance of the experiment by external influences, by technical failures of the recording equipment, or because an animal showed behaviour that deviates in a substantial way from usual. It is advisable to remove such outliers, as they can strongly affect the outcome of an analysis. However, this must be done with care, since very large or small bout lengths may also be legitimate observations. For instance, when bouts have distributions with long tails, such as the exponential distribution, large bout lengths are expected to occur. Therefore it is better to remove suspect data only if there is additional evidence that they may be outliers.

In this phase of inspection of the data it is usually not possible to make strong assumptions about the distribution of bout lengths. Hence, strictly speaking, parametric tests for outliers are inapplicable. Although there have been attempts to derive non-parametric tests (which can be used to judge the evidence in a more objective way), these have not led to operational methods (Barnett and Lewis, 1980, discuss this topic extensively). Although non-parametric methods are not available and the application of parametric methods is in general not justified, it may be helpful to have guidelines for detecting suspect data. To do this we consider outlier tests for the one-para-meter and the two-parameter exponential distribution (see subsection 1.3.1, eqn (1.9)). Tests for lower outliers are only treated for the one-parameter exponential distribution, since the notion of a lower outlier is somewhat strange when there are minimum durations.

Exponential distributions belong to the class of gamma distributions. For the reason mentioned above, it is preferable to test for outliers in a large class of distributions. The considered tests can be regarded as conservative tests for outliers in a gamma distribution, i.e. when they indicate that a bout is an outlier, the corresponding outlier tests for the gamma distribution would in most cases also have given a significant deviation.

Tests for one or two upper outliers in a one-parameter exponential distribution

Data: One sample of bout lengths $x_1,..., x_N$.

Assumptions:

A1: Under H_0 $X_1,...., X_N$ are mutually independent and exponentially distributed with the same (unknown) parameter λ_0.

A2: Under H_1 there are k upper outliers, which are exponentially distributed with (unknown) parameters λ_i. The rest of the observations are identically exponentially distributed with parameter λ. Furthermore, $\lambda_i < \lambda$.

Procedure: We distinguish two cases: $k = 1$ (one upper outlier under the alternative hypothesis H_1) and $k = 2$ (two upper outliers under H_1). Rank all N observations in increasing order (denoted by $x_{(1)},...., x_{(N)}$). The test statistic T_1 for testing the null hypothesis H_0 of no outliers against H_1 of precisely one upper outlier is defined as

$$T_1 = \frac{x_{(N)}}{\sum_{i=1}^{N} x_i},$$
(2.2)

and the test statistic T_2 for testing H_0 against H_1 of two upper outliers is defined as

$$T_2 = \frac{x_{(N)} - x_{(N-2)}}{x_{(N)}}. \tag{2.3}$$

The critical values for T_1, due to Eisenhart *et al.* (1947), corresponding to the levels of significance of 0.05 and 0.01, are given in Table A5 for $N = 2$ (1) 10, 12, 15, 20, 24, 30, 40, 60, 120 and ∞. H_0 is rejected when T_1 exceeds the critical value. See Box 2.2 for an example.

Box 2.2   Test for one upper outlier in a sample of exponentially distributed variables

1. The ordered data are listed in Table 1.

2. We choose $\alpha = 0.05$ as the level of significance.

3. The sum of the bout lengths is 32.72, and the largest value, $x_{(24)}$, is 7.604. Thus from eqn (2.2) it follows that:

$$T_1 = \frac{7.604}{32.72} = 0.2324. \tag{1}$$

4. The sample size, N, is 24. From Table A5 it can be seen that the critical value is 0.2354. Hence, we conclude that the null hypothesis (that the sample does not contain an upper outlier) cannot be rejected.

Table 1 Ordered bout lengths

0.2119	0.2413	0.2472	0.3421	0.3434	0.3665	0.3685	0.3824
0.3969	0.6539	0.7474	0.8378	0.9693	1.011	1.029	1.035
1.501	1.504	2.075	2.174	2.376	2.509	3.792	7.604

The cumulative distribution function of T_2 is:

$$Pr\{T_2 \leq t\} = 1 - N(N-1)(N-2)\left[B\left(\frac{3-2t}{1-t}, N-2\right) - B\left(\frac{3-t}{1-t}, N-2\right)\right], \tag{2.4}$$

where $B(a,b)$ is the beta distribution with parameters a and b (Likeš, 1966). H_0 is rejected when $Pr\{T_2 > t\}$ $(= 1 - Pr\{T_2 \leq t\})$ is smaller than the chosen significance level α. Critical values for $\alpha = 0.05$ and 0.01 are listed in Table A6.

Tests for one or two lower outliers in a one-parameter exponential distribution

Data: One sample of bout lengths $x_1,..., x_N$.

Assumptions:
A1: As in preceding subsection.
A2: Under H_1 there are k lower outliers, which are exponentially distributed with (unknown) parameters λ_i. The rest of the observations are identically exponentially distributed with parameter λ. Furthermore, $\lambda < \lambda_i$.

Procedure: The case of one lower outlier will be treated first. Rank all N observations in increasing order (denoted by $x_{(1)},..., x_{(N)}$). The test statistic T_3 for testing the null hypothesis H_0 of no outliers against H_1 of one lower outlier is defined as

$$T_3 = \frac{x_{(1)}}{\sum_{i=1}^{N} x_i}. \tag{2.5}$$

The null hypothesis is rejected for small values of T_3. Critical values are given in Table A7.

The test statistic T_4 for testing against two lower outliers is defined as

$$T_4 = \frac{x_{(1)} + x_{(2)}}{\sum_{i=1}^{N} x_i}. \tag{2.6}$$

Under H_0 the probability density function of T_4 is as follows:
for $0 < t < 1/(N-1)$:

$$f(t) = \frac{N(N-1)^2}{N-2} \left[\left(1 - \frac{Nt}{2}\right)^{N-2} - \{1 - (N-1)t\}^{N-2} \right]; \tag{2.7}$$

for $1/(N-1) \leq t < 2/N$:

$$f(t) = \frac{N(N-1)^2}{N-2} \left(1 - \frac{Nt}{2}\right)^{N-2}; \tag{2.8}$$

and $f(t) = 0$ for other values of t (Fieller, 1976, Lewis and Fieller, 1979). Critical values are given in Table A8. The null hypothesis is rejected for small values of T_4. See Box 2.3 for an example.

Box 2.3 Test for two lower outliers in a sample of exponentially distributed variables

1. The ordered data are listed in Table 1 in Box 2.2.

2. We choose $\alpha = 0.05$ as the level of significance.

3. As before, the sum of the bout lengths is 32.72. From Table 1, Box 2.2 it can be seen that $x_{(1)}$ is 0.2119 and $x_{(2)}$ is 0.2413. Thus, it follows from eqn (2.6) that

$$T_4 = \frac{0.2119 + 0.2413}{32.72} = 0.01385. \tag{1}$$

4. At $N = 24$, the critical value of T_4 (at $\alpha = 0.05$) lies between 0.000581 and 0.00249 (see Table A8). H_0 is rejected if T_4 is smaller than the critical value. Therefore, we conclude that the hypothesis of two lower outliers is not rejected.

Test for one or two upper outliers in a two-parameter exponential distribution

Data: One sample of bout lengths $x_1,..., x_N$.

Assumptions:
A1: $X_1,..., X_N$ are mutually independent and have a two-parameter exponential distribution (for a definition see eqn (1.9)).
A2: $E X_i = m + 1/\lambda$, $i = 1,..., N$, except for $i = i_1,..., i_k$, in which case $E X_i = m + 1/\lambda_i > m + 1/\lambda$, where λ, all λ_i and m are unknown.

Procedure: We distinguish two cases: $k = 1$ (one upper outlier under the alternative hypothesis H_1) and $k = 2$ (two upper outliers under H_1). Rank all N observations in increasing order (denoted by $x_{(1)},..., x_{(N)}$). The test statistic T_5 for testing H_0 of no outliers against H_1 of precisely one upper outlier is defined as

$$T_5 = \frac{x_{(N)} - x_{(N-1)}}{x_{(N)} - x_{(1)}}, \tag{2.9}$$

and the test statistic T_6 for testing against two upper outliers is defined as

$$T_6 = \frac{x_{(N)} - x_{(N-2)}}{x_{(N)} - x_{(1)}}. \tag{2.10}$$

The critical values for T_5, due to Likeš (1966), corresponding to the levels of significance of 0.05 and 0.01, are tabulated for $N = 3$ (1) 21 in Table A9. Reject H_0 if T_5 exceeds the critical value. The critical value of T_6 for a specific value of N corresponds to the critical value of T_5 at $N - 1$. Thus, Table A9 can also be used for T_6. An example is given in Box 2.4.

Box 2.4 Test for two upper outliers in a sample of two-parameter exponentially distributed variables

1. When we leave out bout lengths in Table 1 (Box 2.2) that are smaller than 0.35 s, the remaining sample has a minimum duration. We will use this sample to illustrate the test based on T_6. Thus, the data set consists of the 19 values from 0.3665 to 7.604.

2. The level of significance, α, is chosen to be 0.05.

3. $x_{(1)} = 0.3665$, $x_{(N-2)} = 2.509$ and $x_{(N)} = 7.604$; hence the test statistic (eqn (2.10)) is

$$T_6 = \frac{7.604 - 2.509}{7.604 - 0.3665} = 0.704. \tag{1}$$

The critical value can be determined from Table A9, since T_6 has the same distribution as T_5, with N replaced by $N - 1$. N is 19, so $N - 1$ is 18 and the critical value at $\alpha = 0.05$ is equal to 0.586. Since T_6 is larger than this value, we conclude that there are two upper outliers in the sample.

Further remarks

Application of the above-mentioned tests during a preliminary inspection of the records may be helpful to judge whether deviations are exceptional or not. The removal of outliers needs care. It is recommended only to take such actions either when there is additional evidence that the bout in question is indeed an abnormal observation or when the p-value of the test statistic is very small. Barnett and Lewis (1980) give a more complete treatment of this topic and many other tests for related situations.

In the tests based on T_1, T_3 and T_5 it is assumed that there is at most one outlier. When it is suspected that there are more outliers, the tests are usually applied successively, i.e. when the test gives a significant result the outlier is removed and the set is tested again. However, this procedure has some disadvantages since the power of a test against one outlier can be affected when the sample contains more outliers. Bendre and Kale (1985) examined this so-called 'masking effect' on T_1. Their results indicate that as long as the number of outliers is smaller than 15% of the sample size the test based on T_1 is not seriously affected. When the test is used in relatively late stages of the analysis a larger proportion of outliers is very unlikely. To our knowledge there are no such results on the test based on T_3. However, it is known that T_5 is very sensitive to masking. When it can be expected that there are more outliers, it is recommended to use the tests based on T_2, T_4 or T_6, since these are less sensitive to masking.

When the (approximate) number of outliers is known, tests for a 'block' of k discordant values can be used (see section 2.4.2 of Barnett and Lewis, 1980). However, this situation does not occur often in practice. Shapiro and Wilk (1972) give a test for an unspecified number of outliers (see also Barnett and Lewis, 1980, p. 88), but the test result does not indicate the number of outliers and it only gives a limited indication about the direction of the outliers (i.e. upper or lower). Therefore, this test is not of much use at the stage where we use outlier tests, i.e. to detect values that should be removed before further analysis.

2.4　EXAMINING INTER-RATER RELIABILITY

In this section we discuss methods for studying the differences between several records of identical behavioural processes. Such records may have been made by the same observer from a videotape of behaviour, or by different observers recording the same behaviour either simultaneously or sequentially from a videotape. The result is a number of records R_{ij} ($i = 1,..., I$ and $j = 1,..., J$), where i denotes the behavioural process and j the rater number (where different 'raters' may in fact be the same person). There are several possible situations:

1. The most common situation in practice is that different degrees of reliability cannot be attached to either of the records. The aim is to get indications of the amount of variability between the records and the effects on behavioural parameters. This is the most difficult situation to handle, since it is not known beforehand which of the records is the most accurate at any specific observation time.

2. There is one perfect record. Other records are compared with this one. Examples are a tutor-student situation, where students have been trained to observe behaviour. The objective is to decide whether the students' records are sufficiently in agreement with the tutor's record.

3. There is a sequence of records of the same behaviour (or several behavioural processes) by one observer. Since the observer learns gradually to record more accurately, the records become more reliable in the course of time. The objective is to decide when the observer has learned to record accurately, i.e. the variability between records is small enough.

Here, we will concentrate on issue 1. Methods for studying issue 2 can be derived relatively simply from the methods for studying 1. It will be indicated how to do this for each method. We will not discuss issue 3. One way to analyse such data, however, is to calculate the sequence of pairwise differences between successive records, with the methods given in subsections 2.4.1–2.4.3. Subsequently, it can be studied whether, and if so, at which rate, differences decrease.

In cases 1 and 2, the goal is to get indications of the effects of the variability between raters on certain statistics, such as mean durations of acts. When the number of different behavioural processes, I (see above), is large, the effects on such statistics can be examined directly by means of analysis of variance (ANOVA). Some of these will be treated in subsection 2.4.4. See also Landis and Koch (1975a,b) for a review. When I is small, however, such methods are not very powerful. In that case, to get an idea of the differences between records, they should be studied in more detail. The theoretical effects on statistics can be considered subsequently. There are, however, several difficulties in comparing different records of the same behavioural process. In subsection 2.4.1 we discuss these problems and give a procedure for preparing two records so that they become comparable. Detailed measures for the differences between two records of one behavioural process are given in subsection 2.4.2. These measures indicate the types and the quantity of differences. Overall measures of differences between two records are treated in subsection 2.4.3. The methods given in 2.4.2 and 2.4.3 are not formal tests. Rather, they should be considered as indicators of whether there are differences between records. There are several problems in connection with comparing such records in a more formal way (see subsection 2.4.1) which are as yet unsolved.

In principle, it is possible to generalize the methods given in subsections 2.4.2 and 2.4.3 for comparing more than two raters and/or records of more behavioural processes. However, these methods have not yet been developed. One possibility is to calculate pairwise differences between all records with the methods given in 2.4.1–2.4.3.

2.4.1 Detailed comparison of behavioural records

Types of differences and their relationships

Two types of differences can be distinguished. Firstly, one observer may assign a certain type of behaviour to a different category than another, or an observer may not be sure about the distinction between two or more behavioural categories. Secondly, there may be differences in the reaction times of observers. In this case, begin and/or end times of acts differ in the records. This may result in different average durations of an act.

If there are many differences in classification, examination of differences in timing is not very meaningful. For instance, mean durations of a certain act may be similar in different records, but at the same time there can be many short bouts and a few long bouts in some records, whereas in other records different numbers of bouts of about the average duration have been recorded. The calculation of an 'overall' measure of similarity of different records, such as the one treated in subsection 2.4.3, is also not recommended in this case, since interpretation of this would be difficult. Therefore, a two-stage procedure is advisable. First check whether there are many differences in classification (by comparing total numbers of acts and transition frequencies, and/or using the methods given in subsection 2.4.2) and, if so, reconsider the definitions of acts in the ethogram (sometimes it might be better to group certain behavioural categories if it is difficult to distinguish between them). When there are few differences in classification left (it may be impossible to remove them all), one can proceed with examining differences in timing and finally apply an overall measure of similarity (subsection 2.4.3).

When measuring differences between records, however, one must be aware that there is a relationship between apparent differences in classification and those in timing. An example is given in Fig. 2.4, where two

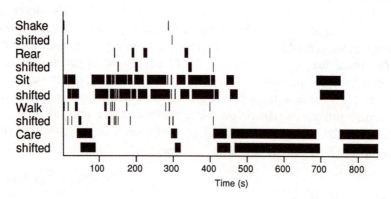

Figure 2.4 Effect of a small shift in timing on the overlap between otherwise identical records.

bar plots of the same record are shown. In the lower half, the bar plot is shifted a little to the right, to demonstrate the effects of a small difference in timing. As a consequence, identical recordings of certain acts no longer overlap. This effect is especially serious for acts that have short mean durations. Therefore, when there are many short-lasting acts, even a small difference in reaction time may have large effects on, for example, the percentage of overlap between records. It is difficult to deal with this problem when there are no systematic differences in timing, but, as will usually occur, one observer is sometimes slower and at other times quicker than the other. In the next subsection we give an *ad hoc* method of taking such time-shifts into account. Certain choices must, however, be made in the course of the process, and each choice has its own pros and cons.

Finding a basis for comparing records

One of the main problems in comparing behavioural records is that the number of recorded acts depends on the decisions that raters have made during observation. For instance, one observer may record a *Groom* bout, whereas during the same time interval, another one records *Groom*, then *Sniff* and then *Groom* again. Furthermore, on top of classification differences, there may be shifts due to timing differences. For instance, in the example given above, the *Groom* bout of the first observer may start at $t = 10$ s and end at $t = 30$ s whereas the sequence '*Groom Sniff Groom*' in the other record starts at $t = 10.5$ s and ends at $t = 31.4$ s. Before measures for differences in classification and timing between records can be applied, a basis for comparing the records must be found. This implies that we must decide which points in the records correspond to each other.

Due to differences in reaction times, the times at which recordings were made are not a good basis for making such decisions. Comparing only the sequences is not possible, since acts in one record may correspond to several acts in another record. Therefore, a combination of both times of recordings and sequences must be used. The procedure for two behavioural records is described below. It is illustrated with parts of records of the same (solitary) rat behaviour, which were made by Dr A. M. M. van Erp and Mrs D.C. Willekens of the Ethopharmacology Group, Leiden University, with CAMERA (see Appendix II). These parts of records are given in Box 2.5. Note that in the course of the procedure neither of the records is altered in any way. All that is done is to seek points of reference in the records that can subsequently be used for calculating measures of similarity between the records.

Box 2.5 Examination of inter-rater reliability: two records of the same behavioural process

Table 1 shows three parts of behavioural records of rat behaviour made by Dr A.M.M. van Erp and Mrs D.C. Willekens (Ethopharmacology Group,

Leiden University). Both observers made the records by means of
CAMERA (Appendix II), from the same videotape.

Table 1 Parts of records of rat behaviour made by two observers from the
same videorecord

a. Part 1

Record 1		Record 2		Correspondence	
Act	Start	Act	Start	Record 1	Record 2
Sniff*	0	Sniff*	0	Sniff	Sniff
Turn	4.68			Turn	
Sniff	5.76			Sniff	
Turn	11.92			Turn	
Care*	13.2				
		Care*	13.52	Care	Care
		Sniff*	19.60		
Turn	19.84			Turn	Sniff
Sniff*	22.16			Sniff	
Turn	23.92			Turn	
Sniff	25.04			Sniff	
Walk*	26.52				
		Walk*	27.56	Walk	Walk
Sniff*	28.56				
		Sniff*	28.76	Sniff	Sniff
Walk	32.76			Walk	
Sniff	33.56			Sniff	
		Care*	34.36		
Care*	34.64			Care	Care
		Walk*	40.64		
Walk*	40.80			Walk	Walk
		Sniff*	42.12		
Sniff*	42.16			Sniff	Sniff
Turn	45.24			Turn	
Walk*	46.16				
		Walk*	46.44	Walk	Walk

Act	Start	Act	Start	Record 1	Record 2
Squat	47.92				
		Sniff	47.96	*Squat*	*Sniff*
*Care**	49.40				
		*Care**	49.88	*Care*	*Care*
		Sniff	52.40		*Sniff*
		Care	54.40		*Care*
*Walk**	56.16				
		*Walk**	56.36	*Walk*	*Walk*

Since the behavioural categories are exclusive, only begin times of the bouts are given. The last column gives the result of the procedure for finding a basis for comparing the records.
* Identical acts that overlap in the two records (each act only counts once, see text).

b. Part 2 (same records)

Record 1		Record 2		Correspondence	
Act	Start	Act	Start	Record 1	Record 2
		*Walk**	324.48	*Walk*	*Walk*
*Walk**	324.68				
*Care**	326.28			*Care*	*Sniff*
		Sniff	326.32	*Sniff*	*Care*
		*Care**	327.08	*Squat*	
Sniff	343.08			*Turn*	
Squat	401.28			*Care*	
Turn	402.44				
Care	403.16				
*Sniff**	408.36	*Sniff**	408.36	*Sniff*	*Sniff*

See remarks below Table 1a.

c. Part 3 (other records, same behaviour)

Record 1 Act	Start	Record 2 Act	Start	Correspondence	
		*Walk**	324.44	*Walk*	*Walk*
*Walk**	324.68				
		*Care**	326.04		
*Care**	326.28			*Care*	*Care*
Squat	343.08			*Squat*	
Care	344.76			*Care*	
Squat	401.28			*Squat*	
Care	403.16			*Care*	
Squat	408.36			*Squat*	
Care	409.96			*Care*	
		Squat	421.64	*Squat*	*Squat*
		Shake	422.80		
		*Walk**	423.24		
Squat	423.36			*Shake*	*Shake*
Shake	424.12				
*Walk**	424.80			*Walk*	*Walk*
		*Sniff**	426.04		
*Sniff**	426.60			*Sniff*	*Sniff*

See remarks below Table 1a.

Data: Two continuous time records of behaviour.

Procedure:

1. Consider the two sequences of acts. For instance, the two records in Box 2.5 (Table 1a) begin with:

record 1: *Sniff Turn Sniff Turn Care Turn Sniff Turn Sniff Walk*
record 2: *Sniff Care Sniff Walk*.

2. Mark all identical acts that overlap in time in the two records, provided that all acts are only counted once:

record 1: *Sniff* Turn Sniff Turn Care* Turn Sniff* Turn Sniff Walk**
record 2: *Sniff* Care* Sniff* Walk**.

Note that, whereas the second *Sniff* bout in record 1 (starting at 5.76 s) overlaps with the first *Sniff* bout in record 2, it is not marked, since the first *Sniff* bout in record 2 also overlaps with the first *Sniff* bout in record 1.

3. Look at what happens in both records between succeeding marked acts. There are several possibilities:

a. No other acts between marked acts in both records. In this case no action needs to be taken.

b. One record contains several acts, the other none. For instance, when we look at the acts between *Sniff** and Care* in Box 2.5, record 1 contains the sequence '*Turn Sniff Turn*', whereas in record 2 there are no acts. Consequently, there is a classification difference between the two records at this point. The problem is how to allocate bouts when there is this difference in classification. In principle there are several possibilities: the first *Sniff* bout is identical in both records and the first *Care* bout in record 2 coincides with '*Turn Sniff Turn Care*' in record 1, or the first sequence '*Sniff Turn*' in record 1 corresponds to *Sniff* in record 2, whereas '*Sniff Turn Care*' in record 1 corresponds to *Care* in record 2, etc. To decide on this, the time recordings of the two records must be considered. The differently classified bouts in record 1 are allocated to the act in record 2 with which they overlap the most. From Box 2.5 it can be seen that '*Turn Sniff Turn*' in record 1 overlaps with *Sniff* in record 2. Consequently, the beginning of the sequences is grouped as follows:
(*Sniff Turn Sniff Turn*) = *Sniff*.
The same situation occurs between *Care** and *Sniff**: record 1 contains a *Turn* bout (starting at 19.84 s), whereas in record 2 there are no acts between *Care** and *Sniff**. Since the *Turn* bout in record 1 overlaps entirely with the *Sniff* bout in record 2, which starts at 19.60 s, the *Turn* bout is allocated to this *Sniff* bout. The same applies to *Turn* (record 1, starting at 23.92 s) and *Sniff* (record 1, starting at 25.04 s), since both overlap entirely with the same *Sniff* bout in record 2. The result is:
record 1: (*Sniff Turn Sniff Turn*) *Care* (*Turn Sniff Turn Sniff*) *Walk*
record 2: *Sniff Care Sniff Walk*.

c. Both records contain one act between two corresponding marked acts. This happens for example with the *Squat* bout in record 1, which starts at 47.92 s, and the *Sniff* bout in record 2, starting at 47.96 s. At this point we have:
record 1: *Walk* Squat Care**
record 2: *Walk* Sniff Care**.
In such cases, it is decided that there is one difference in classification, namely between *Squat* and *Sniff*.

d. Both records contain several acts between two corresponding marked acts. In that case, it must be checked whether the situation may be due to a shift in timing. To do this, shift one of the sequences in such a way that the maximum number of acts correspond in the two sequences. For instance, at observation time 326.28 s in Box 2.5 (Table 1c), record 1 contains the sequence:

Care* Squat Care Squat Care Squat Care Squat Shake Walk*

and record 2 (from 326.04 s onwards):

Care* Squat Shake Walk*.

The two records differ the least if we decide on:

record 1: *Care* Squat Care Squat Care Squat Care Squat* Shake* Walk**

record 2: *Care* Squat* Shake* Walk**.

On the grounds of the overlaps between acts (see situation 'a' above) it is further decided that the sequence '*Squat Care Squat Care Squat Care*' in record 1 is allocated to *Care* in record 2, so we get:

record 1: (*Care Squat Care Squat Care Squat Care*) *Squat Shake Walk*

record 2: *Care Squat Shake Walk*.

Usually, there will not be a large difference in begin times of bouts that correspond according to this method. For instance, the corresponding '*Squat*' bouts in the two records start at 423.36 s and 421.64 s respectively. When required, a maximum allowable difference in begin times can be chosen.

Comments: The procedure results in a correspondence sequence between the two records (see Table 1, Box 2.5, columns 5 and 6), which is used as a basis to calculate measures of similarity (see subsection 2.4.2). In most instances, there will be at most one act in one of the corresponding sequences, e.g. situations where '*Sniff Turn*' corresponds to '*Care Sniff Turn*' will rarely occur. However, on some occasions this may happen. An example is given in Table 1b of Box 2.5, where a sequence of six acts in record 1 corresponds to a sequence of two acts in record 2.

During the procedure several choices have been made which inevitably affect measures for classification as well as timing differences between the records. For instance, if in situation 'd' above we had decided that the grouping of sequences should be:

record 1: *Care (Squat Care Squat Care Squat Care Squat) Shake Walk*

record 2: *Care Squat Shake Walk*,

there would be one less classification difference in *Care* bouts and one extra in *Squat* bouts. Furthermore, the difference in the durations of the *Care* bouts in the two records would be different. Therefore, the measures used for classification and timing differences should complement each other. We will consider such measures in the following subsections.

When there are more than two records, the procedure can be applied sequentially. However, the result may depend on the order in which records are considered. Another possibility is to consider all pairwise combinations of records. A generalization of the procedure, taking several records into account simultaneously, has yet to be developed.

2.4.2 Detailed measures of similarity of two records

Classification

The procedure given in the previous subsection results in a correspondence between the sequences of acts in the two records (see Table 1, Box 2.5, columns 6 and 7). The next step is to derive measures of agreement between the sets of classifications. Several such measures will be considered here.

A measure of agreement based on the contingency table

The situation is comparable with one in which several experts are asked to classify n subjects into K different categories. The 'subjects' are in this case the different acts that an animal has performed during the observation, and the categories are the different possible ethogram units (or their combinations, see below). There are, however, some essential differences: in the case considered here, n is a stochastic variable, whereas in the other situation n is fixed. Furthermore, n depends on the decisions made by both raters as well as the way in which classification differences are scored. For instance, in the example of Box 2.5, the sequence '*Sniff Turn Sniff Turn*' at the start of record 1 corresponds to *Sniff* in record 2. This can be considered in two different ways:

1. Each rater classified four 'subjects'. In two cases the classifications were identical, namely both chose *Sniff*, and in the two other cases one of them chose *Turn* and the other *Sniff*.
2. Each rater classified one 'subject'. One of them classified it as a combination of *Sniff* and *Turn* and the other chose *Sniff*.

Unless it is certain what the animal did at that particular moment (i.e. when there is a standard record), there is no objective way to decide on this issue. We advise the use of method 1, since with this it is possible to use standard methods to get indications of which kind of differences in classifications are made. With this method, the classification categories consist of the ethogram units, and, in principle, all categories can occur together. Alternatively, when method 2 is used, the classification categories consist of all ethogram units plus all combinations of those units. Obviously, this results in a very large set of categories (namely 2^K). Moreover, the procedure of subsection 2.4.1 gives rise to a correspondence sequence where in almost every case at least one of the records contains only one act. Some categories, such as the combination '*Groom Sniff*' with '*Turn Care Sniff*', will therefore rarely occur together. At the moment there are no methods for studying contingencies between classifications under such restrictions.

We outline below how contingencies between classifications in two records are calculated, based on measures that have been developed for the case when *n* is not stochastic. Because of the differences already mentioned, the measures can only be considered as indications of similarity between behavioural records. In particular, it is not certain whether (asymptotic) distributional characteristics of the statistics continue to hold under these conditions.

Data: The correspondence sequence resulting from the procedure given in subsection 2.4.1 (see e.g. Box 2.5).

Procedure: Make a contingency table of the frequencies of occurrence of all possible combinations of different acts. In those cases where one act in one of the records corresponds to two or more acts in the other record, count each combination separately. For instance,

record 1: A

record 2: ABACBA

consists of three 'A,A', two 'A,B' and one 'A,C' combinations. Several measures of agreement have been proposed for such contingency tables. For instance, Cohen (1960) proposed:

$$\kappa = \frac{P_0 - P_e}{1 - P_e}, \tag{2.11}$$

where

$$P_0 = \sum_{k=1}^{K} P_{kk}, \tag{2.12}$$

$$P_e = \sum_{k=1}^{K} P_{k.} P_{.k}, \tag{2.13}$$

and

$$P_{k.} = \sum_{m=1}^{K} P_{km} \quad \text{and} \quad P_{.k} = \sum_{m=1}^{K} P_{mk}. \tag{2.14}$$

In eqns (2.11)–(2.14), p_{km} denotes the relative frequency of combinations of category k in record 1 with category m in record 2 (k, $m = 1,..., K$). When there is complete agreement, $\kappa = 1$. When the raters classify the subjects independently, $\kappa = 0$, and when there is less than chance agreement, $\kappa < 0$. The minimum value of κ is $-p_e/(1 - p_e)$. An example of an application of the procedure described is given in Box 2.6.

Box 2.6 Calculation of classification agreement between two records, based on the contingency table

Table 1 shows the contingency table of classifications for the records given in Table 1a of Box 2.5. According to eqns (2.12) and (2.13):

$$p_0 = \frac{(7 + 4 + 4)}{23} = 0.652 \tag{1}$$

and

$$p_e = \frac{(7 \times 15) + 0 + (5 \times 4) + (5 \times 4) + 0}{23 \times 23} = 0.274. \tag{2}$$

From eqn (2.11) it follows that $\kappa = 0.521$. Thus, the agreement is larger than expected by chance.

Table 1 Contingency table of classifications in the records of Table 1a, Box 2.5

	Record 2					
Record 1	Sniff	Turn	Care	Walk	Squat	Totals
Sniff	7	–	–	–	–	7
Turn	5	–	–	–	–	5
Care	1	–	4	–	–	5
Walk	1	–	–	4	–	5
Squat	1	–	–	–	–	1
Totals:	15	–	4	4	–	23

Further remarks and literature: The measure κ (eqn (2.11)) is only sensitive for deviations from chance agreement (i.e. when the raters classify independently). Krauth (1984) generalized this measure to test for inter-observer bias, i.e. whether raters choose certain categories with different probabilities. Bouza (1987) further improved Krauth's modification of κ.

In a student-tutor situation, the student record can be compared with the standard by means of p_0 (see eqn (2.12)). This case is further discussed by Feinstein (1975) and Light (1971).

Fleiss (1971) gives a measure of agreement for several raters, which can be partitioned into component parts that reflect the agreement for each of the K categories. This method is also described by Landis and Koch (1975b).

When several behavioural processes have been recorded ($I > 1$, see the introduction of section 2.4), measures of agreement can be calculated for each process and then combined. However, the minimum value of κ depends on the expected chance agreement (p_e, see eqn (2.13)), and, accordingly, can differ for the different behavioural processes. The measures given by Krauth (1984) range from -1 to 1, and are therefore easier to combine. The measures can be combined by, for example, (weighted) averages.

Asymptotic means and variances of the measures under the hypothesis of chance agreement have been calculated by Everitt (1968) and Fleiss *et al.* (1969). Krauth (1984) and Bouza (1987) also give the asymptotic characteristics of their (generalized) measures. All these asymptotic properties have been derived, however, for the case of non-stochastic n. It is not certain whether they continue to hold in the case considered here. This makes it difficult to derive an absolute criterion for deciding when records differ too much, on the basis of κ. The measure can, however, be used to compare differences between several pairs of records. Landis and Koch (1975b) review methods for inter-rater agreement (with further references), which can in principle also be used here.

Proportions of similar classifications
Another set of measures that can be used to indicate the similarity with respect to classification are the proportions of cases in which raters agree on classifications of certain acts, e.g. for act 'A':

$$\hat{p}_s(A) = \frac{N_s(A)}{N_s(A) + N_d(A)}, \tag{2.15}$$

where $N_s(A)$ is the number of times that 'A' in one record corresponds to 'A' in the other record and $N_d(A)$ is the number of cases in which 'A' in one of the records corresponds to 'AX' or 'X' or 'AXA' etc. An example is given in Box 2.7.

Box 2.7 Calculating proportions of similar classifications

As an example, we will calculate the proportion of cases that *Care* was classified similarly in the records given in Table 1a, Box 2.5. From the correspondence sequence (columns 5 and 6 in Table 1a, Box 2.5) it can be seen when the records agree on *Care* classifications. There are two such situations: the *Care* bout starting at 13.2 s in record 1 corresponds to the *Care* bout starting at 13.52 s in record 2, and the *Care* bout starting at 34.64 s in record 1 corresponds to the bout starting at 34.36 s in record 2. There is one case in which a disagreement in the records involves *Care* (at 49.40 s), so the proportion of cases that the records agree on *Care* is 2/3.

In the same way it can be calculated that the proportion of agreement on *Squat* in part 3 (Table 1c, Box 2.5) is 1/2: there is one case in which *Squat* bouts agree, and there is one conflict involving *Squat*.

Timing

To study differences in timing, either the begin times or the durations of acts in the different records can be compared. An occurrence of 'A' in one of the records can correspond to 'A' or 'X' or 'AX' or 'XA' or 'AXA' etc. in the other record, where 'X' denotes one or more other acts. To study differences in begin times, only the occurrences of 'A' and 'A' or those of 'A' and 'AX' or 'AXA' etc. are compared, since, in all other instances, classification differences rather than differences in reaction time determine the difference in the begin time of 'A'. Non-parametric tests, such as Wilcoxon's test (Siegel and Castellan, 1988), can indicate whether there are systematic differences in begin times (see Box 2.8 for an example), or the frequency distributions of differences can be studied graphically. One can choose to do such studies for each act separately to see whether differences in reaction time depend on the specific act. Another possibility is to look at the frequency distribution of all timing differences and check whether certain combinations are over-represented in the tails of the distribution.

Box 2.8 Examination of timing differences between two records: comparison of begin times

As an example, we will consider the records in Table 1a, Box 2.5. Since both records begin at $t = 0$, the difference in timing at the beginning is zero. Therefore, we will not include the first begin times in the analysis. As mentioned, begin times are only included when the correspondence between the sequences (columns 5 and 6 in Table 1a, Box 2.5) starts with the same act. The remaining begin times, used for comparison, are listed in Table 1.

To perform Wilcoxon's test, the (signed) differences of the begin times are calculated and ranked. (When there are ties, the average rank is used.) Summation of the ranks with the less frequent sign gives 8.5 as a result. This does not lead to rejection of the hypothesis that the begin times are equal, at a significance level of 0.05 (two-sided test, see e.g. Siegel and Castellan, 1988). Apparently, there is no systematic difference in begin times between the two records.

Table 1 Begin times of Table 1a, Box 2.5, to be used for comparison of the records of the two observers

Act	Record 1 begin times	Record 2 begin times	Difference	Signed rank*
Care	13.20	13.52	-0.32	-6
Walk	26.52	27.56	-1.04	-8
Sniff	28.56	28.76	-0.20	-3.5
Care	34.64	34.36	0.28	5.5
Walk	40.80	40.64	0.16	2
Sniff	42.16	42.12	0.04	1
Walk	46.16	46.44	-0.28	-5.5
Care	49.40	49.88	-0.44	-7
Walk	56.16	56.36	-0.20	-3.5

* The signed ranks are used in the calculation of Wilcoxon's test-statistic.

To study differences in durations which are due to timing, consider all instances in the correspondence sequence in which both records agree on 'A' bouts and compare their durations. Thus, we get two sets of durations '$y_{11},...,y_{1n}$' and '$y_{21},...,y_{2n}$'. See Box 2.9 for an example. The differences between the durations can be studied by comparing frequency distributions or by applying formal tests, such as Wilcoxon's test (see above). Another possibility is to perform one-way layout ANOVA on the sets of durations (see Scheffé, 1959, Landis and Koch, 1975a, and Chapter 7).

Box 2.9 Examination of timing differences between two records: comparison of durations

To study the effect of timing differences on, for example, the durations of *Care* bouts in the records given in Table 1a, Box 2.5, all corresponding *Care* bouts are considered. For instance, the duration of the bout starting at 13.2 s in record 1 is compared with that of the bout starting at 13.52 s in record 2. The bouts starting at 49.40 s (record 1) and 49.88 s (record 2) are not included, since at that point there is a classification difference between the records. Differences in durations between these two bouts are due to the difference in classification rather than timing. Table 1 shows the durations of pairs of *Care* bouts that have been classified similarly in the two records of which parts are shown in Box 2.5.

Application of Wilcoxon's test leads to the conclusion that there are no significant differences between the durations in the two records. (See Siegel and Castellan, 1988, and Box 2.8.)

Table 1 Durations of corresponding *Care* bouts in two records of the same behavioural process

Record 1	Record 2	Difference	Signed rank
6.64	6.08	0.56	3
6.16	6.28	−0.12	−1
11.44	13.48	−2.04	−7
24.12	23.48	0.64	4.5
21.16	19.88	1.28	6
6.30	6.44	−0.14	−2
42.84	42.20	0.64	4.5

2.4.3 Overall measures of similarity of two records

Several measures of similarity of records are used in ethology. We will not treat them all here, since most are very sensitive to timing differences and do not give good insight into classification differences. For instance, a very common method is to calculate the percentage of overlap between records, i.e. the percentage of observation time for which identical acts have been recorded. It can easily be seen that this method is very sensitive to shifts in timing. This is a serious drawback, since usually we are not interested in such small timing differences. One way to correct for systematic timing differences is to calculate the measure for different small shifts. First choose the maximum allowable difference in timing, M. This choice can be based on a combination of the reaction time of observers (somewhere in the order of 0.1 s), the average time needed to recognize an act as such, and the accuracy of the recording equipment. Next, divide the interval $(-M, M)$ into equal parts and shift one of the records backwards or forwards in time according to these time intervals. For instance, if M is 1 second, and the interval $(-M, M)$ is divided into 20, one of the records is shifted -1 s, -0.9 s,..., 1 s. After each shift, calculate the percentage of overlap between the records and use the maximum of these as a measure of similarity. This method does not take into account the possibility that at one time one rater may be slower, whereas at another time he/she is faster than the other rater.

Another method, which we describe here, is based on likelihood theory. It consists of calculating two likelihood ratio test statistics for differences between the two records (the likelihood ratio principle is explained in Box 2.10). It is assumed that the behavioural process can be described by a CTMC. At an early stage of analysis it is usually not known whether this is true. In that case, the result must be viewed as an indication of similarity between the records. An advantage of the method is that it is robust for small random timing differences.

Box 2.10 The likelihood ratio principle

There are a few generally applicable principles to derive tests which possess optimal power properties for large sample sizes. The most important one is the likelihood ratio approach. The majority of the tests we discuss are based on this principle.

Suppose that $X_1,...,X_n$ are independently distributed with unknown parameter (vector) θ and that we want to test the null hypothesis H_0: $\theta \in \Theta_0$ against the alternative hypothesis H_1: $\theta \in \Theta_1$, where $\Theta_0 \subset \bar{\Theta}_1$ (the closure of Θ_1).

The likelihood function $l(x_1,...,x_n;\theta)$ is defined as the product of the probability densities at the observed values $x_1,...,x_n$:

$$l(x_1,...,x_n;\theta) = f(x_1;\theta)f(x_2;\theta)...f(x_n;\theta). \tag{1}$$

The maximum likelihood estimator of θ, $\hat{\theta}$, is found by maximizing (1) over θ (see e.g. Box 1.2, where the maximum likelihood estimator of λ was derived). Note that the likelihoods, and therefore the maximum likelihood estimators, depend on the hypothesis, i.e. in general $\hat{\theta}_0 \neq \hat{\theta}_1$, where $\hat{\theta}_0$ denotes the estimator under H_0 and $\hat{\theta}_1$ that under H_1.

The likelihood ratio test is based on a comparison of the values of the maximum likelihoods under the two hypotheses:

$$\frac{l(x_1,...,x_n;\hat{\theta}_1)}{l(x_1,...,x_n;\hat{\theta}_0)}. \tag{2}$$

The likelihood ratio test statistic is twice the logarithm of (2):

$$\Lambda = 2\{L(x_1,...,x_n;\hat{\theta}_1) - L(x_1,...,x_n;\hat{\theta}_0)\}, \tag{3}$$

where $L(x_1,..., x_n;\theta)$ is log $l(x_1,..., x_n;\theta)$. It can be proved (see e.g. Cox and Hinkley, 1974) that Λ is asymptotically chi-squared distributed (i.e. for large n) with a number of degrees of freedom equal to the difference of the dimension of the parameter spaces Θ_1 and Θ_0.

We will illustrate this approach by means of an example: the two-sample problem for exponentially distributed variables. Suppose that $X_1,..., X_m$ are exponentially distributed with unknown parameter λ, $Y_1,..., Y_n$ with unknown parameter μ, and that the X_i and Y_j are mutually independent. We want to test the null hypothesis H_0: $\lambda = \mu = \eta$ against the alternative hypothesis H_1: $\lambda \neq \mu$.

Under H_0, the likelihood function is equal to

$$l(x_1,...,x_m,y_1,...,y_n;\eta)$$

$$= \eta\exp[-\eta x_1]...\eta\exp[-\eta x_m]\eta\exp[-\eta y_1]...\eta\exp[-\eta y_n]$$

$$= \eta^{m+n}\exp[-\{\eta\sum_{i=1}^{m} x_i\} - \{\eta\sum_{j=1}^{n} y_j\}], \tag{4}$$

and under H_1:

$$l(x_1,...,x_m,y_1,...,y_n;\lambda,\mu)$$

$$= \lambda\exp[-\lambda x_1]...\lambda\exp[-\lambda x_m]\mu\exp[-\mu y_1]...\mu\exp[-\mu y_n]$$

$$= \lambda^m\mu^n\exp[-\{\lambda\sum_{i=1}^{m} x_i\} - \{\mu\sum_{j=1}^{n} y_j\}]. \tag{5}$$

Maximization of (4) and (5) leads to:

$$\hat{\eta} = \frac{m + n}{\sum_{i=1}^{m} x_i + \sum_{i=1}^{n} y_i}, \tag{6}$$

and

$$\hat{\lambda} = \frac{1}{\bar{x}}, \hat{\mu} = \frac{1}{\bar{y}}, \tag{7}$$

(see Box 1.2). $\hat{\eta}$ is substituted in (4), and $\hat{\lambda}$ and $\hat{\mu}$ are substituted in (5). Subsequent substitution of both functions in (3) results in the likelihood ratio test statistic:

$$\Lambda = 2\{(m+n)\log\frac{\sum_{i=1}^{m} x_i + \sum_{i=1}^{n} y_i}{m+n} - m\log\bar{x} - n\log\bar{y}\}. \tag{8}$$

Under H_0 is Λ chi-squared distributed. Under H_0 there is one parameter, η, whereas under H_1 there are two parameters, λ and μ. Therefore, the number of degrees of freedom is equal to one. From Table A2 it follows that the critical value at a level of significance of $\alpha = 0.05$ is 3.84.

Likelihood ratio tests can be derived for all cases in which the likelihoods under the alternative and null hypotheses can be calculated. However, the approximation of the distribution of the test statistic Λ with a chi-squared distribution only holds under certain conditions (see e.g. Cox and Hinkley, 1974). The generalization for the k-sample case is given in subsection 5.3.1.

Data: Two continuous time records of behaviour.

Procedure: Calculate the maximum likelihood estimators of the transition rates in the two records (see eqn (1.6)). Then calculate the log-likelihoods:

$$L(A;B)= \sum_{(i,j)\in D} \{N_{ij}(A)\log\hat{\alpha}_{ij}(B) - S_i(A)\hat{\alpha}_{ij}(B)\}, \tag{2.16}$$

where A and B denote the record numbers, D is the set of transitions that occur in both records, $N_{ij}(A)$ is the number of transitions from behavioural category i to j in record A, $S_i(A)$ is the total duration of act i in record A, and $\hat{\alpha}_{ij}(B)$ is the estimated transition rate from i to j in record B. $L(1;1)$ and $L(2;2)$ are the maximum log-likelihoods of records 1 and 2 respectively. $L(1;2)$ is the log-likelihood of record 1 given the estimated transition rates of record 2 and $L(2;1)$ is the log-likelihood of record 2 given the estimates of record 1. It may happen that some transitions occur in only one of the records, i.e. one or more of the estimators $\hat{\alpha}_{ij}(A)$ is/are zero, whereas the corresponding $N_{ij}(B)$ is/are larger than zero. When that happens, the first term in eqn (2.16) becomes infinite. In calculating $L(A;B)$ the summation is therefore only taken over the set of transitions, D, that occur in both records (see Box 2.11). Subsequently, calculate the differences:

$$\Lambda(A;B) = 2\{L(A;A) - L(A;B)\}. \tag{2.17}$$

$\Lambda(A;B)$ is the likelihood ratio statistic for testing the null hypothesis $\alpha_{ij} = \hat{\alpha}_{ij}(B)$, with the data of record A. If both $\Lambda(1;2)$ and $\Lambda(2;1)$ are less than $\chi^2_m(\alpha)$ (see Table A2), where m is the number of different transitions between behavioural categories that occur in both records (i.e. the number of elements of D), and α is a chosen level of significance, it is decided that the records are similar. The method is illustrated in Box 2.11.

Box 2.11 Calculation of an overall measure of similarity of two records, based on the likelihood ratio principle

As an example, we will use the two records of rat behaviour, parts of which are shown in Table 1a, Box 2.5. To illustrate the calculation of the log-likelihood (eqn (2.16)), we will calculate $L(1;2)$. The total durations of acts in record 1, $S_i(1)$, are shown in Table 1, transition frequencies in record 1, $N_{ij}(1)$, in Table 2, and the estimated transition rates of record 2, $\hat{\alpha}_{ij}(2)$, in Table 3.

Table 1 Total durations of acts in record 1 ($S_i(1)$)

Act	Total duration (s)
Turn	20.76
Shake	4.32
Care	248.36
Yawn	2.12
Squat	27.60
Walk	75.36
Sniff	176.40
Rear	5.08

Table 2 Transition frequencies in record 1 ($(N_{ij}(1))$)

To: From:	*Turn*	*Shake*	*Care*	*Yawn*	*Squat*	*Walk*	*Sniff*	*Rear*
Turn	–	1	4	0	4	1	9	0
Shake	0	–	0	0	0	3	0	0
Care	1	0	–	0	4	4	4	0
Yawn	0	0	0	–	0	1	0	0
Squat	2	1	4	0	–	1	4	0
Walk	3	0	4	0	2	–	22	1
Sniff	12	1	2	1	2	22	–	1
Rear	0	0	0	0	0	0	2	0

Table 3 Estimated transition rates (in s^{-1}) of record 2* ($\hat{\alpha}_{ij}(2)$)

To: From:	Turn	Shake	Care	Squat	Walk	Sniff	Rear
Turn	–	0	0.758	0	0	0	0
Shake	0	–	0	0	0.568	0.189	0
Care	0	0	–	0.0082	0.023	0.0328	0
Squat	0	0	0	–	0.067	0.268	0
Walk	0	0.0407	0.0407	0	–	0.508	0.0203
Sniff	0.0042	0.0083	0.0416	0.0125	0.0957	–	0.0042
Rear	0	0	0	0	0	0.403	–

* *Yawn* did not occur in record 2.

From Tables 2 and 3 it can be seen that several transitions that occur in record 1 do not occur in record 2 and vice versa. Transitions from *Turn* to *Shake*, *Squat*, *Walk* and *Sniff*, for instance, only occur in record 1. As a consequence, these transitions are not included in the set *D*. From eqn (2.16), it follows that

$$L(1;2) = \{(4\log0.758 - 20.76 \times 0.758)\} + \{(3\log0.568 - 4.32 \times 0.568)\}$$

$$+ \{(4\log0.0082 - 248.36 \times 0.0082) + (4\log0.023 - 248.36 \times 0.023)$$

$$+ (4\log0.0328 - 248.36 \times 0.0328)\} + \{(\log0.067 - 27.60 \times 0.067)$$

$$+ (4\log0.268 - 27.60 \times 0.067)\} + \{(4\log0.0407 - 75.36 \times 0.0407)$$

$$+ (22\log0.508 - 75.36 \times 0.508) + (\log0.0203 - 75.36 \times 0.0203)\}$$

$$+ \ ...$$

$$+ \{(2\log0.403 - 5.08 \times 0.403)\} = -352.48. \tag{1}$$

The other log-likelihoods are: $L(1;1) = -312.64$, $L(2;1) = -323.94$, $L(2;2) = -297.42$. Therefore, it follows from eqn (2.17) that the test statistics are

$$\Lambda(1;2) = 2\{(-312.64) - (-352.48)\} = 79.68,$$

$$\Lambda(2;1) = 2\{(-297.42) - (-323.94)\} = 53.04. \tag{2}$$

The number of different transitions that occur in record 1 as well as 2 is 17 (see Tables 2 and 3). The critical value of the chi-squared distribution with 17 degrees of freedom at $\alpha = 0.05$ is equal to 27.6 (Table A2). It thus follows that, according to this measure, the records are dissimilar.

Further remarks and literature: Further details on likelihood methods can be found in, e.g., Kalbfleisch (1979). In principle, it is possible that $\Lambda(1;2)$ is less than the critical value, whereas $\Lambda(2;1)$ is not, or vice versa. In practice, this will rarely happen when the numbers of bouts of different acts in the two records are approximately equal. If it does occur, however, it is best to conclude that the records differ too much. Since the two records cannot be considered as independent observations, the method is not a formal test. Instead, it must be considered as a measure of the degree of similarity. If either of the test statistics $\Lambda(A;B)$ is larger than the critical value, however, it is a strong indication that there are essential differences between the records, since they are based on the same behavioural process.

2.4.4 Methods for comparing records of several behavioural processes

In this subsection we will discuss methods for the situation where J observers have made records of I behavioural processes, where both I and J are larger than or equal to two. In this case, the effect of inter-rater variability on (summary) statistics can be analysed directly.

The effects on mean durations of acts and/or transition rates can be studied through ANOVA methods (see also Chapter 7). Landis and Koch (1975a) give several ANOVA models that can be used in this situation, of which we only mention two. Further details can be found in their paper (see also Scheffé, 1959; Sokal and Rohlf, 1981).

The data are statistics y_{ij} ($i = 1,..., I$, $j = 1,..., J$), for instance mean durations of an act in the different records, or (log-transformed) transition rates between a certain pair of acts. The models are two-way layout models with a random effect of the subjects, i.e. the behavioural processes. Thus, the subjects are assumed to be a sample from a larger population. This is sensible, since we want to infer the effect of inter-rater variability on records of behavioural processes in general, rather than the specific processes that have been observed so far. The models differ with respect to the rater effect, which is either random or fixed. When the raters are considered as a sample from a large population of observers, a model with random observer effect is appropriate:

$$Y_{ij} = \mu + S_i + D_j + \varepsilon_{ij}, \tag{2.18}$$

where the D_j, the observer effects, are assumed to be $N(0,\sigma_d^2)$ distributed, and the S_i, the effects of the behavioural processes, are $N(0,\sigma_s^2)$ distributed.

To study the differences between the specific raters, rather than the variability in any set of raters, a fixed observer effect should be used:

$$Y_{ij} = \mu + S_i + \delta_j + \varepsilon_{ij}, \tag{2.19}$$

where

$$\sum_{j=1}^{d} \delta_j = 0. \tag{2.20}$$

In eqn (2.19), δ_j are the (fixed) observer effects. As before, it is assumed that the S_i, the effects of the behavioural processes, are $N(0, \sigma_s^2)$ distributed.

In addition to the differences in mean durations and/or transition rates, it may be desired to examine the differences in the classification of acts. To do this, calculate the relative frequencies of each act in the different records. Thus, per record, we have a set of frequencies $f_1, \ldots, f_K$, where K is the number of different acts. Note that the sum of the frequencies is equal to one for each record. Rater and subject effects on such frequencies can be studied by means of generalized linear models (McCullagh and Nelder, 1983, Aitkin *et al.*, 1989).

Landis and Koch (1975a) give references for examining inter-rater reliability with multivariate ANOVA on summary statistics. (Multivariate ANOVA is described by e.g. Morrison, 1967, Mardia *et al.*, 1979, Srivastava and Khatri, 1979.) With such methods the inter-rater variability on a set of means of several acts, for instance, can be studied simultaneously. They should be used when it is expected that there will be relatively high correlations between the summary statistics per record. When the processes can be modelled as (semi-)Markov chains, the correlations between the transition rates can be calculated as described in Chapter 7.

3   ANALYSIS OF TIME INHOMOGENEITY

3.1   INTRODUCTORY REMARKS

Frequently, behaviour shows systematic and profound changes during the observation period. Such changes may consist of the appearance and/or disappearance of certain acts (e.g. meals or sleep) or characteristic sequences of acts (e.g. courtship behaviour). Changes may also concern the mean bout length. The day-night cycle of mosquitoes, for example, is accompanied by a change in the mean bout length of flights (Peterson 1980). However, behaviour may also change in a less obvious way, under externally constant circumstances. In that case, unless bouts have a deterministically determined length (like in pacing), changes in the time structure are hard to detect. It is very important to take time inhomogeneities into account since otherwise they may mask treatment effects and make the results of an analysis ambiguous. Furthermore, from an ethological point of view, the occurrence of inhomogeneities can be of importance in itself since it may indicate that there are motivational changes. From a mathematical point of view, to know whether or not the assumption of homogeneity is justified is of great practical importance as homogeneity reduces the complexity of the modelling and the data analysis considerably.

Two types of methods for analysis can be distinguished: visual scanning and formal methods. By applying visual scanning methods a first impression of the main characteristics of a behavioural record can be obtained, i.e. whether or not there is gross periodicity and whether changes are abrupt or gradual. A description of these methods is given in subsections 3.2.1-3.2.4. Formal methods are based on explicitly formulated mathematical models of a behavioural process. These methods are in general more refined and especially appropriate when specific aspects of the data are tested.

For practical reasons formal methods for abrupt and gradual changes are treated separately. Parametric as well as non-parametric estimators and tests for abrupt change points are available and will be treated in section 3.3. Changes can occur in the mean bout length of one behavioural act or simultaneously in more acts (see subsection 3.3.1). Sudden changes can also affect sequences of acts. This will be discussed in subsection 3.3.2. Finally, these types of changes can influence the mean duration of one or more acts as well as the transition probabilities between acts. An example of the general approach to this problem is presented in subsection 3.3.3.

For gradual changes only *ad hoc* methods exist for a few specific situations, and these will be treated in section 3.4.

3.2   VISUAL SCANNING METHODS

Visual scanning is an important step in the analysis of the records which ensures that there are no apparent deviations between the assumptions necessary to apply more refined formal methods (such as tests on the presence of abrupt change points) and the main properties of the data. We recommend the application of all the visual scanning methods treated in the following subsections because each method highlights important, but different, characteristics of the behavioural record.

3.2.1 Bar plots

A simple way of representing the record both completely and concisely is to make plots in which bars represent the occurrence and the duration of acts (see Fig. 3.1 and also Box 1.1). These plots give an impression of whether behaviour is rhythmic or irregularly spread over the observation time, as well as whether there are large variations in mean bout length or in gap length. Figure 3.1 shows a bar plot of mother-infant interaction in

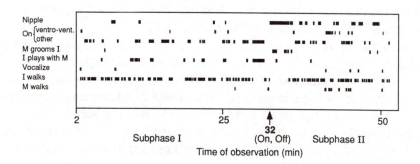

Figure 3.1 Bar plot of an inhomogeneous record of mother-infant interaction of rhesus monkeys (from Haccou *et al.*, 1983, with permission of Academic Press Inc.); 'M' is mother, 'I' is infant. *On ventro-ventral* and *On other* comprise two different forms on *On mother*, where the infant is in close contact with the mother without taking the nipple. During *Nipple* the infant sits on its mother's lap while holding the nipple in its mouth. After about 32 min (see arrow) the infant gets drowsy and *Nipple* as well as *On mother* bouts increase.

rhesus monkeys (taken from Haccou *et al.*, 1983). There are major shift in durations of *Nipple* and *On mother* bouts after about 32 min. Furthermore, the frequency of grooming by the mother increases at this point. Bar plots also reveal whether changes in different acts occur more or less simultaneously, as in Fig. 3.1. Bar plots, however, seldom permit precise decisions about the time structure of the behaviour.

3.2.2 Cumulative bout length plots

In a cumulative bout length plot the total sum of the bout durations of an act or a group of acts is plotted against the sequence number. These graphs are especially useful to detect changes in mean bout length, which are indicated by a consistent change in the slope of the lines. It is easy to obtain an impression about the number, the magnitude and the degree of abruptness of changes. It is recommended to use sequence number, instead of observation time, as the independent variable, thus avoiding the dependency between plots of different acts, since the total duration of the acts occupies 100% of the observation time. Figure 3.2 shows the cumulative duration of *Off mother* bouts in an observation of mother-infant interaction of rhesus monkeys (from Haccou *et al.* 1983), with one apparent, fairly abrupt decrease in the mean bout length after about 30 bouts.

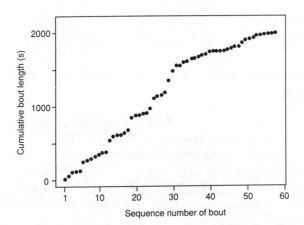

Figure 3.2 Cumulative bout length plot of *Off mother* in an inhomogeneous record of mother-infant interaction of rhesus monkeys (from Haccou *et al.* 1983, with permission of Academic Press Inc.). After about 30 bouts the slope decreases, which indicates a decrease in mean bout length.

3.2.3 Log-bout-length plots

Another method of studying the properties of the time structure of separate acts is to plot the logarithm of the bout lengths against the start times of the bouts for each act (see Fig. 3.3). This provides a visual check on exponentiality, as the vertical width of scattering should be approximately constant, irrespective of the value of the termination rate (see Box 3.1). A sudden change in termination rate appears in these plots as an abrupt vertical shift with an unaltered width of the band of points (Dienske *et al.*, 1980). When the distribution of the bout lengths deviates from exponentiality before or after a change in termination rate, the vertical shift is accompanied by a decrease or an increase in the width of the scattering.

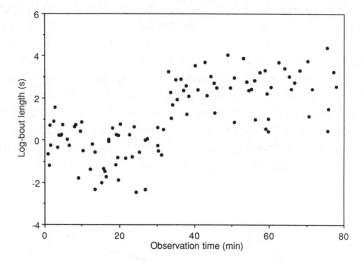

Figure 3.3 Example of a log-bout-length plot. A vertical shift indicates a change in mean bout length.

Box 3.1   Mean and variance of a log-transformed exponentially distributed variable

Let X be an exponentially distributed variable with parameter λ and let Y be log X. Since λ is a scale-parameter, X/λ has an exponential distribution with parameter 1. Furthermore:

$$Y = \log X = -\log\lambda + \log X/\lambda, \tag{1}$$

and thus Y is equal to a constant plus the logarithm of an exponentially distributed variable with parameter 1. From this result it can be derived that the expectation of Y is a linear function of λ and its variance is equal to the variance of a log-transformed exponential(1) distributed variable, which does not depend on λ, and thus is constant. It can be shown that the exact values of the expectation and variance of Y are equal to:

$$E \log Y = -\log\lambda - \psi(1) \approx -\log\lambda - 0.577, \qquad (2)$$

and

$$Var\ Y = \psi'(1) \approx 1.645, \qquad (3)$$

where ψ and ψ' are respectively the digamma and trigamma function, which are tabulated by Abramowitz and Stegun (1972).

3.2.4 Inspection of sequences of acts

Although a bar plot gives a complete representation of the observed behaviour and a global overview of the record, it is not easy to determine from such a plot whether there are global changes in the sequence of acts. Such information includes whether the average frequency per time unit of act A is constant or not, whether the occurrence of act A depends on what happened before and, if so, to what extent, or whether the occurrence of act A is influenced by one or more covariables, e.g. the time of day or the presence or absence of another individual. There are several ways in which a sequence of acts, e.g.

Shake, Walk, Sit, Walk, Sit, Walk, Care, Sit, Walk, ... (3.1)

(see Table 1, Box 1.1), can be represented graphically in order to answer these questions.

Occurrence of one act

If the frequency of an act, say act A, varies in the course of an experiment, the number of acts between two succeeding acts A varies too. A diagram like Fig. 3.4, a bar chart of such 'gaps' between two consecutive acts A, reveals whether these gaps (not to be confused with gap lengths, the durations of the time interval between two acts A) tend to increase, decrease or fluctuate in another way. In the example of Fig. 3.4 the frequency of act A obviously decreases.

Another possible graphical representation is to divide the total observation time into a small number of periods of equal duration and to plot the frequencies of A in these periods in a bar chart as in Fig. 3.5. The number of periods to be chosen must be large enough to detect all fluctuations on the relevant time scale and it must be small enough to average

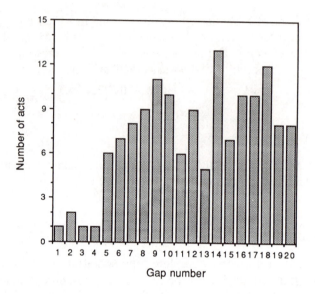

Figure 3.4 Bar chart of the number of acts between two succeeding occurrences of act A. The gaps are numbered consecutively.

out the stochastic nature of the data. As a rule of thumb, it is advisable to choose the number of periods such that the average frequency of occurrences of A is at least 15.

It must be noted that plots such as Fig. 3.5 may give some idea of the presence of fluctuations in the occurrence of an act, but they are of limited value since the observed frequencies may be negatively correlated with the bout lengths due to chance fluctuations: the frequency of A is smaller if the bouts are large (on average) and vice versa.

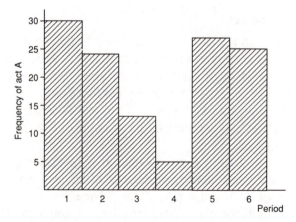

Figure 3.5 Frequency of occurrences of act A in six time intervals of equal length.

The division of the total observation time can sometimes be based on the value of a covariable, for instance the presence or absence of another individual, the availability of food, etc. In that case one must plot the frequency per time unit instead of the absolute number (see Fig. 3.6), since the lengths of the intervals are not necessarily equal. This is less arbitrary than the division of the total observation time into a small number of periods of equal length (which may restrict the possibilities for interpretation of the plot).

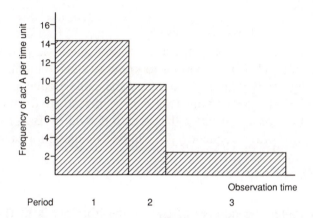

Figure 3.6 Frequency of occurrences of act A per time unit in three periods obtained by dividing the observation time according to a covariable.

This type of bar chart is subject to the same limitations as the one mentioned above in the sense that the durations of acts are not taken into account. This may lead to apparent fluctuations or may obscure them.

We can also make a bar chart of the relative frequency of occurrences of act A from the total number of acts, where the periods can be chosen such that the totals per period are approximately equal or where the periods are related to a covariable. See Fig. 3.7 for an example of this.

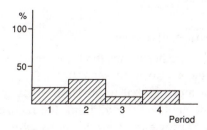

Figure 3.7 Relative frequency of occurrences of act A in four periods.

We would like to emphasize that bar charts like those in Figs 3.4-3.7 are tools to detect the main characteristics of the observed process. For instance, Fig. 3.4 suggests that the frequency of act A is high at the beginning of the experiment and decreases in the course of time. Such plots can be used to judge whether the behavioural process is approximately stationary or not, whether changes are abrupt or gradual, etc. So, graphs like these can be useful for a global inspection, but are not appropriate for a refined analysis.

Transitions between acts

For a sophisticated analysis, based on formal methods derived from a mathematical model, it is important to know if a behavioural act, say A, is followed by act B with approximately the same frequency during the entire observation time or not. This can be investigated graphically by plotting the fraction f_{AB} of cases that A is followed by B in two or more periods. The transition frequency, f_{AB}, can be calculated for each period by

$$f_{AB} = \frac{n_{AB}}{n_A} ,$$

(3.2)

where n_{AB} denotes the number of transitions from act A to B and n_A the total number of occurrences of A in that period (Fig. 3.8).

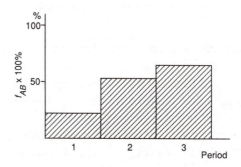

Figure 3.8 Transition frequencies from act A to B in three consecutive periods.

Inspection of transition matrices

It is possible to study whether or not the transition frequencies vary during the experiment for each transition separately. In general, however, it is desirable to obtain a more global impression of the presence of inhomogeneities. This can be achieved by dividing the record and by calculating the transition matrix, *T*, for each period. Let us assume that the ethogram consists of four acts, denoted by A, B, C and D. The matrix is defined by

$$T = \begin{vmatrix} - & f_{AB} & f_{AC} & f_{AD} \\ f_{BA} & - & f_{BC} & f_{BD} \\ f_{CA} & f_{CB} & - & f_{CD} \\ f_{DA} & f_{DB} & f_{DC} & - \end{vmatrix} \qquad (3.3)$$

The transition frequencies are defined according to eqn (3.2). Note that the row sums are exactly equal to 1 and that the transition frequency of an act to 'itself' is zero, since an act cannot be succeeded by itself.

The comparison of transition matrices of different periods is not as straightforward as it may seem; because of the dependence within the rows (the row sums are equal to one) a large frequency of one act implies that the others are small. See Table 3.1 for an example. It might be concluded, for instance, that f_{DA} is considerably smaller in the second matrix, whereas f_{DB} and f_{DC} are larger. However, this is at least partially due to the fact that $f_{DA} + f_{DB} + f_{DC} = 1$. A formal method for detecting changes in matrices while taking such dependencies into account is given in subsection 3.3.2.

Table 3.1 Comparison of two transition matrices

Period 1:			
–	0.21	0.43	0.37
0.02	–	0.57	0.41
0.39	0.26	–	0.35
0.86	0.01	0.13	–
Period 2			
–	0.49	0.35	0.16
0.04	–	0.63	0.33
0.37	0.45	–	0.18
0.52	0.19	0.29	–

3.3 FORMAL METHODS FOR ABRUPT CHANGES

3.3.1 Abrupt changes in mean bout lengths

In this subsection we describe tests for abrupt changes in the bout lengths of an act (see Fig. 3.9). Presumably, behaviour usually does not change abruptly from one instant to another. However, if the transition period from one behavioural phase to another is relatively short compared to the bout lengths of the considered act(s), the assumption of an abrupt change is a

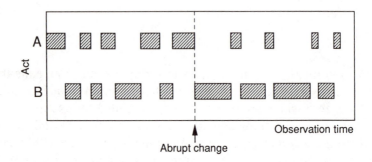

Figure 3.9 Example of a bar plot with an abrupt change (indicated by the arrow) in the mean bout lengths.

good first approximation. This assumption can be checked by means of cumulative plots such as given in Fig. 3.2. In our experience, the assumption appears to be accurate enough in most cases. For instance, we used it for modelling alternations between drowsy and fully awake phases of infant rhesus monkeys (Haccou *et al.* 1983, 1988*c*) and changes in agonistic interactions of male rats (Haccou *et al.*, 1988*a*). We denote the set of bout lengths of the act under consideration (in order of occurrence) by $x_1, ..., x_N$.

A non-parametric (multiple) change point test

Several distribution-free change point tests are described in the statistical literature. However, most of these are tests for cases where it is known in advance that there is at most one change point, e.g. the test described in Pettitt (1979) which is based on Wilcoxon's two-sample test. In ethological research this is usually uncertain. The non-parametric multiple change point test described in this subsection is based on the Kruskal-Wallis test and has been especially developed for this more general case.

Data: One sample of bout lengths $x_1, ..., x_N$.

Assumptions:
A1: $X_1, ..., X_N$ are mutually independent.
A2: $E X_i = \mu_i + \varepsilon_i$, $i = 1, ..., N$, where the ε_i come from the same continuous distribution.

A3: $\mu_1 = ... = \mu_{n_1} = \theta_1,\ \mu_{n_1 + 1} = ... = \mu_{n_1 + n_2} = \theta_2,$

$\mu_{n_1 + n_2 + 1} = ... = \mu_{n_1 + n_2 + n_3} = \theta_3 ,\ \mu_{n_1 + n_2 + n_3 + 1} = ... = \theta_4,$

where n_1, n_2 and n_3 are unknown.

A4:   Under the null hypothesis, H_0, it is assumed that the means are equal: $\theta_1 = \theta_2 = \theta_3 = \theta_4$; and under the alternative hypothesis, H_1, there is at least one inequality.

Procedure: The procedure for testing H_0 against H_1 consists of a sequence of tests which are performed successively. The procedure begins with a test of three against two change points, followed by a test of two against one change points and a test of one against zero change points. The procedure is stopped as soon as a test result is significant, i.e. when the test statistic is larger than the critical value corresponding to the level of significance. If the hypothesis of three against two change points is rejected, it is decided that the number of change points is at least three. If the hypothesis of k ($k = 1$ or 2) against ($k - 1$) change points is rejected, it is decided that the number of change points is equal to k.

Table 3.2 Arrangement of the data and assignment of ranks

Group	1	2	3	4
Data	$x_1,...,x_{n_1}$	$x_{n_1+1},...,x_{n_1+n_2}$	$x_{n_1+n_2+1},...,x_{n_1+n_2+n_3}$	$x_{n_1+n_2+n_3+1},...,x_N$
Ranks	$r_1,...,r_{n_1}$	$r_{n_1+1},...,r_{n_1+n_2}$	$r_{n_1+n_2+1},...,r_{n_1+n_2+n_3}$	$r_{n_1+n_2+n_3+1},...,r_N$
Rank sums	$R_1 = \sum\limits_{i=1}^{n_1} r_i$	$R_2 = \sum\limits_{i=n_1+1}^{n_1+n_2} r_i$	$R_3 = \sum\limits_{i=n_1+n_2+1}^{n_1+n_2+n_3} r_i$	$R_4 = \sum\limits_{i=n_1+n_2+n_3+1}^{N} r_i$

First, rank all N observations in increasing order and let $r_1,..., r_N$ denote the ranks of $x_1,..., x_N$. In the case of tied (= equal) observations use average ranks. If there are three change points, the data set is divided into four groups. Let n_1, n_2, n_3 and n_4 denote the group sizes (see Table 3.2).

Next, compute the sum R_j of the ranks of group j, $j = 1,..., 4$. The Kruskal-Wallis test statistic, K_m, for testing the equality of m mean rank sums is defined as

$$K_m = \left(\frac{12}{N(N+1)} \sum_{j=1}^{m} \frac{R_j^2}{n_j} \right) - 3(N+1) \tag{3.4}$$

in the absence of ties, and otherwise as

$$K_m' = \frac{K_m}{1 - \dfrac{\sum\limits_{k=1}^{g} (t_k^3 - t_k)}{N^3 - N}} , \qquad (3.5)$$

where g is the number of tied groups and t_k is the size of the kth tied group (see e.g. Siegel and Castellan, 1988).

Following this, compute K_4 (or K_4') for all possible choices of $n_1,..., n_4$ and determine the maximum L_3 of K_4 (or K_4') for all possible places of three change points, i.e.

$$L_3 = \max_{\{n_j \geq 1, \, j=1,....,4; \Sigma n_j = N\}} K_4. \qquad (3.6)$$

Then repeat this procedure under the assumption of *two* change points as well as of only *one* change point. The corresponding maximum Kruskal-Wallis statistics are denoted by L_2 and L_1, respectively. The three test statistics for testing three against two, two against one and one against zero change points are defined by $\Lambda_3 = L_3 - L_2$, $\Lambda_2 = L_2 - L_1$ and $\Lambda_1 = L_1$. The critical values are listed in Tables A12, A11 and A10, respectively. It is advisable to choose $\alpha_3 = 0.025$, $\alpha_2 = 0.05$ and $\alpha_1 = 0.10$, where α_i denotes the level of significance corresponding to Λ_i. This choice of an increasing level of significance in the course of the procedure is recommended in order to equalize the risks of wrong decisions for the several types of alternative hypotheses (see Meelis *et al.*, 1991). See Box 3.2 for an example of this procedure.

Box 3.2   The non-parametric (multiple) change point test

1. The data and their ranks are listed in Table 1. First we assume that there are three change points. To calculate L_3, the Kruskal-Wallis test statistic is maximized over all possible locations of the three change points. To illustrate, we calculate the test statistic for the case when there are three changes after bout numbers 5, 12 and 16. Hence there are four groups of size 5, 7, 4, 4 respectively, as indicated in Table 1. The computation of the corresponding Kruskal-Wallis statistic is shown in Table 2.
   Accordingly,

$$K_4 = \frac{12}{20 \times 21} \times 2329.74 - 3 \times 21 = 66.56 - 63 = 3.56.$$

Table 1 Data and ranks

Bout lengths	Ranks
0.0471	2
0.1737	5
0.5769	11
0.9232	15
0.4170	8
0.0163	1
0.1945	6
0.4614	9
0.5109	10
1.7984	16
0.1146	4
2.3442	20
2.2352	18
0.2941	7
0.0771	3
0.8456	14
0.6751	13
2.2373	19
0.5936	12
1.8195	17

Table 2 Calculation of Kruskal-Wallis test statistic

	Group	n_j	R_j	R_j^2	R_j^2/n_j
	1	5	41	1681	336.20
j	2	7	66	4356	622.29
	3	4	42	1764	441.00
	4	4	61	3721	930.25
Total		20	210		2329.74

When K_4 is calculated for all

$$(1/6)(N - 1)(N - 2)(N - 3) = (1/6) \times 19 \times 18 \times 17 = 969$$

possible group sizes it is found that L_3, the maximum K_4-value, is attained for group sizes 11, 2, 2 and 5, see Table 3 for the computations. Hence, the maximum of the K_4-values is equal to

$$L_3 = \frac{12}{20 \times 21} \times 2585.09 - 3 \times 21 = 10.86.$$

Table 3 Calculation of K_4 for group sizes 11, 2, 2, 5

	Group	n_j	R_j	R_j^2	R_j^2/n_j
	1	11	87	7569	688.09
j	2	2	38	1444	722.00
	3	2	10	100	50.00
	4	5	75	5625	1125.00
Total		20	210		2585.09

2. The same computations for the case of two or one change points are summarized in Tables 4 and 5. The maximum of the K_3-values is attained when changes occur after bouts 11 and 13, and is equal to

$$L_2 = \frac{12}{20 \times 21} \times 2442.23 - 3 \times 21 = 6.78.$$

Finally, the maximum of the K_2-statistics is

$$L_1 = \frac{12}{20 \times 21} \times 2369.09 - 3 \times 21 = 4.69.$$

Table 4 Calculation of K_3 for two change points

	Group	n_j	R_j	R_j^2	R_j^2/n_j
	1	11	87	7569	688.09
j	2	2	38	1444	722.00
	3	7	85	7225	1032.14
Total		20	210		2442.23

Table 5 Calculation of K_2 for one change point

	Group	n_j	R_j	R_j^2	R_j^2/n_j
j	1	11	87	7569	688.09
	2	9	123	15129	1681.00
Total		20	210		2369.09

3. The three test statistics are equal to:
$\Lambda_1 = L_1 = 4.69$, $\Lambda_2 = L_2 - L_1 = 2.09$ and $\Lambda_3 = L_3 - L_2 = 4.08$.
The critical value for the test of three against two change points at $\alpha_3 = 0.025$ is 7.86 for $N = 20$ (Table A12), for the test of two against one change points at $\alpha_2 = 0.05$ it is 5.14 (Table A11) and for the test of one against zero change points at $\alpha_1 = 0.10$ it is 5.76 (Table A10). Hence, if we apply the three tests, starting with Λ_3, none of the tests is significant. The conclusion is that there are no significant changes.

4. Note that the estimated change point under the assumption of one change point coincides with the estimates under the assumption of more than one change point, and so on. In general, this need not be the case.

Large-sample approximations: Owing to the extremely low rate of convergence of the distributions of the three test statistics as N goes to infinity there are no useful large-sample approximations available.

Properties: A parametric test is preferable if the corresponding assumptions concerning the distribution of the bout lengths of an act are sufficiently justified. For instance, if the behaviour within homogeneous periods can be described by a CTMC, bout lengths are exponentially distributed. It appears that in these cases the parametric test, treated below, possesses more power than its non-parametric counterpart. See Meelis *et al.* (1991) for details. The non-parametric test is to be preferred if the assumption of a constant termination rate of the act under consideration is inadequate or not satisfactorily supported.

Literature: The non-parametric multiple change point test was first published in an ethological context by Meelis *et al.* (1990). More mathematical details and an extensive discussion on the properties of this procedure are given in Meelis *et al.* (1991).

Likelihood ratio change point test for exponentially distributed bout lengths

If the variables are exponentially distributed the procedure is analogous to the non-parametric case. The main difference is that the procedure starts with a test of $(N - 1)$ against two change points instead of three against two change points. The tests are based on the maximum of the log-likelihood functions under the several hypotheses.

Data: One sample of bout lengths $x_1, ..., x_N$.

Assumptions:
A1: $X_1, ..., X_N$ are mutually independent and exponentially distributed.
A2:   Under H_0 all means are equal, i.e. $E X_1 = ... = E X_N$. Under H_1 there is at least one inequality.

Procedure: The procedure consists of a sequence of at most three tests performed in succession. The procedure starts with a test of $N - 1$ against two change points, followed by a test of two against one change points and a test of one against zero change points. The procedure is stopped as soon as a test result is significant. If the hypothesis of $N - 1$ against two change points is rejected then it is decided that the number of change points is at least three. If the hypothesis of k ($k = 1$ or 2) against $(k - 1)$ change points is rejected then it is decided that this number is equal to k.

The maximum of the log-likelihood functions under the several hypotheses are denoted by L_{N-1}^*, L_2^*, L_1^* and L_0^*. First compute L_{N-1}^*, the maximum of the log-likelihood function if there are $N - 1$ change points, i.e. each group consists of precisely one observation. It is straightforward to derive that

$$L_{N-1}^* = -\left(N + \sum_{i=1}^{N} \log x_i \right). \tag{3.7}$$

Next, compute L_2^* defined by

$$L_2^* = \max_{\{n_1 \geq 1, n_2 \geq 1, n_3 \geq 1 \; ; \; n_1 + n_2 + n_3 = N\}} l_2, \tag{3.8}$$

where l_2 is the log-likelihood when there are two change points and hence three groups of size n_1, n_2 and n_3 respectively,

$$l_2 = \sum_{i=1}^{n_1} \{(\log \hat{\lambda}_1) - x_i \hat{\lambda}_1\} + \sum_{i=n_1+1}^{n_1+n_2} \{(\log \hat{\lambda}_2) - x_i \hat{\lambda}_2\}$$

$$+ \sum_{i=n_1+n_2+1}^{N} \{(\log\hat{\lambda}_3) - x_i\hat{\lambda}_3\}, \tag{3.9}$$

and where the parameters λ_1, λ_2 and λ_3 are estimated by

$$\hat{\lambda}_1 = \frac{n_1}{\sum_{i=1}^{n_1} x_i}, \qquad \hat{\lambda}_2 = \frac{n_2}{\sum_{i=n_1+1}^{n_1+n_2} x_i} \quad \text{and}$$

$$\hat{\lambda}_3 = \frac{n_3}{\sum_{i=n_1+n_2+1}^{N} x_i}. \tag{3.10}$$

L_1^* is defined in an analogous way, assuming one change point and thus two groups. L_0^* is defined more simply:

$$L_0^* = \sum_{i=1}^{N} (\log\hat{\lambda} - x_i\hat{\lambda}), \tag{3.11}$$

where

$$\hat{\lambda} = \frac{N}{\sum_{i=1}^{n} x_i}. \tag{3.12}$$

Then calculate the three statistics Λ_3^*, Λ_2^* and Λ_1^* defined by

$$\Lambda_3^* = 2(L_{N-1}^* - L_2^*)$$

$$\Lambda_2^* = 2(L_2^* - L_1^*) \tag{3.13}$$

$$\Lambda_1^* = 2(L_1^* - L_0^*).$$

Critical values (adapted from Haccou and Meelis, 1988) are listed in Tables A13, A14 and A15. As in the non-parametric case, it is advisable to choose $\alpha_3 = 0.025$, $\alpha_2 = 0.05$ and $\alpha_1 = 0.10$, in order to equalize the risks of wrong decisions for the several alternative hypotheses. A hypothesis should be rejected if the corresponding test statistic exceeds the critical value. See Box 3.3 for an example.

Box 3.3   (Multiple) change point test for the exponential case

1. The data are listed in Table 1 of Box 3.2. Substitution of the data into eqn (3.7) results in: $L_{N-1}^* = -3.10$.

2. The maximum of l_2 is determined as follows. Calculate l_2 as defined by eqn (3.9) for all possible group sizes: $(n_1, n_2, n_3) = (1, 1, 18), (1, 2, 17), ...,$ $(2, 1, 17), (2, 2, 16), ..., (17, 1, 2), (17, 2, 1), (18, 1, 1)$. It turns out that the maximum is attained for $n_1 = 1$, $n_2 = 8$ and $n_3 = 11$, and
$L_2^* = \max l_2 = 2.06 - 0.85 - 12.87 = -11.66$.
The estimated $\hat{\lambda}$ for the three groups are equal to, respectively,

$\hat{\lambda}_1 = 21.24$, $\hat{\lambda}_2 = 2.44$ and $\hat{\lambda}_3 = 0.84$.

3. The maximum of l_1 is determined in the same way: calculate l_1 for all possible group sizes n_1 and n_2, where $n_1 + n_2 = 20$. It is attained for $n_1 = 9$ and $n_2 = 11$, and $L_1^* = \max l_1 = -0.03 - 12.86 = -12.89$.
In this case the estimated $\hat{\lambda}$s are

$\hat{\lambda}_1 = 2.71$ and $\hat{\lambda}_2 = 0.84$.

and, finally, it follows by substitution of the data into eqn (3.11) that
$L_0^* = -15.98$, with $\hat{\lambda} = 1.22$.

4. The three test statistics are:
$\Lambda_1^* = 2(L_1^* - L_0^*) = 6.16$
$\Lambda_2^* = 2(L_2^* - L_1^*) = 2.46$
$\Lambda_3^* = 2(L_{N-1}^* - L_2^*) = 17.12$.
The three critical values at $\alpha_3 = 0.025$ for Λ_3^*, at $\alpha_2 = 0.05$ for Λ_2^* and at $\alpha_1 = 0.10$ for Λ_1^* are 25.21, 8.61 and 6.99 respectively. Hence, we conclude that there are no significant change points.

Note that, just as in the non-parametric case, the estimated change points need not coincide.

Large-sample approximations: Haccou *et al.* (1988*b*) proved that the distribution of Λ_1^* tends to the so-called extreme value distribution as N goes to infinity. Analogous results for the other two statistics are not yet available. However, owing to the very low rate of convergence of the distribution of Λ_1^*, the large-sample approximation is not useful (Haccou and Meelis, 1988).

Properties: This parametric procedure is preferable to the non-parametric counterpart if the assumption of exponentiality is sufficiently justified (Meelis *et al.* 1991).

Literature: The test for one change point is treated in Haccou *et al.* (1983). The multiple change point test is described in Haccou and Meelis (1988). Meelis *et al.* (1991) give a review of parametric and non parametric multiple change point tests.

Comments: If the bouts appear to have a common, but unknown, minimum duration or time lag, i.e. if the bout lengths follow the two-parameter exponential distribution (see eqn (1.9)), the same procedure can be applied as an (approximate) likelihood ratio method. The estimated minimum bout length should be subtracted from each bout length, i.e. the procedure is applied to

$$x_i' = x_i - x_{\min}, \text{ where } x_{\min} = \min_i x_i \text{ , } i = 1,...,n. \tag{3.14}$$

Note that the transformed bout length(s) x_i' of length zero should be omitted.

Likelihood ratio change point test for (log-)normally distributed bout lengths

If the (log-transformed) durations of an act have an (approximately) normal distribution, a test for an abrupt shift in the mean bout length which takes this into account is preferable. We assume that the variances are equal and unknown, as will usually be the case in practice. This model applies, for instance, if a bout length is constant plus an error term caused by relatively small internal or external fluctuations or by inaccurate registration.

Data: One sample of (log-transformed) bout lengths $x_1,..., x_n$.

Assumptions:
A1: $X_1,..., X_n$ are mutually independent and normally distributed.
A2: $E X_i = \mu_i$, $Var X_i = \sigma^2$, $i = 1,..., n$, σ^2 unknown.
A3: $\mu_1 = ... = \mu_k = \theta_1$ and $\mu_{k+1} = ... = \mu_n = \theta_2$, where θ_1, θ_2 and k are unknown.
A4:   Under H_0 there is no change point, i.e. $\theta_1 = \theta_2$, and under H_1 there is one change point.

Procedure: The mean of the first k observations is denoted by $\bar{x}_k$ and that of the last $n - k$ observations by $\bar{x}_k'$. The within-groups sum of squares is denoted by

$$S_k^2 = \sum_{i=1}^{k} (x_i - \bar{x}_k)^2 + \sum_{i=k+1}^{n} (x_i - \bar{x}_k')^2 \tag{3.15}$$

and Student's two-sample statistic for the two-sided test is

$$W_k = \frac{|\bar{x}_k - \bar{x}_k'|}{S_k} \sqrt{\frac{k(n-k)}{n}} \sqrt{n-2}. \qquad (3.16)$$

The likelihood ratio statistic W is defined as the maximum of W_k over all possible values of k, i.e. $k = 1, 2, ..., n - 1$:

$$W = \max_k W_k. \qquad (3.17)$$

The critical values are listed in Table A16. H_0 is rejected if W exceeds the critical value.

Large-sample approximations: Gombay and Horvath (1990) proved that the limit distribution of W is the extreme value distribution. Owing to the low rate of convergence to this limit distribution it does not provide a useful large-sample approximation.

Properties: James *et al.* (1987) compare the power properties of a few change point tests for the normal case.

Literature: The test is described by Worsley (1979).

Combined estimate of an abrupt change point for two or more acts: the exponential case

Especially motivational changes will usually imply shifts in the characteristics of more than one act. When rhesus infants get drowsy, for instance, the mean bout lengths of *Nipple* and *On mother* usually increase, whereas the durations of *Off mother* bouts decrease (Haccou *et al.*, 1983). In this subsection we consider the problem of how to estimate a simultaneous change point in the mean bout lengths of two or more different acts (see Fig. 3.10 for an illustration). This might be done by estimating change points separately for each of the acts and subsequently taking the average. However, the variance of the change point estimator appears to be very large. A combined maximum likelihood estimator is more accurate. A significance test for simultaneous change points is not yet available.

Data: A sample of bout lengths $x_1,..., x_m$ from act A with the corresponding start times of the bouts $s_1,..., s_m$ and a sample of bout lengths $y_1,..., y_n$ from act B with the corresponding start times $t_1,..., t_n$.

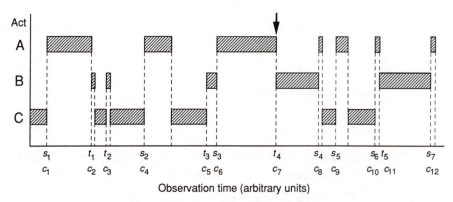

Figure 3.10 Illustration of the simultaneous occurrence of an abrupt change point (indicated by the arrow) in the mean bout lengths of acts A and B.

Assumptions:

A1: $X_1,..., X_m$ and $Y_1,..., Y_n$ are mutually independent and exponentially distributed.

A2: $E X_i = \lambda_1$ for $i = 1,..., \tau_1$ and $E X_i = \lambda_2$ for $i = \tau_1+1,..., m$, λ_1, λ_2 and τ_1 unknown.

A3: $E Y_i = \mu_1$ for $i = 1,..., \tau_2$ and $E Y_i = \mu_2$ for $i = \tau_2+1,..., m$, μ_1, μ_2 and τ_2 unknown.

Procedure: Consider the start times of bouts A and B pooled together. Denote the ordered set by $c_1,..., c_{m+n}$. Denote the number of bouts of act A that ended before c_j by k_j, and the corresponding number of act B by l_j. The maximum likelihood estimate of the time at which the change occurs is that value of c for which the maximum of the log-likelihood is attained:

$$\max_j \left\{ k_j \log \hat{\lambda}_1 - \hat{\lambda}_1 \sum_{i=1}^{k_j} x_i + (m-k_j)\log\hat{\lambda}_2 - \hat{\lambda}_2 \sum_{i=k_j+1}^{m} x_j \right.$$

$$\left. + l_j\log\hat{\mu}_1 - \hat{\mu}_1 \sum_{i=1}^{l_j} y_i + (n-l_j)\log\hat{\mu}_2 - \hat{\mu}_2 \sum_{i=l_j+1}^{n} y_i \right\} \qquad (3.18)$$

where the $\hat{\lambda}$ and $\hat{\mu}$ are defined by

$$\hat{\lambda}_1 = \frac{k_j}{\displaystyle\sum_{i=1}^{k_j} x_i}, \quad \hat{\lambda}_2 = \frac{m - k_j}{\displaystyle\sum_{i=k_j+1}^{m} x_i},$$

$$\hat{\mu}_1 = \frac{l_j}{\displaystyle\sum_{i=1}^{l_j} y_i} \; , \; \hat{\mu}_2 = \frac{m - l_j}{\displaystyle\sum_{i=l_j+1}^{n} y_i} \; . \tag{3.19}$$

See Box 3.4 for a worked example.

Box 3.4 Maximum likelihood estimation of a change point occurring simultaneously in the bout lengths of two acts

1. For computational purposes eqn (3.18) can be simplified by determining the maximum over j from $\{LX_j + LY_j\}$, where

$$LX_j = k_j \log k_j + (m - k_j)\log(m - k_j)$$

$$- k_j \log\left(\sum_{i=1}^{k_j} x_i\right) - (m - k_j)\log\left(\sum_{i=k_j+1}^{m} x_i\right) - m,$$

$$LY_j = l_j \log l_j + (n - l_j)\log(n - l_j)$$

$$- l_j \log\left(\sum_{i=1}^{l_j} y_i\right) - (n - l_j)\log\left(\sum_{i=l_j+1}^{n} y_i\right) - n.$$

2. The start times of the bouts of acts A and B and the bout lengths are listed in Table 1. The calculation of the LX_j and the LY_j are summarized in Tables 2 and 3. The sum of the two components is listed in Table 1. Obviously the maximum is attained after the second bout of act B. Hence, after $26.278 + 0.974 = 27.252$ min. there is possibly a change in the mean bout length of act A as well as act B.

Comment: A formal test procedure can be performed by applying a change point test for each act separately. The test results can subsequently be combined by one of the combination procedures described in section 6.2, if the test results are independent, or, if there is no sufficient evidence for this assumption, by the improved Bonferroni procedure for dependent test statistics given in subsection 6.3.2.

Table 1 Data and summary of the calculation

j	c_j	Bout lengths act A	act B	LX_j	LY_j	Sum
1	0.665	0.807	–	−10.40	–	–
2	4.189	0.178	–	− 9.92	–	–
3	7.239	1.645	–	−10.38	–	–
4	10.144	–	0.728	−10.38	−10.49	−20.87
5	10.991	2.968	–	−10.06	−10.49	−20.55
6	14.490	1.399	–	− 9.82	−10.49	−20.31
7	21.458	2.165	–	− 7.99	−10.49	−18.48
8	26.278	–	0.974	− 7.99	−10.11	−18.10
9	30.415	0.158	–	− 9.06	−10.11	−19.17
10	32.884	0.063	–	−10.01	−10.11	−20.12
11	34.927	–	1.664	−10.01	−10.09	−20.10
12	41.250	–	7.202	−10.01	−10.86	−20.87
13	61.020	0.137	–	−10.43	−10.86	−21.29
14	68.239	–	2.451	−10.43	−10.56	−20.99
15	72.157	–	0.798	−10.43	−11.00	−21.43
16	73.074	0.925	–	–	–	–

Table 2 Calculation of the *LX*

I*	II*	III*	IV*	m	LX
0	19.775	−0.214	20.391	10	−10.40
1.386	16.636	−0.0296	17.976	10	−9.92
3.296	13.621	−2.901	14.393	10	−10.38
5.545	10.751	−6.889	9.471	10	−10.06
8.047	8.047	−9.727	6.190	10	−9.82
10.751	5.545	13.290	0.998	10	−7.99
13.621	3.296	15.625	0.354	10	−9.06
16.636	1.386	17.911	0.120	10	−10.01
19.775	0	20.281	−0.078	10	−10.43
23.026	–	23.461	–	10	–

*I $= k_j \log k_j$
II $= (m-k_j)\log(m-k_j)$
III $= k_j \log \Sigma\, x_i$
IV $= (m-k_j)\log \Sigma\, x_i$

Table 3 Calculation of the *LY*

I*	II*	III*	IV*	n	LY
0	8.047	-0.317	12.859	6	-10.49
1.386	5.545	1.064	9.978	6	-10.11
3.296	3.296	3.642	7.040	6	-10.09
5.545	1.386	9.431	2.357	6	-10.86
8.047	0	12.832	-0.225	6	-10.56
10.751	–	15.756	–	6	–

$$
\begin{aligned}
*I &= l_j \log l_j \\
II &= (m - l_j) \log(m - l_j) \\
III &= l_j \log \Sigma\, y_i \\
IV &= (m - l_j) \log \Sigma\, y_i
\end{aligned}
$$

3. The estimated termination rates before and after the change point are equal to

$$\hat{\lambda}_{1,A} = 0.655, \quad \hat{\lambda}_{2,A} = 3.118,$$

$$\hat{\lambda}_{1,B} = 0.615, \quad \hat{\lambda}_{2,B} = 0.378.$$

The use of the estimation as well as the test procedure is only justified if the assumption of simultaneous occurrence of change points is sufficiently supported.

Literature: Haccou *et al.* (1983) indicate the idea behind this method for the more general case of two or more acts and/or changes in the transition rate.

3.3.2　Abrupt changes in the properties of sequences of acts

In practice, major behavioural changes will usually comprise shifts in the mean bout lengths as well as in characteristics of sequences of acts. However, it is possible that only the sequence changes, while the mean bout lengths stay the same or show only slight shifts. An obvious example is the sudden emergence or disappearance of (an) act(s) from the behavioural repertoire. Such changes are easily detected. Other sequential properties, however, such as relative frequencies of acts, are less obvious, but may nevertheless also change without accompanying changes in the mean bout lengths. Therefore, methods are needed to detect changes in sequential properties without making assumptions about accompanying changes in the mean bout lengths. Unfortunately, changes in, for example, the expected relative frequency of an act, are usually difficult to detect, even with formal methods. Strictly speaking, detection is only possible if the record contains large numbers of acts. In an analysis of the homogeneity of sequences of acts one can choose either a top-down approach, by application of likelihood methods, or a bottom-up approach, by considering each act or transition between pairs of acts separately. The disadvantage of a top-down approach is that there are no formal tests available, only an estimation procedure. The bottom-up approach suffers from poor power properties if several acts have to be taken into account.

A sudden reduction in the relative frequency of occurrence of an act, say A, means that the average number of acts between two succeeding occurrences of A increases. A test on an abrupt change in the relative frequency can be based on this number of acts between two occurrences of A. The non-parametric change-point test (subsection 3.3.1) might be applied as an approximate test. There is some violation of the assumptions since the numbers of acts between two succeeding occurrences of A are integers instead of realizations of continuous variables; the tie correction (eqn (3.5)) can be applied if necessary. Another possibility is to apply this test to the durations of gaps between occurrences of A. In that case, however, significant results may be due to changes in the mean durations of the intermediate acts rather than the frequency of occurrence of A.

The same procedures could be applied for all other acts. It could also be applied to the occurrence of transitions, for instance the transition from act A to B. If the relative frequency of this transition is suddenly reduced, it results in a change in the average number of acts between two succeeding occurrences of a transition from A to B. Another possibility is to apply it to the occurrence of so-called triplets (e.g. act A, followed by act B, followed in turn by act C), and so on. Obviously these test results are strongly correlated. For instance, a reduction of the relative frequency of transitions to act A implies an increase of the relative frequency of transitions to one or more other acts. In such cases the conclusion of the occurrence of a change point is wrongly confirmed due to the correlation of the test results.

The test results could be combined by, for example, the improved Bonferroni method (described in subsection 6.3.2) if p-values were available. Unfortunately, this is not yet the case. Moreover, the fact that the results are dependent would imply a serious loss of power (i.e. a reduction of the probability of detecting a change point when it is indeed present).

The alternative approach for detecting abrupt changes in the properties of sequences, the top-down approach, could be based on the likelihood ratio test. Unfortunately, critical values are not yet available. This method thus merely provides a means of estimating the change point, rather than a test of the significance of changes.

Maximum likelihood estimate of a change point in a transition matrix

Data: One sequence of acts A_1, A_2,..., A_{n-1}, A_n.

Assumptions:

A1:   Under H_0 the sequence of acts is a first-order discrete time Markov chain.

A2:   Under H_1 there are two sequences, A_1,..., A_τ and $A_{\tau+1}$,..., A_n, which are both first-order discrete time Markov chains with different transition matrices. The two matrices as well as the change point index τ are unknown.

Procedure: First, denote the number of occurrences of acts A, B, C, ... by n_A, n_B, n_C, ... and the number of transitions from one act to another by n_{AB}, n_{AC}, n_{BC}, ... Then calculate the log-likelihood L from

$$L = \sum_A \left\{ \sum_{B:\, B \neq A} n_{AB} \log n_{AB} \right\} - \sum_A n_A \log n_A. \qquad (3.20)$$

(The summations are taken over all acts A and B, such that $B \neq A$.) Next, calculate the log-likelihoods $L(\tau)$ and $L^*(\tau)$ for the sequence of the first τ and the last $n - \tau$ acts respectively. The maximum likelihood estimate τ of a change point is the value of τ which maximizes the likelihood ratio test statistic $\Lambda(\tau)$, defined by

$$\Lambda(\tau) = 2\{L(\tau) + L^*(\tau) - L\}. \qquad (3.21)$$

Since the distributional properties of the maximum of $\Lambda(\tau)$ are unknown, it is not yet possible to use it as a test statistic. Furthermore, it is advisable to restrict the range of admitted τ-values, for instance to the interval $(0.1n, 0.9n)$, in order to avoid undesirable effects for small or large values of τ caused by the discrete character of the data. See Box 3.5 for an example of the estimation procedure.

Box 3.5   Maximum likelihood estimate of a change point in a transition matrix

1. We illustrate the procedure with the data from Table 1 and Fig. 1 in Box 1.1.

2. The total number of acts, n, is equal to 100.

3. We calculate $\Lambda(\tau)$ defined by eqn (3.21) for $\tau = 10, 11, ..., 90$. The maximum is attained for $\tau = 26$. As an example we give the results for this value of τ.

4. The matrix for the first 26 acts is

To: From:	Care	Walk	Sit	Rear	Shake
Care	0	0	1	0	0
Walk	1	0	8	0	0
Sit	0	8	0	3	0
Rear	0	0	3	0	0
Shake	0	1	0	0	0

The matrix for the last 74 acts is

To: From:	Care	Walk	Sit	Rear	Shake
Care	0	0	8	0	4
Walk	1	0	10	3	0
Sit	10	11	0	4	5
Rear	0	0	7	0	0
Shake	1	4	5	0	0

And the matrix for the whole record is

To: From:	Care	Walk	Sit	Rear	Shake
Care	0	0	9	0	4
Walk	2	0	18	3	0
Sit	10	19	0	7	6
Rear	0	0	10	0	0
Shake	1	5	5	0	0

Note that one transition is omitted if it is assumed that there is one change point. In this case the transition from act 26 to act 27: from *Sit* to *Shake*. Therefore the sum of the matrix elements (3,5) of the first and the second matrix is one smaller than that of the third matrix.

5. The corresponding log-likelihoods are respectively equal to

$L(26) =$ 0 log 0 + 1 log 1 + 0 log 0 + 0 log 0 + 1 log 1
+ 8 log 8 + 0 log 0 + 0 log 0 + 0 log 0 + 8 log 8
+ 3 log 3 + 0 log 0 + 0 log 0 + 0 log 0 + 3 log 3
+ 0 log 0 + 0 log 0 + 1 log 1 + 0 log 0 + 0 log 0
− 1 log 1 − 9 log 9 − 11 log 11 − 3 log 3 − 1 log 1
$=$ (0 + 0 + 0 + 0) + (0 + 16.636 + 0 + 0)
+ (0 + 16.636 + 3.2958 +0) + (0 + 0 + 3.2958 + 0)
+ (0 + 0 + 0 + 0) − (0 + 19.775 + 26.377 + 3.2958 + 0)
$= − 9.585.$

(Note that 0 log 0 = 0.) It follows that
$L^*(26) = − 66.738$ and $L = − 87.354$; hence $\Lambda(26) = 22.063$.

6. The maximum of $\Lambda(10), \Lambda(11), ..., \Lambda(89), \Lambda(90)$ is attained for $\Lambda(26)$. Thus the estimated change point based on a change in the transition matrix is after the 26th act.

Comment: In this procedure it is assumed that the sequence of acts is a discrete time Markov chain with not more than one change point in the matrix of transition probabilities. However, in the explorative phase of research it may be desirable to drop the Markovity assumption and/or restrict attention to one single act, say act A. The following suggestion may be helpful: divide the sequence into two, the first τ acts and the last $n - \tau$, and compare the relative frequency of occurrence of A in both sequences. As a measure for the difference the well-known chi-squared test statistic for a 2 × 2 contingency table can be used (Siegel and Castellan, 1988). Calculate this statistic for all possible values of τ between $0.1n$ and $0.9n$. The τ-value at which the maximum is attained is an estimate of the change point in the occurrence of A in the sequence. There are no critical values available for the test statistic and properties of the estimator, such as confidence intervals, are unknown. Presumably, the estimator is not very accurate unless n is very large.

3.3.3 Abrupt changes in bout lengths and transition matrix: the exponential case

The occurrence of abrupt changes in behaviour need not be restricted to either the mean bout lengths of one or more acts or the matrix of transition probabilities. It is far more likely that a combination of both occurs. In this subsection we give the maximum likelihood estimator of a change point if the data satisfy the CTMC assumptions. There are no critical values available for testing the significance of the estimated change point.

Data: A sequence of n acts with the corresponding bout durations: $(a_1, x_1), (a_2, x_2), ..., (a_n, x_n)$.

Assumptions:

A1:   Under H_0 the sequence of acts and their durations is a first-order CTMC.

A2:   Under H_1 there are two sequences: $(a_1, x_1), ..., (a_\tau, x_\tau)$ and $(a_{\tau+1}, x_{\tau+1}), ..., (a_n, x_n)$; both are CTMCs, but with at least one different transition rate. Furthermore, the change point index τ is unknown.

Procedure: We assume that the ethogram consists of k acts, indexed by $i = 1, ..., k$, and we denote the frequency of act i by n_i, the number of transitions from act i to act j by n_{ij} and the mean duration of act i by $\bar{x}_i$.

First, calculate the log-likelihood L of the complete record of observed data, defined by

$$L = \sum_{i=1}^{k} n_i \log \bar{x}_i + \sum_{i=1}^{k} \sum_{j \neq i}^{k} n_{ij} \log \frac{n_{ij}}{n_i} . \qquad (3.22)$$

Then divide the record into two parts: one with the first τ acts with corresponding bout lengths and the other with the last $n - \tau$. Next, calculate the log-likelihood for each part, denoted by $L(\tau)$ and $L^*(\tau)$, respectively. The change point $\hat{\tau}$ is that value τ which maximizes the likelihood ratio statistic $\Lambda(\tau)$ defined by

$$\Lambda(\tau) = 2\{L(\tau) + L^*(\tau) - L\}, \qquad (3.23)$$

by varying τ. Since the distributional properties of the maximum of $\Lambda(\hat{\tau})$ are unknown, it is not yet possible to base a test on it. Furthermore it is advisable to restrict the range of admitted τ-values, for instance to the interval $(0.1n, 0.9n)$, in order to avoid undesirable effects caused by the inaccuracy of the parameter estimates of the smallest sample. See Box 3.6 for an example of the estimation procedure.

Box 3.6   Estimation of a change point in the bout lengths and the transition matrix

1. The procedure is illustrated with the data from Table 1 and Fig. 1 in Box 1.1.

2. The total number of acts with recorded bout lengths, n, is 100.

3. The calculation of the statistic $\Lambda(\tau)$, eqn (3.23), is given for $\tau = 26$.

4. The matrices with the numbers of transitions between the five different acts are given in Box 3.5, under 4.

5. The mean bout lengths, calculated over the first 26 bouts, the last 74 bouts and the whole record, for the five acts (in s) are respectively equal to

$$
\begin{array}{llll}
x_{Care} & = 40.290 & 47.369 & 46.825 \\
x_{Walk} & = 2.774 & 2.259 & 2.453 \\
x_{Sit}   & = 16.933 & 22.576 & 20.964 \\
x_{Rear} & = 6.433 & 2.914 & 3.970 \\
x_{Shake} & = 0.450 & 0.450 & 0.450
\end{array}
$$

6. The statistics $L(26)$, $L^*(26)$ and $L(26)$ are calculated as follows:

$$
\begin{aligned}
L(26) = & - \{1 \log 40.290 + 9 \log 2.774 + 11 \log 16.933 \\
        & + 3 \log 6.433 + 1 \log 0.45\} + 0 \log 0/1 + 1 \log 1/1 \\
        & + 0 \log 0/1 + 0 \log 0/1 + 1 \log 1/9 + 8 \log 8/9 \\
        & + 0 \log 0/9 + 0 \log 0/9 + 0 \log 0/11 + 8 \log 8/11 \\
        & + 3 \log 3/11 + 0 \log 0/11 + 0 \log 0/3 + 0 \log 0/3 \\
        & + 3 \log 3/3 + 0 \log 0/3 + 0 \log 0/1 + 0 \log 1/1 \\
        & + 0 \log 0/1 + 0 \log 0/1 \\
      = & - \{\log 40.290 + 9 \log 2.774 + 11 \log 16.933 \\
        & + 3 \log 6.433 + \log 0.45\} + 0 + 0 + 0 + 0 + \log 1/9 \\
        & + 8 \log 8/9 + 0 + 0 + 0 + 8 \log 8/11 + 3 \log 3/11 \\
        & + 0 + 0 + 0 + 0 + 0 + 0 + 0 + 0 + 0 \\
      = & - (3.696 + 9.183 + 31.122 + 5.584 - 0.799) \\
        & + (-2.197 - 0.942 - 2.548 - 3.898) \\
      = & - 58.37.
\end{aligned}
$$

In the same way it can be calculated that $L^*(26) = -217.45$ and that $L = -290.79$. Hence, $L(26) = 29.93$ and this value appears to be the maximum value for all possible τ, such that $10 \le \tau \le 90$.

7. The estimated change point based on a change in the transition matrices as well as on the bout lengths is after the 26th act.

3.4   GRADUAL CHANGES: A FEW EXAMPLES

Gradual changes in behaviour can occur in many different ways. A multitude of possibilities could be modelled. We will restrict ourselves, however, to a few examples: periodic changes, monotonically increasing or decreasing mean bout lengths and changes in behaviour related to an external (co)variable, for instance temperature, concentration of a drug, etc.

We suppose, furthermore, that the changes concern the mean bout length of only one act. Hence, we maintain most CTMC assumptions, except those of the bout length distribution of one single act. In the following subsections we illustrate how to test for the presence or absence of a gradual change in mean bout length.

3.4.1 Periodicities

The case when the periodicity is related to the time of day is treated as an example. An example of such periodic changes in behaviour is given by e.g. Gibson (1980), who found variations in the durations of swimming and eating bouts of juvenile plaice corresponding to tidal changes. Whereas in that case there is an external covariable (the tide) correlating with the changes, we will here consider a method for examining periodic changes in the absence of such covariables. The mean bout length is assumed to be

$$\frac{1}{\lambda(t)} = \frac{1}{\lambda}\exp\left[a\sin\left(\frac{2\pi t + b}{T}\right)\right], \tag{3.24}$$

where $1/\lambda$ is the mean when there is no time dependency, a is the amplitude of the fluctuation, b is the phase shift in time and T is the duration of a period. In general, these four parameters will be unknown and must be estimated by a numerical optimization procedure. Exponentially distributed data have a large variance and this results in poor convergence properties of these methods. Hence, it is strongly advised to restrict the number of unknown parameters as much as possible. In the example of eqn (3.24) it is possible to set T equal to 1440 min (= 24 h) and to derive a graphical estimate of the phase shift b. See Fig. 3.11 for a visual representation of such a parameter reduction.

Data: One sample of bout lengths $x_1,..., x_n$ with the corresponding start times $t_1,..., t_n$.

Assumptions:
A1: $X_1,..., X_n$ are mutually independent and exponentially distributed.
A2:   Under H_0, $E\,X_i = 1/\lambda$, where λ is unknown.
A3:   Under H_1, $E\,X_i = 1/\{\lambda\exp[a\,\sin(2\pi t_i)]\}$, where λ and a are unknown. The time t_i is the time at the onset of the act. The time unit is the fraction of 24 hours.

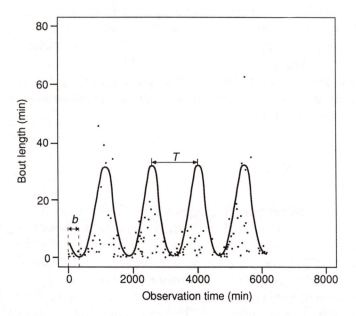

Figure 3.11 Periodic changes in the mean bout length with the phase shift b and period T.

Procedure: Under H_0 the unknown parameter λ is estimated by the reciprocal of the mean bout length: see eqn (11), Box 1.2.

Under H_1 there is no explicit solution for the two likelihood equations (3.27) and (3.28). The log-likelihood L_1 is equal to

$$L_1 = \sum_{i=1}^{n} \log \lambda_1 \exp[-\lambda_1 x_1], \qquad (3.25)$$

where

$$\lambda_1 = \lambda \exp[a \sin(2\pi t_i)]. \qquad (3.26)$$

The partial derivatives with respect to λ and a are equal to

$$\frac{\partial \log L_1}{\partial \lambda} = \frac{n}{\lambda} - \sum_{i=1}^{n} x_i \exp[a \sin(2\pi t_i)], \qquad (3.27)$$

$$\frac{\partial \log L_1}{\partial a} = \sum_{i=1}^{n} \sin(2\pi t_i) - \lambda \sum_{i=1}^{n} x_i \sin(2\pi t_i) \exp[a \sin(2\pi t_i)]. \qquad (3.28)$$

By equating the two derivatives to zero one can find the solutions $\hat{\lambda}$ and $\hat{a}$. Using eqn (3.25) it is possible to express $\hat{\lambda}$ in $\hat{a}$ as follows:

$$\hat{\lambda} = \frac{n}{\displaystyle\sum_{i=1}^{n} x_i \exp[\hat{a}\sin(2\pi t_i)]} \; . \qquad (3.29)$$

Substitution of eqn (3.29) into eqn (3.28) leads to an equation which must be solved numerically, for instance by the Newton-Raphson method (see e.g. Stoer and Bulirsch, 1980).

The likelihood ratio test statistic Λ is defined by

$$\Lambda = 2(L_1 - L_0), \qquad (3.30)$$

where L_0 and L_1 are defined by

$$L_0 = n\log n - n\log \sum_{i=1}^{n} x_i - n, \qquad (3.31)$$

$$L_1 = \sum_{i=1}^{n} \hat{\lambda}_i - \sum_{i=1}^{n} \hat{\lambda}_i x_i, \qquad (3.32)$$

and

$$\hat{\lambda}_i = \hat{\lambda}\exp[\hat{a}\sin(2\pi t_i)]. \qquad (3.33)$$

There are no small-sample critical values available.

Large-sample approximation: For large sample sizes Λ has approximately a chi-squared distribution with one degree of freedom. H_0 (absence of periodicity) should be rejected if Λ exceeds $\chi^2_1(\alpha)$, see Table A2.

Comment: Under the alternative hypothesis, conditional on the start time t_i, the bout length X_i is independent of the preceding bout lengths. Hence, t_i is a renewal point and apparent correlation between succeeding bout lengths is due to the fact that the expected ith bout length is a function of t_i, the sum of all preceding bouts of all different acts.

3.4.2 Monotonic changes

As an example of a monotonic change we treat the case when the mean bout length is a linear function of the observation time, i.e. $EX = (1+at)/\lambda$, where $1/\lambda$ is the mean in the absence of any time dependency and the regression coefficient a can be positive as well as negative. The two parameters have to estimated by a numerical optimization procedure.

Data: One sample of bout lengths $x_1,..., x_n$ with the corresponding start times of the bouts $t_1,..., t_n$.

Assumptions:

A1: $X_1,..., X_n$ are mutually independent and exponentially distributed.

A2:   Under H_0, $E\,X_i = 1/\lambda$, where λ is unknown.

A3:   Under H_1, $E\,X_i = (1 + at_i)/\lambda$, where λ and a are unknown. The time t_i is the time at the onset of the act.

Procedure: Under H_0 the unknown parameter λ is estimated by the reciprocal of the mean bout length: see eqn (11), Box 1.2.

Under H_1 there is no explicit solution of the two likelihood equations (3.36) and (3.37). The log-likelihood L_1 is equal to

$$L_1 = \sum_{i=1}^{n} \log \lambda_i \exp[-\lambda_i x_i],\tag{3.34}$$

where

$$\lambda_i = \frac{\lambda}{1 + at_i}.\tag{3.35}$$

The partial derivatives with respect to λ and a are equal to

$$\frac{\partial \log L_1}{\partial \lambda} = \frac{n}{\lambda} - \sum_{i=1}^{n} \frac{x_i}{1 + at_i},\tag{3.36}$$

$$\frac{\partial \log L_1}{\partial a} = \sum_{i=1}^{n} -\frac{1}{1 + at_i} + \sum_{i=1}^{n} \frac{x_i t_i}{(1 + a_i t_i)^2}.\tag{3.37}$$

By equating the two derivatives to zero one can find the solutions $\hat{\lambda}$ and $\hat{a}$. Using eqn (3.32) it is possible to express $\hat{\lambda}$ in $\hat{a}$ as follows:

$$\hat{\lambda} = \frac{n}{\displaystyle\sum_{i=n}^{n} \frac{x_i}{1 + \hat{a}t_i}}.\tag{3.38}$$

Substitution of eqn (3.38) into eqn (3.37) leads to an equation which must be solved numerically, for instance by the Newton-Raphson method.

The likelihood ratio test statistic Λ is defined by

$$\Lambda = 2(L_1 - L_0),\tag{3.39}$$

where L_0 and L_1 are given by eqns (3.31) and (3.32); however,

$$\hat{\lambda}_i = \frac{\hat{\lambda}}{1 + \hat{a}t_i}.\tag{3.40}$$

There are no small-sample critical values available.

Large-sample approximation: For large sample sizes, Λ approximately follows a chi-squared distribution with one degree of freedom. H_0 (absence of periodicity) should be rejected if Λ exceeds $\chi^2_1(\alpha)$, see Table A2.

Comment: See our comment in subsection 3.4.1.

3.4.3 Changes depending on internal or external covariables

In many cases the mean bout length can be considered as a function y of measurable internal or external covariables. Previously mentioned examples are, for instance, the variation in mosquito flight activity with the day-night cycle (Peterson, 1980) and the changes in swimming and eating behaviour of juvenile plaice due to tidal changes (Gibson, 1980). Here, we will consider a model for the effect of a drug on bout lengths of one act. The function y is a covariable which is parametrized by an unknown parameter α (which may be a vector), and t_i is the time at the onset of the ith act. Figure 3.12 shows an example where the mean is influenced by the concentration, α_1, of a drug and the moment, α_2, of administration. The largest effect appears shortly after the drug enters the bloodstream and the effect diminishes in the course of time. The maximum of the expected mean bout length is apparently determined by the dose.

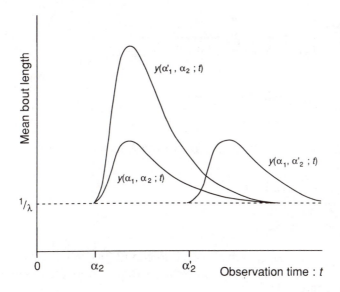

Figure 3.12 Example of the influence of a drug on the mean bout length of an act: α_1 and α_1' are two different concentrations; α_2 and α_2' are different times of administration of the drug.

Data: One sample of bout lengths $x_1,..., x_n$ with the corresponding start times $t_1,..., t_n$.

Assumptions:

A1: $X_1,..., X_n$ are mutually independent and exponentially distributed.

A2:   Under H_0, $E X_i = 1/\lambda$, where λ is unknown.

A3:   Under H_1, $E X_i = y(\alpha,t_i)/\lambda$, where $y(\alpha,t_i)$ is specified up to the unknown parameter vector α. The time t_i is the time at the onset of the act.

Procedure: Under H_0 the unknown parameter λ is estimated by the reciprocal of the mean bout length (see eqn (11), Box 1.2). Under H_1 there is in general no explicit solution of the likelihood equations. The log-likelihood L_1 is equal to

$$L_1 = \sum_{i=1}^{n} \log \lambda_i \exp[-\lambda_i x_i], \tag{3.41}$$

where $\lambda_i = y(\alpha,t_i)/\lambda$.

The partial derivatives with respect to λ and the vector $\alpha = (\alpha_1,..., \alpha_k)$ have to be equated to zero and solved:

$$\left[\frac{\partial \log L_1}{\partial \lambda}\right]_{\hat{\lambda},\hat{\alpha}} = \frac{n}{\hat{\lambda}} - \sum_{i=1}^{n} \frac{x_i}{y(\hat{\alpha},t_i)} = 0, \tag{3.42}$$

$$\left[\frac{\partial \log L_1}{\partial \alpha_j}\right]_{\hat{\lambda},\hat{\alpha}} = \sum_{i=1}^{n} \frac{\hat{\lambda} x_i - y(\hat{\alpha},t_i)}{\{y(\hat{\alpha},t_i)\}^2} = 0, \; j = 1,...,k. \tag{3.43}$$

Using eqn (3.38) it is possible to express $\hat{\lambda}$ in the vector $\hat{\alpha}$ as follows:

$$\hat{\lambda} = \frac{n}{\displaystyle\sum_{i=1}^{n} \frac{x_i}{y(\hat{\alpha},t_i)}}. \tag{3.44}$$

The estimated $\hat{\lambda}$ can be substituted into eqn (3.39), which has to be solved numerically, for instance by the Newton-Raphson method (see Stoer and Bulirsch, 1980).

The likelihood ratio test statistic Λ is defined by

$$\Lambda = 2(L_1 - L_0), \tag{3.45}$$

where L_0 and L_1 are given by eqns (3.31) and (3.32) with

$$\hat{\lambda}_1 = \frac{\hat{\lambda}}{y(\hat{\alpha}, t_i)} . \tag{3.46}$$

There are no small-sample critical values available.

Large-sample approximation: For large sample sizes Λ follows approximately a chi-squared distribution with k degrees of freedom. H_0 is rejected if Λ exceeds $\chi^2_k(\alpha)$, see Table A2.

Comment: See our comment in subsection 3.4.1.

4 TESTS FOR EXPONENTIALITY

4.1 INTRODUCTORY REMARKS

In this chapter we discuss methods for analysing the bout length distributions of the recorded behavioural categories. Since the CTMC is used as the basic model, these distributions are exponential under the null hypothesis (see Box 1.3). We therefore focus on detecting deviations from exponentiality. Our strategy is to investigate this for each act separately. In a CTMC or a semi-Markov chain the process of alternation between one state and the group of other states is a so-called alternating renewal process (e.g. Cox and Lewis, 1978). Thus, we give tests for the exponentiality of the residence times in one state of an alternating renewal process, irrespective of the residence times in the other state. We restrict our attention to so-called scale-invariant tests which do not depend on the recording time unit. Furthermore, we have chosen procedures that can easily be applied, since either the exact null distribution or the asymptotic distribution is known. The (approximate) critical values are listed in the tables in Appendix I.

Although some of the methods treated can indicate whether there is time inhomogeneity (see subsection 4.2.1), we assume here that the data are more-or-less homogeneous and concentrate on other types of deviations from the basic model. The analysis of time inhomogeneity is treated separately in Chapter 3. At a relatively early stage of analysis there may also be outliers (extremely large bout durations, due to some rare external or internal disturbance, or very short durations). Outlier tests are treated in subsection 2.2.3. In this chapter it is assumed that outliers have previously been removed.

In homogeneous records without outliers, deviations from exponentiality imply that the basic model should be adjusted, i.e. behaviour categories have to be redefined or a generalization of a CTMC (subsection 1.3) should be used. The types of departures indicate which adjustments should be made. Guidelines are given in section 4.2 and subsection 4.8.2. Note, however, that small deviations from exponentiality, caused by, for example, slight 'mood-changes' of an animal, may not be of interest. Therefore it should also be considered whether the order of magnitude of the deviations is indeed relevant.

The tests can be classified according to their sensitivity to specific types of departures. Sections 4.6 and 4.7 are arranged according to this classification. In section 4.3 we explain this classification further.

Testing for exponentiality is usually more complicated than one would expect. This is due to the fact that behavioural data are often censored. This means that not all the begin or the end times of acts are recorded. In section 4.4 we give a few examples of censoring in ethological observations. To prevent wrong conclusions, it is essential to use adjusted methods when there is censoring (see Bressers *et al.*, 1991). Section 4.6 contains tests for uncensored observations while censor-adjusted methods are described in section 4.7.

Section 4.5, which may be omitted at first reading, contains some theoretical considerations concerning the fact that the sample size is stochastic when the exponentiality of the bout lengths of a selected act is tested. This poses a formal problem since the properties of test statistics are usually derived under the assumption that the sample size is fixed. It can be shown, however, that the properties of tests discussed in this chapter (which were derived in the first instance for the fixed sample size case) continue to hold.

We end this chapter with a few remarks on the performance of the tests and a strategy for their application in practice.

4.2   TYPES OF DEVIATIONS FROM EXPONENTIALITY AND THEIR IMPLICATIONS FOR MODELLING

As already mentioned, specific types of departures from exponentiality indicate that certain adjustments must be made to the basic model. We consider here ethologically relevant types of departures which will be interpreted within the framework of the generalizations of the CTMC model (section 1.3). Thus, departures are used as indications that either a semi-Markov model, or a function of a Markov model, applies. In the initial stages of analysis we are often especially interested in deviations in the direction of a function of a Markov chain, since these indicate that the behavioural category under consideration should, if possible, be split up. Thus, if such deviations do occur, the ethogram (see Box 1.1) should be adjusted. Within this context, two types of distributions are of particular interest, namely mixtures and convolutions of exponentials.

4.2.1 Mixtures of exponentials

A mixture of m exponential distributions is defined by

$$f(x) = \sum_{k=1}^{m} w_k \lambda_k \exp[-\lambda_k x], \quad \lambda_k > 0, \quad \sum_{k=1}^{m} w_k = 1, \tag{4.1}$$

where w_k ($k = 1,..., m$) are weight coefficients. Here, we consider mixtures in the strict sense, with positive weight coefficients. When the bout length distribution of a behavioural category is a mixture of exponentials, this can mean either that there is time inhomogeneity or that the category consists of subcategories which differ with respect to their time structure. In the latter case, the category can be considered as a group of lumped Markov states, i.e. the process can be modelled by a function of a Markov chain (Chapter 7). Methods for investigating the first possibility further are given in Chapter 3. Here we consider the second option.

Let C be a group of Markov states that are not distinguished. Furthermore, suppose that the subcategories in C do not necessarily always occur together. Then the residence times of C are distributed as a mixture of exponentials. For instance, if C consists of two states A and B; then one occurrence of C can mean that either A occurred, or B, or AB, or BA. In that case the residence-time distribution of C is a mixture of two exponential distributions.

When the bout length distribution of a behavioural category appears to be a mixture of exponentials, one should try to split the category up according to observable criteria. For instance, in their study of mother-infant body contact in rhesus monkeys, Dienske and Metz (1977) initially distinguished only two states, namely *On mother* and *Off mother*. Since the bout length distribution of *On mother* appeared to be a mixture of exponentials, they split this category up according to whether the infant had the mother's nipple in its mouth or not. In the resulting model there were three behavioural categories: *On mother without nipple*, *On nipple* and *Off mother*. Bout length distributions of these three categories did not deviate significantly from exponentiality. This is illustrated in Box 4.1.

Box 4.1 Example of splitting up a behavioural category: mother-infant interaction in rhesus monkeys

Dienske and Metz (1977) studied body-contact alternations in mother and infant rhesus monkeys. At first they only distinguished two behavioural categories: *On mother* and *Off mother*. An example of a record of such an observation is given in Table 1. The resulting process could not be described by a Markov chain, since the bout length distribution of *On mother* bouts is a mixture rather than an exponential distribution. Therefore, they split up *On mother* bouts into *On mother without nipple* and *On nipple* (see Table 1). Both *On mother* (s.s.) bouts and *On nipple* bouts are exponentially distributed. The *On nipple* bouts have a much lower termination rate, since the mother's as well as the infant's tendencies to stop *On nipple* bouts are much lower than their tendencies to stop *On mother* bouts. Thus, apparently, by taking the nipple into its mouth, the infant can prolong body contact with the mother.

Table 1 Records of mother-infant interaction in rhesus monkeys: 1, without distinction whether or not the infant had the nipple in its mouth; and 2, with *On mother* bouts split up

Record 1		Record 2	
Time	Act	Time	Act
0	*On mother*	0	*On mother s.s.*
10	end *On mother*	10	end *On mother s.s.*
10	*Off mother*	10	*Off mother*
43	end *Off mother*	43	end *Off mother*
43	*On mother*	43	*On mother s.s.*
		51	end *On mother s.s.*
		51	*On nipple*
65	end *On mother*	65	end *On nipple*
65	*Off mother*	65	*Off mother*
83	end *Off mother*	83	end *Off mother*
83	*On mother*	83	*On nipple*
90	end *On mother*	90	end *On nipple*

4.2.2 Convolutions of exponentials

A convolution is the distribution of the sum of (independent) random variables. Bout length distributions that are convolutions indicate that the behavioural category under consideration consists of a series of acts that are performed subsequently. For instance, in many species self-grooming proceeds in a certain order, from head to tail. When a bout length distribution is a convolution of exponentials, the behavioural process can be described by a special type of function of a Markov chain, where an occurrence of the lumped state implies that all subcategories in the state are visited. For example, if state C is such a group, consisting of two states A and B, an occurrence of C means that either (AB) or (BA) has occurred. In that case the distribution of the residence times of C is a convolution of two exponentials

$$f(x) = \frac{\lambda_A \lambda_B \{\exp[-\lambda_A x] - \exp[-\lambda_B x]\}}{\lambda_B - \lambda_A}, \quad \lambda_A \neq \lambda_B, \tag{4.2}$$

where λ_A and λ_B are the termination rates of the two states.

When the termination rates are equal and the number of lumped states is k ($k \geq 2$), the residence times have a gamma distribution with discrete 'shape parameter' k:

$$f(x) = \frac{\lambda(\lambda x)^{k-1}\exp[-\lambda x]}{\Gamma(k)}, \tag{4.3}$$

where $\Gamma(k)$ denotes the gamma function which is equal to $(k - 1)!$ when k is a positive integer (see Box 4.2).

Box 4.2   The gamma distribution

1. Properties of the gamma distribution

The probability density function $f(x)$ ($0 < x < \infty$) of a random variable X which has a gamma distribution with parameters k and λ is defined by eqn (4.3). Examples of such probability densities are given in Fig. 1. If $k = 1$, $f(x)$ reduces to an exponential distribution. The cumulative distribution function $F(x)$, defined by $F(x) = Pr\{X \leq x\}$, is called an incomplete gamma function and is extensively tabulated. See Johnson and Kotz (1970) for references. The chi-squared distribution is a special case of the gamma distribution: if $\lambda = \frac{1}{2}$ and $k = \frac{1}{2}n$, (4.3) is the probability density function of the chi-squared distribution with n degrees of freedom.

The mathematical expectation of X is equal to

$$EX = \int_0^\infty \frac{x\lambda(\lambda x)^{k-1}\exp[-\lambda x]\,dx}{\Gamma(k)}$$

$$= \frac{1}{\lambda\Gamma(k)}\int_0^\infty (\lambda x)^{(k+1)-1}\exp[-\lambda x]\,d(\lambda x) = \frac{\Gamma(k+1)}{\lambda\Gamma(k)} = \frac{k}{\lambda}. \tag{1}$$

In an analogous way it can be shown that

$$VarX = \frac{k}{\lambda^2}. \tag{2}$$

2. Maximum likelihood estimators (MLEs) of λ and k

The MLEs of λ and k are the solutions of the equations

$$\bar{x} = \hat{k}\hat{\lambda} \tag{3}$$

and

$$\frac{\sum_{i=1}^{n} \log x_i}{n} - \log \bar{x} = \log \hat{\lambda} + \psi(\hat{k}), \tag{4}$$

where $\psi(k) = \dfrac{d\log\Gamma(k)}{dk}$.

Eqns (3) and (4) have to be solved numerically; see Johnson and Kotz (1970) for further details or Bowman and Shenton (1988) for a monograph on this topic. Bowman and Shenton also give an approximation of the joint critical region for k and λ. Simulations show that their approximation is sufficiently accurate.

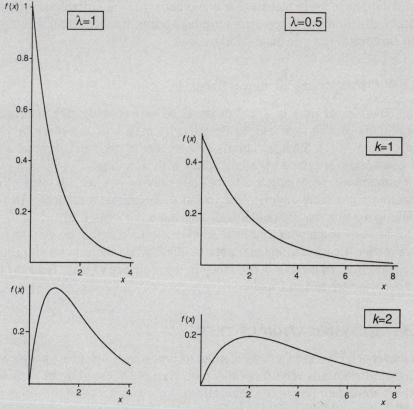

Figure 1 Examples of Gamma probability densities for several values of k and λ.

When the bout lengths of a behavioural category have this type or distribution, one can try to split up the behavioural category according to observable criteria. However, with this type of deviation it is usually also possible to use a semi-Markov model. This can be done if tests for first-order dependency (given in Chapter 5) do not give significant results. The model which is chosen depends on whether the transitions between sub-categories are of interest. For instance, in the case of grooming behaviour, one might either be interested in the tendency to start or stop grooming (in which case grooming can be modelled as one category), or in the specific elements during grooming, for instance, whether certain experimental conditions affect the relative proportions of time spent grooming different body parts.

4.2.3 Other convolutions

Time-lags (subsection 1.3.1) can cause deviations in the direction of convolutions of an exponential and another type of distribution, such as a normal distribution. When such time-lags occur, the behavioural category can be considered as a semi-Markov state.

4.2.4 Other types of departures

Combinations of time-lags and mixtures of exponentials may also occur. Deviations of this type can be detected by means of log-survivor plots (subsection 4.6.1). Such deviations indicate that behavioural categories can be considered as lumped Markov states with time-lags.

Furthermore, bout length distributions may be neither mixtures of exponentials nor convolutions. For instance, an animal's tendency to stop walking may increase monotonically during a *Walk* bout. In such cases, a semi-Markov chain with a suitable residence-time distribution may be used as a model. However, this can only be done when there is no significant deviation from first-order dependency in the sequence of acts. Tests for this property are treated in Chapter 5.

4.3 CLASSIFICATION OF TESTS

In this chapter we treat a large number of procedures for investigating bout length distributions. The procedures are classified according to their sensitivity to certain (types of) alternatives.

We start with graphical procedures, so-called visual scanning methods, which are useful as a device to indicate the type of departure(s).

Next we treat a few goodness-of-fit tests that are sensitive against all kinds of deviations from exponentiality. These are generalizations of non-parametric goodness-of-fit tests such as the Cramér-von Mises test, the

Kolmogorov-Smirnov test and the well-known chi-squared test. These tests have less discriminating power against specific alternative hypotheses than tests that are specially designed to detect those departures.

We subsequently consider tests for detecting broad classes of alternatives that are in some sense restricted. These tests are formulated in terms of the termination rate (in our terminology), as defined in eqn (1.2). In the statistical literature they are called 'failure rate' (or 'hazard rate') tests. As we have proved in Box 1.3, in a Markov chain the termination rates are constant. In a more general (semi-Markov) model the termination rates may depend on the residence time in a state, x. Tests of exponentiality can thus be considered as tests of the constancy of the termination rate $\lambda(x)$. The different classes of alternatives considered by the tests we describe are the classes of increasing or decreasing failure (= termination) rates (IFR or DFR). An increasing termination rate indicates deviations in the direction of a convolution, whereas a decreasing termination rate indicates a mixture. We treat the 'total time on test' statistic proposed by Barlow *et al.* (1972).

Finally, we treat tests for specific alternatives. We have included a test for the Weibull distribution, which is sensitive to departures in the direction of monotonic increasing or decreasing termination rates. When bout lengths have a Weibull distribution, the termination rate has the form:

$$\lambda(x) = \lambda\rho\,(\lambda x)^{\rho-1} \tag{4.4}$$

with λ and ρ larger than zero (see Fig. 4.1). This termination rate is monotonically decreasing when ρ is less than one and monotonically increasing when ρ is larger than one. It is to be expected that one-sided tests for Weibull alternatives ($\rho < 1$ and $\rho > 1$) also perform fairly well

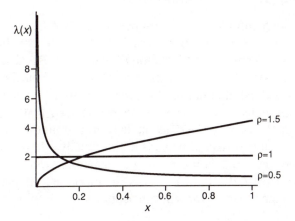

Figure 4.1 Termination rate corresponding to the Weibull distribution for three different values of ρ.

against mixtures and convolutions of exponentially distributed bout lengths respectively.

Instead of a gradual change in the termination rate, there may be an abrupt change during a bout. For instance, a bird's tendency to stop singing may change abruptly after it has sung for 5 minutes. In this case, the form of the termination rate is as given in Fig. 4.2. In subsection 4.6.4 we discuss a test against such alternatives which was developed by Matthews and Farewell (1982). We also present tests that are especially sensitive against mixtures of exponentials and gamma distributions.

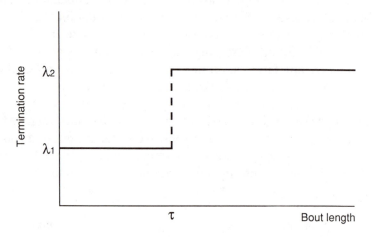

Figure 4.2 Example of a termination rate which changes abruptly during a bout.

4.4   OCCURRENCE OF CENSORING

In section 1.5 we mentioned the occurrence of censors: bouts for which the beginning or end has not been recorded. In the case considered there, censoring occurs because the observation time is fixed. Because of this, the beginning of the first bout and/or the end of the last bout is usually not observed. This is illustrated in Fig. 4.3, where the first *Groom* bout and the last *Lie* bout are censored. In section 1.5 we advised that such censored observations should be discarded, as the loss of information is slight if the number of completely observed bouts is not too small. Alternatively, as also mentioned, the censored bout(s) can be treated as randomly censored observation(s). (See Gill, 1980.) This is of importance if the number of observations is small or if, for some reason, there are more censors (see below).

Although censored data are quite common in behavioural research, they are often not recognized as such. However, applying methods which are

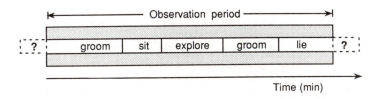

Figure 4.3 Graphical representation of the behaviour of a rat. Only events within the shaded area are observed. The first (*Groom*) bout began before observation started and is therefore left-censored. The last (*Lie*) bout ends after the end of the observation period and is right-censored.

not adjusted for censors often leads to erroneous conclusions. See Bressers *et al.* (1991) for a number of striking examples. In this subsection we will discuss the occurrence of censors. In section 4.7 we treat several methods for testing for exponentiality when there are censors.

The main reason why we treat this subject extensively here is that tests on exponentiality can be strongly influenced by the presence of censors. Incorrect application of unadjusted tests can lead to rejection of the hypothesis of exponentiality in favour of a much more complicated model. Although we discuss censors especially here, censor-adjusted methods for other types of tests are mentioned in other chapters where they are available and of relevance.

To give an impression of the situations in which censors occur, we give a few examples.

Example 1: Observation ends before the occurrence of a certain event

Many examples of this kind of censoring can be found in studies of latencies. Consider, for instance, the latency of the aggressive response of rats during 10 seconds of electrical brain stimulation. Both stimulation and observation end after 10 seconds. Figure 4.4 shows the record of two stimulated rats. The first rat has a latency of 6 seconds. The second rat did not attack within 10 seconds. Therefore, its attack latency cannot be determined: we only know that it exceeds 10 seconds. Hence, a censored observation of 10 seconds is recorded. The occurrence of this type of censored observation clearly depends on the length of the observation period. In theory, the number of censors can be reduced by prolonged observation. In practice, however, this is not always possible, because there may be interfering side-effects. In this example, for instance, prolonged stimulation may cause brain damage.

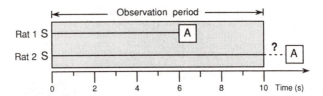

Figure 4.4 Graphical representation of the behaviour of two rats in the experiment of Example 1. Only events within the shaded region are observed (S = start of stimulus, A = attack).

Example 2: Restricted observation area

Observation is often restricted to a limited area for practical or technical reasons. This occurs in the field, e.g. in observations on foraging birds, as well as in laboratory experiments, e.g. in studies on parasitoids laying eggs in an artificial patch of hosts. Censored observations occur when the animals enter or leave the area.

Example 3: Interactive behaviour

In interactions between mother and infant rhesus monkeys, body contact is an important behavioural category. A fictitious example is given in Fig. 4.5. If one studies how quickly the mother terminates contact with her infant, X_{mother} is clearly an uncensored bout length. The second bout of body contact is, however, terminated by the infant. In that case, we only know that the mother did not break contact during the interval X_{infant}. The observations on lengths of bouts of body contact ended by the mother are censored by the infant's acts. Similarly, when bouts of body contact ended

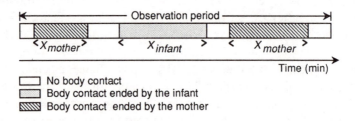

Figure 4.5 Graphical representation of the behavioural record of a mother-infant pair of monkeys (X = lengths of bouts of body contact; the subscript indicates which individual broke the contact).

by the infant are studied, X_{mother} is a censored observation. This censoring mechanism, inherent to social interactions, shows up in many situations. Returning for instance to Example 1 in which latencies to the first attack are studied, suppose that two types of attack can be distinguished: a violent one (A) and a mild one (B). Observation is stopped after the first attack, and the type of attack is recorded. When type A attack latencies are studied, every bout ended by a type A attack is an uncensored observation. However, when a rat shows a type B attack, the observed latency is a censored observation.

Thus, in general, an observation on a bout length X is censored at c if the only available information is that $X > c$. As should be clear from the examples, censoring occurs often in behavioural research and a substantial part of the data may be censored. Adjusted methods have been developed in the engineering and biomedical sciences (known as 'survival analysis') for the analysis of 'failure time' data. A bout length (in a broad sense: the duration of an act, or the time interval between two acts, or a latency) can be regarded as a failure time. Therefore, techniques originally developed in the field of survival analysis can also be used in ethology.

Censoring can be modelled as follows. Let Y denote the hypothetical bout length and let C be the time until censoring. Then the minimum of Y and C, denoted by X, is the observed bout length. X is censored when X equals C and uncensored when X equals Y. In general it is assumed that Y and C are stochastically independent. The distribution of C can have various forms, e.g. exponential, normal, uniform or degenerate (in which case the censoring time is a fixed constant).

The censoring mechanism treated in this section is called type I censoring (e.g. Kalbfleisch and Prentice, 1980, Miller, 1981*b*). If the censoring times depend on previous termination or censoring times, the mechanism is called type II censoring. For instance, suppose that the experiment described in example 1 is performed simultaneously with 20 rats and observation is stopped after five rats have attacked; then the other attack latencies are type II censored. Throughout this book we restrict our attention to type I censoring.

Within the context of behavioural records such as considered in this book, the most obvious type of censoring occurs at the beginning and/or the end of observation. A more obscure type of censoring occurs due to the fact that a bout of a certain act can be followed by several other acts. For instance, when we study the bout lengths of act A followed by B, all bouts of A that were followed by other acts are censored observations. In the CTMC model and its generalizations, this censoring is taken implicitly into account, since the transition rate from A to B can be interpreted as the chance per time unit of a transition from A to B if the other acts were excluded. Note that censoring occurring in social interactions due to changes in behaviour of the other animal(s) is a special case of this, since A and B can be combinations of acts of different individuals (see section 1.2 and Box 1.5).

In 'survival analysis' terms, the transition rates, as defined by eqn (1.1), or the termination tendencies are called 'cause-specific hazard rates' (e.g. Kalbfleisch and Prentice, 1980, Chapter 7). The above interpretation of the transition rates implies that bout terminations are assumed to be due to independent 'competing risks', where the 'risks' are the different possible following acts. It can be proved that the observations from any process with possibly dependent competing risks can also be generated by a process with independent risks (e.g. David and Moeschberger, 1978, section 4.3). Therefore, it is not possible to determine on the basis of the observed process of alternations between acts whether or not the 'risks' are indeed independent. In solitary behaviour there is no basis for a model phrased in terms of either dependent or independent risks, unless there is detailed information about the underlying internal processes. In this case, a model with independent risks is thus the most parsimonious. In social interactions, dependencies between the terminations of different individuals would imply that they exchange signals. When the individuals can observe such signals, a really good human observer should also be able to detect them (although, admittedly, this might be difficult, e.g. in the case of ultrasonic sounds, or scents). Such signals should then be included in the definitions of behavioural categories. Conditional on the process of signal values the bout terminations are independent.

When the hypothetical bout length Y as well as the censor time C have an exponential distribution, the resulting distribution of X is also exponential. Tests for exponentiality can therefore in first instance be applied to the whole set of bout lengths, to find out whether both Y and C are exponential. If such tests give significant results, tests adjusted for censoring can be used subsequently to find out which of the two distributions deviate from exponentiality (see also Box 4.14).

4.5 SOME PRELIMINARY THEORETICAL CONSIDERATIONS

The tests are carried out conditionally on $N(T) = n$ (see section 1.5), where $N(T)$ denotes the number of bouts that were observed in a period of T time units. In testing for exponentiality of $X_1,..., X_{N(T)}$ only scale-invariant procedures are relevant. Therefore, inference is made conditionally on the sum of the X_i, since for $N(T) = n$ this is a sufficient statistic for the nuisance parameter (see eqn (1.3)). Let $y_1,..., y_{n(T)}$ denote the lengths of the gaps between the bouts of the studied act (see Box 1.1 and Fig. 5.8). Conditional on the observed values of $\Sigma\, X_i$ and $N(T)$, the statistic $(Y_1,\ldots, Y_{N(T)})$ does not contain information about the distribution of $X_1,\ldots, X_{N(T)}$ and is therefore an ancillary statistic (e.g. Cox and Hinkley, 1974, Chapter 2). The tests are thus also performed conditionally on the observed values of $Y_1,\ldots, Y_{N(T)}$.

From the discussion in section 1.5 it can be inferred that the asymptotic results based on non-random sample sizes continue to hold for the test statistics given here.

Alternatively, note the following: let T be fixed and conditional on

$$\sum_{i=1}^{N(T)} X_i = S \quad\text{and}\quad N(T) = n. \tag{4.5}$$

Then the joint conditional distribution of $X_1,\ldots, X_{N(T)}$ is the same as that of a sample $X_1,\ldots, X_n$ conditional on $\Sigma\, X_i$. Thus, the conditional distribution of test statistics is the same as in the standard case, under the null hypothesis as well as under the alternative, and small-sample results concerning the power continue to hold conditionally on the observed value of $N(T)$. Moreover, certain optimum properties also hold unconditionally; when for each fixed value of n a test has maximal power against certain alternatives, the expected value of its power as a function of $N(T)$ at those alternatives is also maximal. Thus, a test that is, for example, uniformly most powerful or locally most powerful in the standard case remains so in the situation considered here, albeit within the class of tests that disregard the censored observation(s) at the beginning and/or end of a record.

Strictly speaking we need multivariate versions of the theorems given in section 1.5 when tests are carried out for several acts. However, these readily follow from the theorems described, since in a (semi-)Markov chain the residence times of different states are independent.

4.6 TESTS FOR UNCENSORED OBSERVATIONS

4.6.1 A visual scanning method: the log-survivor plot

Most graphical procedures are based on the cumulative distribution function (see Box 1.2):

$$F(x;\theta) = Pr\{X \le x\}, \tag{4.6}$$

or on the survivor function:

$$\bar{F}(x;\theta) = 1 - F(x;\theta), \tag{4.7}$$

where θ denotes a (possibly unknown) parameter, which may be a vector depending on one or more covariables. In Box 1.3 we derived the relation between the survivor function and the termination rate $\lambda(s;\theta)$ (eqn (2), Box 1.3) which, by taking logarithms, transforms to

$$\log \bar{F}(x;\theta) = -\int_0^x \lambda(s;\theta)\,ds. \tag{4.8}$$

When the distribution of X is exponential (4.8) reduces to

$$\log \bar{F}(x;\lambda) = -\lambda x. \tag{4.9}$$

Accordingly, many procedures are based on so-called 'log-survivor' plots, where an estimate of $\bar{F}(x;\theta)$ is plotted against x on a semi-log scale. When the distribution of X is exponential, such a plot is approximately a straight line through the origin. A convex log-survivor plot indicates a decreasing termination rate, e.g. a mixture of exponentials, whereas a concave plot indicates an increasing termination rate, e.g. a convolution of exponentials. Time-lags are indicated when the plot is nearly horizontal for small x and tends to a straight line for larger x. Examples are given in Fig. 4.6. It may also occur that the log-survivor plot is nearly horizontal initially and becomes convex later. This indicates a combination of a time-lag with a decreasing termination rate later on during the bouts. Note that in survival

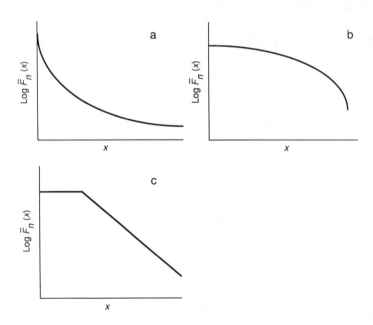

Figure 4.6 Schematic examples of different types of log-survivor plots:
a. a convex log-survivor; this indicates a mixture of exponentials
b. a concave log-survivor; this indicates a convolution
c. a log-survivor with a 'shoulder'; this indicates time-lags.

analysis it is conventional to plot $-\log \bar{F}(x;\theta)$ against x instead of $\log \bar{F}(x;\theta)$, since, according to eqn (4.8), this is equal to the so-called cumulative hazard rate:

$$\int_0^x \lambda(s)\,ds. \tag{4.10}$$

Since in ethology and other biological disciplines it is usual to make log-survivor plots, we follow this convention here.

When there are no censors the distribution function $F(x)$ at point x can be estimated by

$$F_n(x) = J/n, \tag{4.11}$$

where J is the number of observed values smaller than or equal to x. The survivor function is estimated by

$$\bar{F}_n(x) = 1 - F_n(x). \tag{4.12}$$

Note the distinction in notation between $F(x)$, the theoretical cumulative distribution function, and $F_n(x)$, its empirical counterpart and similarly between $\bar{F}(x)$ and $\bar{F}_n(x)$.

For every n and x, $F_n(x)$ has a binomial distribution with parameters n and $p = F(x)$. Thus, the standard deviation of $F_n(x)$ as well as that of $\bar{F}_n(x)$ is proportional to $1/\sqrt{n}$. A heuristic test procedure can be based on the confidence bands derived from this deviation around the empirical distribution or survivor function. A $(1 - \alpha)$ confidence interval of $\log \bar{F}(x)$ that can also be applied when there are censors is given by (e.g. Kalbfleisch and Prentice, 1980)

$$\log \bar{F}_n(x)\exp[z_{1-\frac{1}{2}\alpha}s(x)] < \log \bar{F}(x) < \log \bar{F}_n(x)\exp[z_{\frac{1}{2}\alpha}s(x)],$$

where $\tag{4.13}$

$$(s(x))^2 = \frac{\displaystyle\sum_{\{j\,|\,x_j<x\}} \frac{d_j}{n_j(n_j - d_j)}}{\left\{\displaystyle\sum_{\{j\,|\,x_j<x\}} \log\left(\frac{n_j - d_j}{n_j}\right)\right\}^2}, \tag{4.14}$$

and $z_{1-\frac{1}{2}\alpha}$ and $z_{\frac{1}{2}\alpha}$ are respectively the $1 - \frac{1}{2}\alpha$ and $\frac{1}{2}\alpha$ critical values of the standard normal distribution (see Table A1). The y_j in eqn (4.14) correspond to the ordered *distinct* values of bout lengths (see Box 4.3). n is the total number of observations, d_j is the number of bouts equal to y_j and n_j the number of bouts larger than or equal to y_j. The summations in eqn (4.14) are taken over all values of j for which y_j is smaller than x. An example is given in Box 4.3.

Box 4.3 Calculation of confidence intervals for the log-survivor function

The bout lengths are given in Table 1. There are $n = 22$ bouts, but only 10 distinct bout lengths. The ordered values, y_j, and the corresponding d_j and n_j are given in Table 2a for $j = 1$ to 9. Since at $j = 10$ the estimated log-survivor function is infinite, this point is not included. The calculation of the 95% confidence intervals for log $\bar{F}_n(y)$ is illustrated in Tables 2a and 2b. Note that $\bar{F}_n(y)$ is discontinuous at the points y_j. In the tables we give the values of

$$\bar{F}_n(y_j) = \lim_{\delta \downarrow 0} \bar{F}_n(y_j + \delta) \tag{1}$$

with corresponding confidence intervals. Define:

$$a_j = \frac{d_j}{n_j(n_j - d_j)} . \tag{2}$$

These values are given in Column 3 of Table 2a. The nominator of $(s(y_j))^2$ is equal to the sum of a_k for $k = 1$ to j (see column 4 of Table 2a). The denominator equals:

$$\left\{ \sum_{k=1}^{j} \log\left(\frac{n_k - d_k}{n_k}\right) \right\}^2 = \left\{ \log\left(\frac{n_j - d_j}{n}\right) \right\}^2 \tag{3}$$

(column 5 of Table 2a). From Table A1 we find that $z_{0.975}$ and $z_{0.025}$ are respectively 1.96 and -1.96. These values are substituted in eqn (4.13) to determine the 95% confidence intervals. For example, the estimated log-survivor function at $y_j = 2.4$ is -1.705. Thus, the 95% confidence interval for log $\bar{F}(2.4)$ is equal to

$$\left(-1.705 \exp\left[1.96\sqrt{0.0704} \right], -1.705 \exp\left[-1.96\sqrt{0.0704} \right] \right) \tag{4}$$
$$= (-2.867, -1.014).$$

Table 1 The ordered data

0.1	0.1	0.1	0.1	0.1	0.2	0.2	0.2	0.2
0.3	0.3	0.3	0.5	0.5	1.0	2.0	2.4	2.4
3.5	4.6	5.0	5.0					

Table 2a Calculation of confidence intervals for the log-survivor function, according to eqn (4.13)

j	y_j	d_j	n_j	a_j	$\sum a_k$	$(\log(n_j\text{-}d_j)/n)^2$
1	0.1	5	22	0.01337	0.01337	0.06648
2	0.2	4	17	0.0181	0.03147	0.2768
3	0.3	3	13	0.02308	0.05455	0.6217
4	0.5	2	10	0.025	0.07955	1.023
5	1.0	1	8	0.01786	0.0974	1.311
6	2.0	1	7	0.02381	0.1212	1.688
7	2.4	2	6	0.08333	0.2045	2.906
8	3.5	1	4	0.08333	0.2879	3.97
9	4.6	1	3	0.1667	0.4545	5.75

Table 2b Calculation of confidence intervals for the log-survivor function, according to eqn (4.13)

j	$(s(y_j))^2$	$\bar{F}_n(y_j)$	$\log \bar{F}_n(y_j)$	Confidence interval	
1	0.2011	0.77	−0.258	−0.621	−0.107
2	0.1137	0.59	−0.526	−1.019	−0.272
3	0.0877	0.45	−0.788	−1.409	−0.441
4	0.0777	0.36	−1.012	−1.747	−0.586
5	0.0743	0.31	−1.145	−1.954	−0.671
6	0.0718	0.27	−1.299	−2.197	−0.768
7	0.0704	0.18	−1.705	−2.867	−1.014
8	0.0725	0.14	−1.992	−3.378	−1.175
9	0.0791	0.09	−2.398	−4.161	−1.382

Some remarks on the properties of log-survivor plots

Figure 4.7a shows the probability densities of two exponentials (f_1 and f_2) with $\lambda_1 = 0.5$, $\lambda_2 = 4.0$ and of the mixture of these exponentials, with weights $w_1 = 0.37$, $w_2 = 0.63$ (f_3, see eqn (4.1)). The probability density of the mixture can hardly be distinguished from an exponential curve. However, the log-survivor plot, given in Fig. 4.7b, makes the distinction clear. There is an apparent deviation from a straight line, which illustrates the usefulness of making such plots.

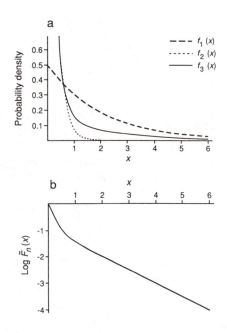

Figure 4.7
a. Example of a mixture of two exponentials:
$f_1(x) = 0.5 \exp[- 0.5x]$
$f_2(x) = 4.0 \exp[- 4.0x]$
$f_3(x) = 0.37\, f_1(x) + 0.63\, f_2(x)$
b. Log-survivor of the mixture of two exponentials given in 4.7a.

Figure 4.8a shows a plot of the probability density of a convolution of two exponentials (see eqn (4.2)). The difference from an exponential density is obvious. However, the log-survivor plot (Fig. 4.8b) does not deviate clearly from a straight line. Apparently, the log-survivor plot is less useful for the detection of this type of deviation from exponentiality.

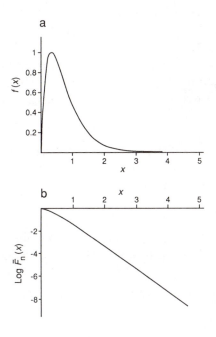

Figure 4.8
a. Probability density function of a convolution of two exponentials with parameters:
$\lambda_A = 2$ and $\lambda_B = 4$
b. Log-survivor plot of the convolution.

4.6.2 Unspecified alternative hypotheses

Besides graphical procedures, non-parametric tests can be used to test against unspecified deviations from an exponential distribution. We give the most commonly used tests plus a test developed by Shapiro and Wilk (1972) which can be used when there are minimum bout durations under the null hypothesis. Other 'overall' tests for exponentiality can be found in, for example, Stephens (1976), O'Reilly and Stephens (1982) and Kimber (1985). The test statistics given here are easy to compute and their asymptotic null distribution is known. Further references on tests for exponentiality can be found in Spurrier (1984).

The chi-squared test for goodness of fit

This test can be applied to all kinds of continuous distribution functions which are specified up to a finite number of unknown parameters. When X is exponentially distributed with unknown parameter λ, the test is as follows (see Box 4.4 for a detailed example).

Data: One sample of bout lengths $x_1,..., x_n$.

Assumptions:

A1:   Under H_0, $X_1,..., X_n$ are mutually independent exponentially distributed with common and unknown parameter λ.

A2:   Under the alternative hypothesis H_1 there are unspecified deviations
of A1.

Procedure: First choose k distinct classes of values of x. It is recommended
to choose k of the order of magnitude of $\sqrt{n}$. Let $0 = I_0, I_1,..., I_{k-1}$ be the
class bounds and let x_i belong to class j when

$$I_{j-1} \leq x_i < I_j, \quad j = 1,..., k - 1 \tag{4.15}$$

and to class k when $x \geq I_{k-1}$, $i = 1,..., n$. The number of observations in
class j is denoted by m_j. The probability of an observation in class j is equal
to

$$p_j = \exp[-\lambda I_{j-1}] - \exp[-\lambda I_j], \quad \text{for } j = 1,..., k - 1 \quad \text{and}$$

$$p_k = \exp[-\lambda I_{k-1}], \tag{4.16}$$

where the termination rate λ is unknown. This parameter has to be estimated by numerically minimizing

$$\chi^2(\lambda) = \sum_{j=1}^{k} \frac{(m_j - np_j)^2}{np_j} \tag{4.17}$$

as a function of λ.
The value $\hat{\lambda}$ for which the minimum of (4.17) is attained,

$$\chi^2(\hat{\lambda}) = \min_{\lambda} \chi^2(\lambda), \tag{4.18}$$

is called the chi-squared minimum estimate. The null hypothesis of ex-
ponentiality is rejected if $\chi^2(\hat{\lambda}) \geq c_\alpha$. (No critical values are available for
the small sample case.)

Large-sample approximation: As n tends to infinity $\chi^2(\hat{\lambda})$ has a chi-squared
distribution with $k - 2$ degrees of freedom under H_0 (i.e. $c_\alpha = \chi^2_{k-2}(\alpha)$).
Critical values are given in Table A2. The approximation is sufficiently
accurate if the expected numbers of observations in each class is larger than
one and at least 80% of these expectations are greater than, or equal to,
five.

Multiple comparison methods: On some occasions it may be of interest to
examine whether for one or more specified classes there are departures

from the expected class frequencies. For instance, to determine whether
there are time-lags (see subsection 1.3.1), deviations in the first classes can
be studied.

To test whether the class frequency in the jth class, m_j, differs significantly from its expected value, p_j, the statistic

$$Z = \frac{m_j - n\hat{p}_j}{\sqrt{n\hat{p}_j\left\{1 - \hat{p}_j - \frac{\hat{\lambda}^2}{\hat{p}_j}\left(\frac{\partial p_j}{\partial \lambda}\right)_{\hat{\lambda}}^2\right\}}}$$

(4.19)

can be used, where $\hat{p}_j$, respectively $(\partial p_j/\partial \lambda)_{\hat{\lambda}}$ denotes that the chi-squared
minimum estimate is substituted for λ (Rao, 1973). The test statistic Z has
an asymptotically standard normal distribution. H_0 is rejected if $Z < z_{1-\frac{1}{2}\alpha}$
or if $Z > z_{\frac{1}{2}\alpha}$, see Table A1 for the critical values and Box 4.4 for a
detailed example.

Box 4.4 The chi-squared test for goodness of fit for the exponential distribution with unknown λ

1. We apply the test on the data listed in Table 1. We distinguish seven
classes (which is approximately equal to $\sqrt{n}$, since n, the number of observations, is 50), with class bounds as indicated in Table 2. The class bounds
can be chosen as desired. However, the expected frequencies per class
should be larger than one and, in at least 80% of the classes, they should
be larger than or equal to five.

Table 1 The ordered data

0.0302	0.0401	0.0604	0.0863	0.160	0.226	0.252	0.326	0.354
0.382	0.387	0.388	0.397	0.505	0.516	0.560	0.641	0.665
0.762	0.876	0.895	0.905	0.935	0.976	1.06	1.08	1.20
1.24	1.28	1.34	1.40	1.44	1.50	1.57	1.60	1.63
1.69	1.69	1.71	1.72	1.83	1.98	2.16	2.26	2.34
2.62	2.69	2.77	2.82	3.47				

2. Next we numerically minimize the function χ^2, which is defined by eqns
(4.16) and (4.17). We therefore apply the Newton–Raphson method for
determining the solution of the equation $f(\lambda) = 0$, where the function f is
defined by

$$f(\lambda) = \frac{\mathrm{d}}{\mathrm{d}\lambda}\chi^2(\lambda).$$

(1)

Table 2 Frequency table

Class no.	Class bounds	Frequency (m)
1	0.00–0.25	6
2	0.25–0.50	7
3	0.50–0.75	5
4	0.75–1.00	6
5	1.00–1.50	9
6	1.50–2.50	12
7	2.50–∞	5

As starting value λ_0 we choose the maximum likelihood estimate (i.e. the inverse of the mean bout length, see eqn (15) in Box 1.2), which is equal to 0.841 in this example. The successive iteration values λ_1, λ_2, ... are calculated by applying

$$\lambda_i = \lambda_{i-1} - \frac{f(\lambda_{i-1})}{f'(\lambda_{i-1})}, \quad \text{for } i = 1, 2, \ldots . \tag{2}$$

f and f' are rather complicated functions of λ. By straightforward calculations it can be shown that

$$f(\lambda) = \sum_{j=1}^{k} \frac{\mathrm{d}p_j}{\mathrm{d}\lambda}\left(n - \frac{m_j^2}{np_j^2}\right) \tag{3}$$

and

$$f'(\lambda) = \sum_{j=1}^{k} \frac{2m_j^2}{np_j^3}\left(\frac{\mathrm{d}p_j}{\mathrm{d}\lambda}\right)^2 + \left(n - \frac{m_j^2}{np_j^2}\right)\frac{\mathrm{d}^2 p_j}{\mathrm{d}\lambda^2}, \tag{4}$$

where the m_j denote the class frequencies of Table 2 and

$$\frac{\mathrm{d}p_j}{\mathrm{d}\lambda} = I_j\exp[-\lambda I_j] - I_{j-1}\exp[-\lambda I_{j-1}], \text{ for } j = 1,\ldots,k-1,$$

$$= -I_{j-1}\exp[-\lambda I_{j-1}], \text{ for } j = k, \tag{5}$$

$$\frac{\mathrm{d}^2 p_j}{\mathrm{d}\lambda} = I_j^2\exp[-\lambda I_j] - I_{j-1}{}^2\exp[-\lambda I_{j-1}], \text{ for } j = 1,\ldots,k-1,$$

$$= -I_{j-1}{}^2\exp[-\lambda I_{j-1}], \text{ for } j = k. \tag{6}$$

First, (5) and (6) are calculated for $\lambda = \lambda_0$. The results are substituted in (3) and (4). Subsequently, (3) and (4) are substituted in (2). Thus, one obtains a new iteration value λ_1. This is repeated to get λ_2, etc. The procedure is repeated until $|d\chi^2/d\lambda|$ is sufficiently small. See Table 3 for the result.

Table 3 Results of the iteration process

Step	λ	$\chi^2(\lambda)$	$|d\chi^2/d\lambda|$
0	0.841	4.2398	10.1
1	0.753	3.8094	0.58
2	0.75744	3.8081491	2.4×10^{-3}
3	0.75746046	3.80814909	4.1×10^{-8}
4	0.757460463	3.80814909	5.0×10^{-15}

3. Substituting $\lambda_4 = 0.757460463$ in (4.16) and multiplying p_j by n results in the estimated expected class frequencies listed in Table 4.

The minimum χ^2 value is 3.808 (Table 3, column 3, row 4). Since $k = 7$, the number of degrees of freedom is $7 - 2 = 5$. From Table A2 it can be seen that the corresponding p-value lies between 0.75 and 0.50. Hence, the null hypothesis that the data are exponentially distributed is not rejected.

Table 4 Estimated class frequencies

Class no.	Class bounds	Observed frequency	Estimated expected frequency
1	0.00–0.25	6	8.626
2	0.25–0.50	7	7.138
3	0.50–0.75	5	5.906
4	0.75–1.00	6	4.887
5	1.00–1.50	9	7.391
6	1.50–2.50	12	8.526
7	2.50–∞	5	7.526

4. Application of the multiple comparison test based on Z (eqn (4.19)) on the frequency of the first class leads to the following result:

$$m_1 = 6, \quad np_1 = 8.626,$$

$$\frac{dp_1}{d\lambda} = 0.25 \exp[-0.7575 \times 0.25] - 0 = 0.25 \times 0.8275 = 0.2069.$$

Hence Z is equal to

$$\frac{6 - 8.626}{\sqrt{8.626\left(1 - 0.1725 - \dfrac{0.7575^2}{0.1725}0.2069^2\right)}} = \frac{-2.626}{2.43} = -1.08$$

and from Table A1 it can be concluded that the observed class frequency does not differ significantly from the expected estimated frequency under the hypothesis of exponentiality. This is to be expected, since the overall test does not give a significant result.

It is assumed for simplicity (without loss of generality) that deviations in the first r class frequencies, $p_1,..., p_r$, $1 < r < k - 2$ are of interest. Consider the vector $d = (d_1,..., d_r)'$, where

$$d_i = \frac{m_i - n\hat{p}_i}{\sqrt{n\hat{p}_i}} , \tag{4.20}$$

the matrix $B = (b_{ij})$, with

$$b_{ii} = 1 - \hat{p}_i \quad \text{and} \quad b_{ij} = -\sqrt{\hat{p}_i\hat{p}_j} \tag{4.21}$$

and the matrix $C = (c_{ij})$, with

$$c_{ii} = 1 - \frac{\dfrac{1}{\hat{p}_i}\left(\dfrac{\partial p_i}{\partial \lambda}\right)_{\hat{\lambda}}^2}{\displaystyle\sum_{l=1}^{k} \dfrac{1}{\hat{p}_l}\left(\dfrac{\partial p_l}{\partial \lambda}\right)_{\hat{\lambda}}^2}$$

$$c_{ij} = - \frac{\dfrac{1}{\sqrt{\hat{p}_i\hat{p}_j}}\left(\dfrac{\partial p_i}{\partial \lambda}\right)_{\hat{\lambda}}\left(\dfrac{\partial p_j}{\partial \lambda}\right)_{\hat{\lambda}}}{\displaystyle\sum_{l=1}^{k} \dfrac{1}{\hat{p}_l}\left(\dfrac{\partial p_l}{\partial \lambda}\right)_{\hat{\lambda}}^2} , \quad i \neq j, \tag{4.22}$$

$i = 1,..., k$ and $j = 1,..., k$. The matrix D_r is defined as the left upper $r \times r$ partition of the matrix $D = C\,B\,C$. See e.g. Table 4.1. For large n, the test statistic

$$\chi^2(\hat{\lambda};r) = d'D_r^{-1}d \tag{4.23}$$

Table 4.1 Left upper r-partition of a $k \times k$ matrix

M_{ij} = the ijth element of the matrix.

has approximately a chi-squared distribution with r degrees of freedom. H_0 is rejected if $\chi^2(\hat{\lambda};r) > \chi^2_r(\alpha)$. See Table A2 for the critical values (no critical values are available for the small sample-case).

Comments: Note that the chi-squared tests described above are asymptotically non-parametric. This means that the asymptotic null hypothesis distribution does not depend on the distribution of the X_i. The chi-squared test for goodness of fit is not very sensitive to deviations for large values of x (the upper tail of the distribution) and can only be used for relatively large sample sizes. An advantage of this test is the availability of a multiple comparison method to trace the cause of deviations from exponentiality.

Many introductory textbooks treat the chi-squared test for goodness of fit, for instance, Dixon and Massey (1969), Siegel and Castellan (1988), Sokal and Rohlf (1981). For a discussion on the accuracy of the approximations and for further references, see Horn (1977).

The Kolmogorov-Smirnov test

In the standard Kolmogorov-Smirnov test it is assumed that the distribution under the null hypothesis is completely specified. Durbin (1975) gives a generalization of this test, where the maximum likelihood estimator is substituted for λ in calculating the test statistic. This test is asymptotically non-parametric.

Data and assumptions: See the chi-squared test for goodness of fit.

Procedure: λ is estimated by the maximum likelihood estimator

$$\hat{\lambda} = \frac{1}{\bar{x}} = \frac{n}{\sum_{i=1}^{n} x_i} \ . \tag{4.24}$$

In the exponential case, the two-sided Kolmogorov-Smirnov test statistic is defined by

$$D_n = \max \{D_n^+, D_n^-\}, \text{ where}$$
$$D_n^+ = \max_{y_1,\ldots,y_k} \{\exp[-\hat{\lambda}y_i] - \bar{F}_n(y_i)\} \tag{4.25}$$
$$D_n^- = \max_{y_1,\ldots,y_k} \{\bar{F}_n(y_{i-1}) - \exp[-\hat{\lambda}y_i]\}$$

and $y_1, \ldots, x_k$ are the ordered *distinct* bout lengths. D_n can be considered as the maximum distance between the empirical survivor function and the theoretical survivor function for $\lambda = \hat{\lambda}$. See Box 4.5 for a detailed example. H_0 is rejected if $\sqrt{n}\, D_n > c_\alpha(n)$. The critical values $c_\alpha(n)$ are given in Table A17 for sample sizes $n = 2$ (1) 10 (2) 30 (5) 50 (10) 100.

Box 4.5   The Kolmogorov-Smirnov test for the exponential distribution with unknown λ

1. We use the same data as in Box 4.4. Since there are no identical bout lengths, the y_i correspond to the ordered bout lengths, $y_{(1)},\ldots, y_{(50)}$ which are listed in the second column of Table 1.

2. Column 3 gives the values of the estimated survivor function $\bar{F}_n(y_{(i)})$ (see also Fig. 1.) and column 4 those of $\bar{F}_n(y_{(i-1)})$, with $\bar{F}_n(y_{(0)}) = 1$.

3. The MLE of λ, $\hat{\lambda}$, is 0.841 (eqn (4.24)). This value is substituted in the function $\exp[-\lambda y_{(i)}]$ ($i = 1,\ldots, k$), to obtain the estimate of the survivor function under H_0. The results are given in the fifth column and in Fig. 1.

4. To calculate D_n^+ subtract the values in column 3 from those in column 5 (results are given in column 6). The maximum difference is attained at $y_{(20)}$ and $D_n^+ = 0.1415$. To calculate D_n^-, the values in column 5 are subtracted from those in column 4 (see column 7). The maximum is attained at $y_{(49)}$ and D_n^- equals 0.0732. It follows that $D_n = 0.1415$.

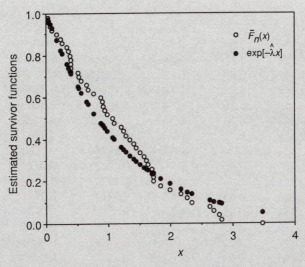

Figure 1 Plot of the two estimated survivor functions of the data from Table 1.

5. The test statistic $\sqrt{n}\,D_n = \sqrt{50} \times 0.1415 = 1.001$. Since the critical value at a level of significance of 0.05 is equal to 1.0668 (Table A17), the null hypothesis of exponentiality is not rejected.

6. The large-sample statistic is (eqn (4.26)):
$$T_n = (0.1415 - 0.2/50)(\sqrt{50} + 0.26 + 0.5/\sqrt{50}) = 1.0177,$$
which leads to the same result (from Table A18 it can be seen that the critical value is 1.094). However, we advise to restrict the application of the large-sample approximation to cases where $n > 100$.

Table 1 Calculation of the Kolmogorov-Smirnov test statistic

i	$y_{(i)}$	$\bar{F}_n(y_{(i)})$	$\bar{F}_n(y_{(i-1)})$	$\exp[-\hat{\lambda}y_{(i)}]$	D^+	D^-
1	0.0302	0.98	1.00	0.9749	−0.0051	0.0251
2	0.0401	0.96	0.98	0.9668	0.0068	0.0132
3	0.0604	0.94	0.96	0.9504	0.0104	0.0096
4	0.0863	0.92	0.94	0.9299	0.0099	0.0101
5	0.160	0.90	0.92	0.8740	−0.0260	0.0460
6	0.226	0.88	0.90	0.8268	−0.0532	0.0732
7	0.252	0.86	0.88	0.8089	−0.0511	0.0711

8	0.326	0.84	0.86	0.7601	−0.0799	0.0999
9	0.354	0.82	0.84	0.7424	−0.0776	0.0976
10	0.382	0.80	0.82	0.7251	−0.0749	0.0949
11	0.387	0.78	0.80	0.7220	−0.0580	0.0780
12	0.388	0.76	0.78	0.7214	−0.0386	0.0586
13	0.397	0.74	0.76	0.7160	−0.0240	0.0440
14	0.505	0.72	0.74	0.6538	−0.0662	0.0862
15	0.516	0.70	0.72	0.6478	−0.0522	0.0722
16	0.560	0.68	0.70	0.6242	−0.0558	0.0758
17	0.641	0.66	0.68	0.5831	−0.0769	0.0969
18	0.665	0.64	0.66	0.5714	−0.0686	0.0886
19	0.762	0.62	0.64	0.5266	−0.0934	0.1134
20	0.876	0.60	0.62	0.4785	−0.1215	0.1415*
21	0.895	0.58	0.60	0.4709	−0.1091	0.1291
22	0.905	0.56	0.58	0.4669	−0.0931	0.1131
23	0.935	0.54	0.56	0.4553	−0.0847	0.1047
24	0.976	0.52	0.54	0.4398	−0.0802	0.1002
25	1.06	0.50	0.52	0.4098	−0.0902	0.1102
26	1.08	0.48	0.50	0.4030	−0.0770	0.0970
27	1.20	0.46	0.48	0.3643	−0.0957	0.1157
28	1.24	0.44	0.46	0.3522	−0.0878	0.1078
29	1.28	0.42	0.44	0.3406	−0.0794	0.0994
30	1.34	0.40	0.42	0.3238	−0.0762	0.0962
31	1.40	0.38	0.40	0.3078	−0.0722	0.0922
32	1.44	0.36	0.38	0.2977	−0.0623	0.0823
33	1.50	0.34	0.36	0.2830	−0.0570	0.0770
34	1.57	0.32	0.34	0.2668	−0.0532	0.0732
35	1.60	0.30	0.32	0.2602	−0.0398	0.0598
36	1.63	0.28	0.30	0.2537	−0.0263	0.0463
37	1.69	0.26	0.28	0.2412	−0.0188	0.0388
38	1.69	0.24	0.26	0.2372	−0.0028	0.0228
39	1.71	0.22	0.24	0.2352	0.0152	−0.0048
40	1.72	0.20	0.22	0.2412	0.0412	−0.0212
41	1.83	0.18	0.20	0.2144	0.0344	−0.0144
42	1.98	0.16	0.18	0.1890	0.0290	−0.0090
43	2.16	0.14	0.16	0.1624	0.0224	−0.0024
44	2.26	0.12	0.14	0.1493	0.0293	−0.0093
45	2.34	0.10	0.12	0.1396	0.0396	−0.0196
46	2.62	0.08	0.10	0.1103	0.0303	−0.0103
47	2.69	0.06	0.08	0.1040	0.0440	−0.0240
48	2.77	0.04	0.06	0.09719	0.0572	−0.0372
49	2.82	0.02	0.04	0.09319	0.0732*	−0.0532
50	3.47	0	0.02	0.05393	0.0539	−0.0339

Large-sample approximation: Although the asymptotic distribution of D_n is known (e.g. Hájek and Šidák, 1967, p. 189), sample sizes are usually too small to use the corresponding critical values (given by e.g. Sokal and Rohlf, 1981). It is better to use the transformed test statistic T_n given by Pearson and Hartley (1972):

$$T_n = (D_n - 0.2/n)(\sqrt{n} + 0.26 + 0.5/\sqrt{n}). \tag{4.26}$$

H_0 is rejected if $T_n > c^*_\alpha$, where the c^*_α are listed in Table A18 for the usual values of α.

Comments: Durbin (1975) gives tables of the critical values of the one-sided analogue of this test. The Kolmogorov-Smirnov test is not very sensitive to deviations for either small or large values of x. This is a serious drawback, since mixtures of exponentials give deviations in the right tail of the distribution, whereas time-lags give deviations for small bout lengths. Kimber (1985) gives results on the power against gamma distributions with discrete parameter k larger than one (subsection 1.3.1, eqn (1.17), and eqn (4.3)) and against Weibull alternatives (eqn (4.4)). The power against Weibull alternatives gives a good indication of how a test performs against mixtures or convolutions of exponentials (see section 4.2).

The Cramér-von Mises test

A third well-known goodness-of-fit test against unspecified alternatives is the Cramér-von Mises test. It gives more weight to deviations in the tails of a distribution. Therefore this test is more sensitive to departures of exponentiality for very small or very large bout lengths.

Data and assumptions: See the chi-squared test for goodness of fit.

Procedure: The unknown parameter λ is estimated by the maximum likelihood estimator $\hat{\lambda}$ (see eqn (4.24)). We denote the ordered bout lengths by $x_{(1)}, x_{(2)}, ..., x_{(n)}$. The cumulative distribution function

$$F(x;\lambda) = 1 - \exp[-\lambda x], \tag{4.27}$$

(see eqn (2), Box 1.2) is estimated by

$$F(x;\hat{\lambda}) = 1 - \exp[-\hat{\lambda} x]. \tag{4.28}$$

The Cramér-von Mises test statistic W^2 is defined by:

$$W^2 = \frac{1}{12n} + \sum_{i=1}^{n} \left\{ F(x_{(i)};\hat{\lambda}) - \frac{2i-1}{2n} \right\}^2. \tag{4.29}$$

Hence, the test statistic is equal to the sum of the squared differences between the expected values of the empirical distribution function and the (estimated) exponential distribution function at the points $x_{(1)}, x_{(2)}, ..., x_{(n)}$. H_0 is rejected if W^2 exceeds the critical value $W^2(\alpha)$. See Box 4.6 for a worked example.

Box 4.6 The Cramér-von Mises test for the exponential distribution with unknown λ

1. The ordered data $x_{(1)}, ..., x_{(n)}$ ($n = 50$) are listed in the second column of Table 1.

Table 1 Calculation of the Cramér-von Mises test statistic

i	$x_{(i)}$	$F(x_{(i)};\hat{\lambda})$	$\dfrac{2i-1}{2n}$	$\left\{F(x_{(i)};\hat{\lambda}) - \dfrac{2i-1}{2n}\right\}^2$
1	0.0302	0.02509	0.01	0.00023
2	0.0401	0.03318	0.03	0.00001
3	0.0604	0.04956	0.05	0.00000
4	0.0863	0.07005	0.07	0.00000
5	0.160	0.1260	0.09	0.00129
6	0.226	0.1732	0.11	0.00399
7	0.252	0.1911	0.13	0.00373
8	0.326	0.2399	0.15	0.00809
9	0.354	0.2576	0.17	0.00768
10	0.382	0.2749	0.19	0.00721
11	0.387	0.2780	0.21	0.00462
12	0.388	0.2786	0.23	0.00236
13	0.397	0.2840	0.25	0.00116
14	0.505	0.3462	0.27	0.00581
15	0.516	0.3522	0.29	0.00387
16	0.560	0.3758	0.31	0.00433
17	0.641	0.4169	0.33	0.00756
18	0.665	0.4286	0.35	0.00617
19	0.762	0.4734	0.37	0.01068
20	0.876	0.5215	0.39	0.01730
21	0.895	0.5291	0.41	0.01419

22	0.905	0.5331	0.43	0.01062
23	0.935	0.5447	0.45	0.00897
24	0.976	0.5602	0.47	0.00813
25	1.06	0.5902	0.49	0.01004
26	1.08	0.5970	0.51	0.00757
27	1.20	0.6357	0.53	0.01118
28	1.24	0.6478	0.55	0.00956
29	1.28	0.6594	0.57	0.00800
30	1.34	0.6762	0.59	0.00743
31	1.40	0.6922	0.61	0.00674
32	1.44	0.7023	0.63	0.00523
33	1.50	0.7170	0.65	0.00449
34	1.57	0.7332	0.67	0.00399
35	1.60	0.7398	0.69	0.00248
36	1.63	0.7463	0.71	0.00132
37	1.69	0.7588	0.73	0.00083
38	1.69	0.7588	0.75	0.00008
39	1.71	0.7628	0.77	0.00005
40	1.72	0.7648	0.79	0.00063
41	1.83	0.7856	0.81	0.00059
42	1.98	0.8110	0.83	0.00036
43	2.16	0.8376	0.85	0.00015
44	2.26	0.8507	0.87	0.00037
45	2.34	0.8604	0.89	0.00087
46	2.62	0.8897	0.91	0.00041
47	2.69	0.8960	0.93	0.00115
48	2.77	0.9028	0.95	0.00223
49	2.82	0.9068	0.97	0.00399
50	3.47	0.9461	0.99	0.00193

2. Calculate the MLE of λ (eqn (4.24), which gives $\hat{\lambda} = 0.841$) and substitute this value in the cumulative distribution function $F_n(x_{(i)};\lambda) = 1 - \exp[-\lambda x_{(i)}]$ ($i = 1,..., n$) to obtain the third column of Table 1.

3. Calculate the expected values from the cumulative distribution function $F_n(x_{(i)})$ for $i = 1,..., n$. (See fourth column of Table 1.)

4. Calculate the sum of the squared differences between these two estimates (fifth column) and determine the value of W^2 (see eqn (4.29)): $W^2 = 0.231$.

5. The correction proposed by Stephens according to eqn (4.30) results in $(W^2)' = 0.232$; hence $0.025 < p < 0.05$, and the null hypothesis of exponentiality is rejected (see Table A19).

Large-sample approximation: Stephens (1976) gives asymptotic critical values. These are listed in Table A19. The transformation

$$(W^2)' = W^2(1 + 0.16/n)$$ (4.30)

improves the accuracy considerably. Stephens (1974) showed by Monte Carlo methods that with this transformation the large-sample critical values can already be used at sample sizes of 10. Our simulation results show that even for smaller sample sizes the approximation is satisfactory, provided that the level of significance is not too small. See Table 4.2.

Table 4.2 Levels of significance of the test based on $(W^2)'$ at small sample sizes, when the large-sample critical values are used

Sample size		Estimated level of significance			
3	0.169	0.113	0.040	0.008	0.000
5	0.148	0.098	0.042	0.021	0.006
10	0.161	0.108	0.056	0.025	0.009
50	0.141	0.096	0.044	0.025	0.010
∞ (exact)	0.150	0.100	0.050	0.025	0.010

The data are based on simulation runs of length 1000.

Comments: The Cramér-von Mises test usually performs better than the Kolmogorov-Smirnov test, since it puts more weight on the tails. Stephens (1974) gives simulation results. These indicate that in tests on normality with unknown μ and σ^2, the Cramér-von Mises test does much better than the Kolmogorov-Smirnov test. However, there are no power comparisons for the exponential case.

The Shapiro-Wilk test

The Shapiro-Wilk test is especially designed for testing exponentiality if there may be minimum bout durations (time-lags). Models for time-lags were treated in subsection 1.3.1. Thus, this test differs from the others given in this chapter in that its null hypothesis distribution is not necessarily the one-parameter exponential distribution. Instead, there may be a fixed minimum duration under the null hypothesis.

Data: One sample of bout lengths $x_1, ..., x_n$.

Assumptions:
A1:   Under H_0, $X_1, ..., X_n$ are mutually independent and identically distributed according to a two-parameter exponential distribution.
A2:   Under the alternative hypothesis H_1 there are unspecified deviations of A1.

Procedure: The n ordered bout lengths are denoted by $x_{(1)}, x_{(2)}, ..., x_{(n)}$, where $x_{(1)}$ is the smallest and $x_{(n)}$ the largest observation. Calculate the statistics W and U, defined by

$$W = \frac{n(\bar{x} - x_{(1)})^2}{(n-1)S^2}, \tag{4.31}$$

$$U = \frac{x_{(1)}}{\bar{x} - x_{(1)}}, \tag{4.32}$$

where $S^2 = \sum_{i=1}^{n}(x_i - \bar{x})^2$ and $\bar{x} = \frac{1}{n}\sum_{i=1}^{n} x_i$.

W is sensitive to departures from exponentiality irrespective of the presence of a minimum bout duration. U can be used for testing whether the minimum bout duration is zero. The p-value of W, p_W, can be obtained from Table A20 for $n = 3$ (1) 100. The p-value of U is equal to $(1 + U)^{-(n-1)}$. Since W and U are stochastically independent, the p-values can be combined by Fisher's omnibus procedure in order to test the two hypotheses of exponentiality and minimum bout duration simultaneously (see also Chapter 6). The corresponding test statistic is

$$T_1 = -2\log p_w p_u. \tag{4.33}$$

Under the null hypothesis of exponentiality, T_1 is chi-squared distributed with four degrees of freedom. H_0 is rejected if T_1 exceeds the critical value $\chi^2_4(\alpha)$ listed in Table A2. See Box 4.7 for an example.

Box 4.7   The Shapiro–Wilk test

1. We illustrate the application of this test using the data listed in Table 1 of Box 4.4.

2. Calculate the test statistic W. The smallest observation $x_{(1)} = 0.0302$, the mean $\bar{x} = 1.1883$ and $S^2 = 35.416$. Substitution in (4.31) yields

$$W = \frac{50 \times (1.1883 - 0.0302)^2}{49 \times 35.416} = 0.0386.$$

It follows from Table A20 that the approximate p-value p_W lies between 0.01 and 0.005. By linear interpolation we arrive at an approximate value of 0.0062. It follows that the null hypothesis of exponentiality is rejected.

3. Calculate U defined by (4.32):

$$U = \frac{0.0302}{1.1883 - 0.0302} = 0.02608;$$

hence, the p-value is equal to

$$p_U = (1 + 0.02608)^{-(50-1)} = 0.2836,$$

which is in accordance with the null hypothesis of the absence of a minimum bout duration.

4. To combine these two test results we calculate T_1, which gives:

$$T_1 = -2\log p_W p_U = -2\log(0.2836 \times 0.0062) = 12.69.$$

This means that the overall hypothesis is also rejected, since T_1 is larger than the critical value of a chi-squared distribution with four degrees of freedom (= 9.49 at a level of significance of 0.05, see Table A2).

Large-sample approximation: Large-sample approximations for the distribution of W for $n > 100$ are not available. It can be shown, however, that the asymptotic distribution of $(1/W) - 1$ is equal to that of Darling's test statistic (see Darling, 1953, and eqns (4.42) and (4.43)), with n replaced by $n - 1$.

Multiple comparison methods: It is possible to proceed in two stages: first test whether there are deviations from an exponential distribution, with T_1. If this gives a significant result, perform a multiple comparison by means of W and U to see whether deviations are due to a minimum duration (i.e. U gives a significant result, W does not) or to other causes (W gives a significant result).

Comments: The Shapiro-Wilk test performs well against skew alternatives and against distributions with either short or long tails. This test differs from the others treated in this section in that there may be time-lags under the null hypothesis. The asymptotic equivalence of $(1/W) - 1$ and K_n (eqn (4.42)) indicates that the test based on W will have good local power properties against mixtures of exponentials (see also subsection 4.6.4).

4.6.3 Broad classes of alternative hypotheses: Barlow's 'total time on test' statistic

Barlow *et al.* (1972) describe a test against alternative hypotheses formulated in terms of the termination rate, based on the so-called 'cumulative total time on test' statistic.

Data: One sample of bout lengths $x_1,..., x_n$.

Assumptions:
A1: $X_1,..., X_n$ are mutually independent and identically distributed.
A2:   Under H_0, the X_i are exponentially distributed with unknown parameter λ: $f(x) = \lambda \exp[-\lambda x]$, $0 < x < \infty$.
A3:   Under the alternative hypothesis H_1 the termination rate $\lambda(x)$ is a monotonically increasing or decreasing function of x.

Procedure: We denote the ordered bout lengths by $x_{(1)}, x_{(2)},..., x_{(n)}$. The 'cumulative total time on test' statistic V_n is defined by

$$V_n = \frac{2}{n\bar{x}} \sum_{i=1}^{n-1} (n - i)x_{(i)} . \tag{4.34}$$

Large values indicate an increasing termination rate, small values a decreasing one. The upper critical values for $n = 3$ (1) 26 are listed in Table A21, which is partially adapted from Barlow *et al.* (1972, p. 269). The lower critical values can also be calculated from this table, since the distribution of the test statistic is symmetric around $(n - 1)/2$. Thus, if c_α is the upper critical value of level α, the lower critical value of level α is equal to $(n - 1) - c_\alpha$.

Large-sample approximation: For large n the distribution of

$$Z = \left(\frac{V_n}{n - 1} - \frac{1}{2} \right) \sqrt{12(n - 1)} \tag{4.35}$$

tends to a standard normal distribution. For a right-sided test reject H_0 if Z exceeds the critical value $z_{\frac{1}{2}\alpha}$ listed in Table A1. See Box 4.8 for a detailed example.

Box 4.8   Barlow's test

1. The data are given in Table 1. They are ordered according to increasing values. The total number of observations, n, is 25. The mean value, $\bar{x}$, is 0.2853

Table 1 Ordered data

0.003913	0.006023	0.05354	0.06414	0.06664
0.07449	0.1497	0.1609	0.1655	0.2060
0.2202	0.2286	0.2410	0.2420	0.2442
0.2461	0.2571	0.2924	0.3707	0.3743
0.4539	0.5214	0.6078	0.9158	0.9653

2. Compute the statistic V_n (eqn (4.34)):

$$V_n = 2 \times \frac{24 \times 0.003913 + 23 \times 0.006023 + \dots + 1 \times 0.9158 + 0}{25 \times 0.2853} = 13.08.$$

3. The upper critical value for the two-sided test at a level of significance $\alpha = 0.05$ ($\frac{1}{2}\alpha = 0.025$) is equal to 14.767 (see Table A21; note that $n - 1 = 24$) and the null hypothesis of a constant termination rate cannot be rejected.

The results of Chen (1984) indicate that the asymptotic approximation holds well for $n \geq 30$. Table 4.3 shows that it can also be applied for smaller values of n, provided that the level of significance is not too small.

Table 4.3 Exact levels of significance of the test based on Z (eqn (4.35)) when the normal approximation is used

	α				
$n-1$	0.10	0.05	0.025	0.01	0.005
5	0.103	0.0506	0.0239	0.0083	0.0034
10	0.102	0.0503	0.0245	0.0092	0.0043
15	0.101	0.0502	0.0247	0.0095	0.0045
20	0.101	0.0501	0.0247	0.0096	0.0046
25	0.101	0.0501	0.0248	0.0097	0.0047

Properties: The one-sided tests based on V_n have asymptotically the highest minimum power that a test can have against alternatives with, in some sense, equally strong increasing (or decreasing) termination rates (Barlow *et al.*, 1972, section 6.3). Note that DFR (decreasing failure rate) tests are also sensitive to deviations in the direction of mixtures of exponentials, whereas IFR (increasing failure rate) tests are sensitive to convolutions of exponentials or to time-lags (section 4.3.)

Although the test was originally developed for one-sided alternatives, a two-sided analogue can also be used: reject H_0 if V_n is either smaller than the critical value for $1 - \frac{1}{2}\alpha$ or larger than the critical value for $\frac{1}{2}\alpha$. The results of Chen (1984) and Deshpande (1983) indicate that the two-sided test based on V_n has good power against Weibull alternatives (see eqn (4.4)) with increasing or decreasing termination rates.

4.6.4 Specific alternative hypotheses

Most tests treated in this subsection are illustrated with the data given in Box 4.9.

Box 4.9 Data used in Boxes 4.10, 4.12 and 4.13

i	x_i	$(x_i - \bar{x})^2$	$\log x_i$	$x_i \log x_i$	$x_i(\log \hat{\lambda} x_i)^2$
1	0.108	0.905	-2.22	-0.241	0.564
2	0.0266	1.07	-3.63	-0.0964	0.361
3	0.229	0.691	-1.48	-0.337	0.538
4	0.295	0.585	-1.22	-0.36	0.482
5	0.142	0.842	-1.95	-0.277	0.574
6	0.0317	1.06	-3.45	-0.109	0.39
7	0.0813	0.958	-2.51	-0.204	0.536
8	1.95	0.797	0.669	1.31	0.729
9	0.402	0.433	-0.911	-0.366	0.378
10	0.738	0.104	-0.304	-0.224	0.0967
11	2.68	2.62	0.985	2.64	2.3
12	0.942	0.0139	-0.0595	-0.0561	0.0131
13	0.142	0.843	-1.95	-0.277	0.574
14	0.181	0.772	-1.71	-0.31	0.565
15	0.27	0.624	-1.31	-0.354	0.505
16	0.0823	0.956	-2.5	-0.205	0.537
17	1.87	0.656	0.626	1.17	0.602
18	0.319	0.548	-1.14	-0.365	0.46

Box 4.9 Continued

i	x_i	$(x_i - \bar{x})^2$	$\log x_i$	$x_i \log x_i$	$x_i(\log \hat{\lambda} x_i)^2$
19	0.0124	1.1	−4.39	−0.0543	0.245
20	0.0633	0.993	−2.76	−0.175	0.503
21	5.54	20.1	1.71	9.48	15.1
22	0.0335	1.05	−3.4	−0.114	0.4
23	5.39	18.7	1.68	9.07	14.2
24	0.0607	0.999	−2.8	−0.17	0.497
25	0.53	0.281	−0.634	−0.336	0.254
26	0.0234	1.07	−3.75	−0.0879	0.34
27	0.173	0.787	−1.76	−0.303	0.569
28	0.336	0.524	−1.09	−0.366	0.443
29	2.16	1.22	0.772	1.67	1.1
30	1.01	0.00251	0.00986	0.00996	0.00237
31	0.147	0.833	−1.91	−0.282	0.574
32	0.6	0.212	−0.512	−0.307	0.195
33	0.0233	1.07	−3.76	−0.0876	0.34
34	0.0545	1.01	−2.91	−0.159	0.48
35	1.75	0.479	0.561	0.983	0.442
36	4.68	13.1	1.54	7.23	10.3
37	0.793	0.0711	−0.231	−0.184	0.0666
38	0.176	0.781	−1.74	−0.306	0.567
39	0.0192	1.08	−3.95	−0.0758	0.309
40	1.12	0.00354	0.113	0.126	0.00333
41	0.444	0.38	−0.813	−0.361	0.337
42	0.201	0.738	−1.6	−0.322	0.556
43	6.94	34.6	1.94	13.4	24.5
44	2.48	2.03	0.91	2.26	1.8
45	0.144	0.84	−1.94	−0.279	0.574
46	0.414	0.418	−0.883	−0.365	0.366
47	0.436	0.39	−0.83	−0.362	0.344
48	3.2	4.59	1.16	3.73	3.92
49	0.651	0.167	−0.429	−0.279	0.154
50	2.89	3.35	1.06	3.07	2.91
Totals:	53	127	−54.7	47.4	92.7

The likelihood ratio test against Weibull alternatives

One possible alternative hypothesis corresponding to a monotonically changing termination rate of a specific form is the Weibull distribution (see also section 4.3).

Data: One sample of bout lengths $x_1, ..., x_n$.

Assumptions:

A1: $X_1,..., X_n$ are mutually independent and identically distributed.

A2: Under H_0, the X_i have an exponential distribution with unknown termination rate: $f(x) = \lambda \exp[-\lambda x]$, $0 < x < \infty$.

A3: Under the alternative hypothesis H_1 the termination rate is $\lambda(x) = \lambda \rho (\lambda x)^{\rho-1}$, with λ and ρ unspecified.

Procedure: The test statistic is

$$t = \frac{n + \sum \log x_i - n \dfrac{\sum x_i \log x_i}{\sum x_i}}{I_{\rho\rho} - \dfrac{I_{\rho\lambda}^2}{I_{\lambda\lambda}}}, \tag{4.36}$$

where the elements of the information matrix I are equal to

$$\begin{aligned} I_{\rho\rho} &= n + \sum (\hat{\lambda} x_i)\{\log(\hat{\lambda} x_i)\}^2, \\ I_{\rho\lambda} &= \sum x_i \log(\hat{\lambda} x_i), \\ I_{\lambda\lambda} &= \frac{n}{\hat{\lambda}^2} \end{aligned} \tag{4.37}$$

and $\hat{\lambda}$ is the MLE of λ under the null hypothesis, i.e. $1/\bar{x}$. Positive values of t indicate an increasing termination rate, negative values a decreasing one. There are no critical values available for the small-sample case. See Box 4.10 for an example.

Box 4.10   The likelihood ratio test against Weibull alternatives

1. The data are given in Box 4.9.

2. $\hat{\lambda}$ is $1/\bar{x} = 0.94$. Thus:
   $I_{\rho\rho} = 50 + 0.94 \times 92.7 = 137.1$
   $I_{\rho\lambda} = 47.4 + 53 \log 0.94 = 44.1$
   $I_{\lambda\lambda} = 50/0.94^2 = 56.59$
and the denominator in t equals $137.1 - 44.1^2/56.59 = 103.1$ The test statistic (eqn (4.36)) is $(50 - 54.7 - 50 \times 47.4/53)/103.1 = -0.479$. Accordingly, at a significance level of 0.05 the null hypothesis is not rejected by this test, whether it is applied one or two sided. (See Table A1 for the critical values.)

Large-sample approximation: As n tends to infinity, the test statistic t has a standard normal distribution. See Table A1 for the appropriate critical values. In the right-sided case H_0 is rejected in favour of an increasing termination rate if $t > z_\alpha$ and in the left-sided case H_0 is rejected in favour of a decreasing termination rate if $t < z_{1-\alpha}$. In the two-sided case, H_0 is rejected if $t < z_{1-\frac{1}{2}\alpha}$ or if $t > z_{\frac{1}{2}\alpha}$.

Comments: The test is described by Cox and Oakes (1984).

Matthews and Farewell's test for an abrupt change in the termination rate

Instead of a gradual change in the termination rate, there may be an abrupt change, see for example Fig. 4.2 and Box 4.11. The test proposed by Matthews and Farewell (1982) is especially sensitive to such deviations. We will discuss the improved version of this test, due to Worsley (1988).

Data: One sample of bout lengths $x_1, ..., x_n$.

Assumptions:
A1: $X_1, ..., X_n$ are mutually independent and identically distributed.
A2:   Under H_0 the X_i have an exponential distribution with unknown constant termination rate, i.e. $\lambda(x) = \lambda$. The probability density of the X_i, $f(x) = \lambda \exp[-\lambda x]$, $0 < x < \infty$.
A3:   Under H_1 the termination rate is
$$\lambda(x) = \lambda \quad \text{for } x \leq \tau$$
$$= \rho\lambda \quad \text{for } x > \tau,$$
where λ, ρ and τ are unknown.

Procedure: The n ordered bout lengths are denoted by $x_{(1)}, x_{(2)}, ..., x_{(n)}$, where $x_{(1)}$ is the smallest and $x_{(n)}$ the largest observation. The test statistic is

$$LR_n = \max_t 2\{n\log\bar{x} - s\log\bar{x}_s - (n - s)\log\bar{x}_{n-s}\}, \tag{4.38}$$

where s is the number of the x_i smaller than or equal to t,

$$\bar{x}_s = \frac{1}{s} \sum_{i=1}^{s} x_{(i)} \tag{4.39}$$

and

$$\bar{x}_t = \frac{1}{n-s} \sum_{i=s+1}^{n} x_{(i)} . \tag{4.40}$$

The estimated change point τ is that value of t for which (4.38) attains its maximum, where the maximum is taken over $t = x_{(1)}, ..., x_{(n-1)}$. Reject H_0 if LR_n exceeds the critical value. See Box 4.11 for an example. In Table A22 critical values for $n = 10$ (10) 100 are listed, based on the results of Worsley (1988).

Box 4.11 Matthews and Farewell's test

Presumably an abrupt change of the termination rate of an act will not often occur during a bout. However, there are some occasions when such a model may apply. For instance, ultrasonic sounds made by rats consist of bursts of short 'beeps'. It is often difficult to determine when a vocalization bout has ended. To do this, we must find a way to distinguish short interruptions between subsequent 'beeps' within one vocalization bout from gaps between bouts. One possibility is to assume that the tendency to start another 'beep' increases abruptly after a certain time τ. Pauses within vocalization bouts are shorter than τ, whereas gaps are longer than τ. The log-survivor of the times between 'beeps' will then look like Fig. 1. In such cases Matthews and Farewell's test may be used to test whether there is indeed

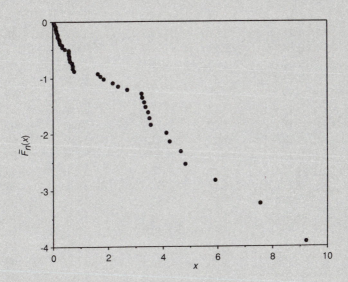

Figure 1 Log-survivor plot of the data given in Table 1. There appears to be an abrupt change in the termination rate during bouts.

a significant change. Furthermore, the value of t at which the likelihood ratio (cf. eqn (4.38)) attains its maximum is the maximum likelihood estimator of τ. This can be used to define vocalization bouts: a bout ends when the interruption lasts longer than τ. (Note that in this example we are investigating the time intervals between 'beeps', so we are looking at gaps between the occurrence of an act rather than at the durations of the act itself.) The calculation of LR_n is illustrated in Table 1.

Table 1 Calculation of LR_n (eqn (4.38))

Bout lengths	$\bar{x}_s$	$\bar{x}_{n-s}$	$s\log \bar{x}_s$	$(n-s)\log \bar{x}_{n-s}$	$LR(t)$
0.0294	0.0294	1.99	−3.53	33.7	6.44
0.0495	0.0395	2.03	−6.47	33.9	11.92
0.0767	0.0519	2.07	−8.88	34.2	16.14
0.0932	0.0622	2.11	−11.1	34.4	20.18
0.0937	0.0685	2.16	−13.4	34.6	24.38
0.106	0.0748	2.2	−15.6	34.8	28.38
0.113	0.0803	2.25	−17.7	34.9	32.38
0.127	0.0861	2.3	−19.6	35	35.98
0.134	0.0914	2.36	−21.5	35.1	39.58
0.159	0.0982	2.41	−23.2	35.2	42.78
0.174	0.105	2.47	−24.8	35.2	45.98
0.175	0.111	2.53	−26.4	35.3	48.98
0.218	0.119	2.59	−27.7	35.2	51.78
0.227	0.127	2.66	−28.9	35.2	54.18
0.243	0.135	2.73	−30.1	35.1	56.78
0.253	0.142	2.8	−31.2	35	59.18
0.325	0.153	2.87	−31.9	34.8	60.98
0.329	0.163	2.95	−32.7	34.7	62.78
0.434	0.177	3.03	−32.9	34.4	63.78
0.567	0.196	3.12	−32.6	34.1	63.78
0.569	0.214	3.2	−32.4	33.8	63.98
0.575	0.231	3.3	−32.3	33.4	64.58
0.59	0.246	3.4	−32.2	33	65.18
0.594	0.261	3.51	−32.3	32.6	66.18
0.629	0.275	3.62	−32.2	32.2	66.78
0.695	0.291	3.74	−32.1	31.7	67.58
0.705	0.307	3.88	−31.9	31.2	68.18
0.725	0.322	4.02	−31.8	30.6	69.18
0.76	0.337	4.17	−31.6	30	69.98*

* Maximum value of $LR(t)$

Table 1 Continued

1.61	0.379	4.3	−29.1	29.2	66.58
1.72	0.423	4.44	−26.7	28.3	63.58
1.83	0.466	4.58	−24.4	27.4	60.78
2.16	0.518	4.73	−21.7	26.4	57.38
2.36	0.572	4.87	−19	25.3	54.18
2.69	0.633	5.02	−16	24.2	50.38
3.22	0.704	5.15	−12.6	22.9	46.18
3.23	0.773	5.3	−9.54	21.7	42.46
3.31	0.839	5.46	−6.65	20.4	39.28
3.36	0.904	5.65	−3.94	19.1	36.46
3.45	0.968	5.87	−1.32	17.7	34.02
3.49	1.03	6.14	1.17	16.3	31.84
3.55	1.09	6.46	3.58	14.9	29.82
4.13	1.16	6.79	6.37	13.4	27.24
4.24	1.23	7.22	9.1	11.9	24.78
4.64	1.31	7.74	12	10.2	22.38
4.8	1.38	8.47	14.9	8.55	19.88
5.9	1.48	9.33	18.4	6.7	16.58
7.54	1.6	10.2	22.7	4.65	12.08
9.22	1.76	11.2	27.7	2.42	6.54
11.2	0	0	0	0	0

1. Arrange the bout lengths in increasing order (column 1).

2. For each bout length t calculate the mean of the bout lengths that are smaller than or equal to t, $\bar{x}_s$ (column 2), and the mean of the bout lengths larger than t, $\bar{x}_{n-s}$ (column 3).

3. For each t calculate:
$$LR(t) = 2\{n\log\bar{x} - s\log\bar{x}_s - (n-s)\log\bar{x}_{n-s}\}. \tag{1}$$

The results of this calculation are given in column 6. (Note that $\bar{x}$ is equal to 1.95 in this case.)

4. From Table 1 it follows that in the example the maximum of $LR(t)$ is equal to 69.98 and is attained at $t = 0.76$. The sample size, n, is 50, so it follows from Table A22 that the critical value at $\alpha = 0.05$ lies between 10.308 and 10.424. Thus, the null hypothesis is rejected at a significance level of 5%. The estimated change point τ is equal to 0.76.

When the null hypothesis is rejected and the maximum likelihood estimate of ρ

$$\hat{\rho} = \frac{\bar{x}_s}{\bar{x}_{n-s}} \tag{4.41}$$

is larger than one, there may be time-lags. This can be checked by inspection of the log-survivor plot (cf. subsection 4.6.1).

Properties: Matthews and Farewell (1982) give results on the power of the test for several ρ and τ. Worsley (1988) improved the test and gave exact results on the critical values. Further adjustments of the test are given by Henderson (1990). Presumably the test is also sensitive to time-lag alternatives ($\rho > 1$), although the termination rate has a different form when there are in fact time-lags.

Darling's test against mixtures of exponentials

A test against mixtures of exponentials with unknown mixing proportions and termination rates is given by Darling (1953).

Data: One sample of bout lengths $x_1, ..., x_n$.

Assumptions:
A1: $X_1, ..., X_n$ are mutually independent with probability density $f(x)$.
A2:   The probability density is a mixture of exponential distributions with weights according to $g(s)$ ($0 < s < \infty$):

$$f(x) = \int_0^\infty s \exp[-sx] g(s) ds.$$

A3:   The mixing function $g(s)$ can be considered as a probability density with mean λ and variance $\xi\lambda$, i.e.

$$\int_0^\infty s g(s) ds = \lambda \quad \text{and} \quad \int_0^\infty (s - \lambda)^2 g(s) ds = \xi\lambda.$$

A4:   Under H_0, $\xi = 0$, and hence the X_i are exponentially distributed with unknown parameter λ.

Procedure: The test statistic K_n is defined as follows:

$$K_n = \sum_{i=1}^n \left(\frac{x_i - \bar{x}}{\bar{x}}\right)^2. \tag{4.42}$$

The null hypothesis is rejected for large values of K_n.

Large-sample approximations: For large n, the standardized test statistic

$$\frac{K_n - \mu}{\sigma} \, ,$$

where   (4.43)

$$\mu = \frac{n(n - 1)}{n + 1} \quad \text{and} \quad \sigma^2 = \frac{4n^4(n - 1)}{(n + 1)^2(n + 2)(n + 3)} \, ,$$

is approximately N(0,1) distributed. The null hypothesis is rejected for large values of the standardized test statistic given in eqn (4.43). (See Table A1 for the critical values and Box 4.12 for a worked example.)

Box 4.12 Darling's test against mixtures of exponentials

1. The data are listed in Box 4.9. The sample size $n = 50$.

2. Compute the mean $\bar{x}$ and $\Sigma(x_i - \bar{x})^2$ and substitute these values, equal to 1.06 and 127 respectively, in K_n defined by (4.42). Hence, $K_n = 113.03$.

3. The expected mean of this test statistic $\mu = 50 \times 49/51 = 48.04$ and the variance $\sigma^2 = 4 \times 50^5/(51^2 \times 52 \times 53) = 174.38$. Hence, $\sigma = 13.2$.

4. We perform a right-sided test, i.e. we test for deviations in the direction of a mixture. The test statistic $(K_n - \mu)/\sigma$ is equal to 4.92. Since under H_0 this statistic has a standard normal distribution, it follows from Table A1 that, at a significance level of 0.05, the data deviate significantly from exponentiality.

Properties: This test is locally most powerful against a mixture of exponential distributions, where the variance of the mixing distribution tends to zero, irrespective of the value of λ. One might argue that in ethology small differences are not of special interest, because they may be caused by time inhomogeneity (for instance, small 'mood-changes'). Therefore, one must distinguish significant and relevant results. Especially when tests such as the one treated here are applied, which are optimal against local alternatives, it should be considered whether the order of magnitude of deviations is indeed relevant. Note that good local power properties do not imply that a test has less good discriminating power against alternatives which are not 'nearby'. On the contrary, most locally most powerful tests have favourable properties in a large part of the parameter space.

A two-sided analogue of the test can also be used. In this case small values indicate an increasing termination rate.

Moran's test against gamma distributions

The left-sided version of Moran's test is optimal against gamma distributions (see eqn (4.3)). It is also expected to have good power properties against related types of alternatives, for instance convolutions of exponentials with unequal termination rates (as e.g. in eqn (4.2)). The two-sided form of the test is also sensitive to deviations in the direction of decreasing termination rates (e.g. mixtures of exponentials).

Data: One sample of bout lengths $x_1,..., x_n$.

Assumptions:
A1: $X_1,..., X_n$ are mutually independent and identically distributed.
A2: Under H_0 the X_i are exponentially distributed with unknown parameter λ: $f(x) = \lambda \exp[-\lambda x]$, $0 < x < \infty$.
A3: Under H_1 the X_i follow a gamma distribution with unknown parameters k and λ:
$f(x) = \lambda \{(\lambda x)^{k-1} \exp[-\lambda x]\}/\{\Gamma(k)\}$, $0 < x < \infty$.

Procedure: The test statistic L_n is defined by

$$L_n = 2\left(n\log\bar{x} - \sum_{i=1}^{n}\log x_i\right). \tag{4.44}$$

To test against a decreasing termination rate, the null hypothesis is rejected for large values of L_n. Small values of L_n indicate an increasing termination rate.

Large-sample approximation: For large n, L_n has approximately a chi-squared distribution with $n - 1$ degrees of freedom. Critical values are given in Table A2. The approximation is improved by the transformation

$$L_n' = \frac{L_n}{1 + \dfrac{n + 1}{6n}} \tag{4.45}$$

(Cox and Lewis, 1978).

Dienske and Metz (1977) give an improved large-sample approximation for the critical values, based on:

$$L_n^* = 3\frac{\mu}{\sigma}\left\{\left(\frac{L_n}{\mu}\right)^{1/3} - 1 + \frac{\sigma^2}{9\mu^2}\right\}, \tag{4.46}$$

where

$$\mu = 1.15443\,n - 1 - \frac{1}{6n}$$

and

$$\sigma^2 = 2.57974\,n - 2 - \frac{2}{3n}. \tag{4.48}$$

For large n, L_n^* is approximately $N(0,1)$ distributed. Critical values are given in Table A1. See Box 4.13 for an example.

Box 4.13 Moran's test against gamma distributions

1. The data are given in Box 4.9.

2. Since $\bar{x} = 1.06$, $\log\bar{x} = 0.058$. The sum over $\log x_i$ is equal to -54.7 (see Box 4.9). Thus the statistic $L_n = 115.2$ (see eqn (4.44)).

3. To apply the large-sample approximation given in eqn (4.41), we calculate:
   $\mu = 1.15443\times50 - 1 - 1/300 = 56.7$ and
   $\sigma^2 = 2.57974\times50 - 2 - 2/150 = 127$; hence,
   $L_n^* = 3(56.7/11.3)\{(115.2/56.7)^{1/3} - 1 + 127/(9\times56.7^2)\} = 4.08$.
When we apply a two-sided test at a significance level of 0.05, the critical values are -1.96 and 1.96, since L_n^* has approximately a standard normal distribution under H_0 (see Table A1). The null hypothesis is thus rejected. The fact that L_n^* is large indicates a deviation in the direction of a decreasing termination rate, i.e. a mixture of exponentials.

Properties: This test is uniformly most powerful unbiased against deviations in the family of gamma distributions (Shorack, 1972), irrespective of the true value of λ. Although the test was originally developed as a left-sided test, it can also be applied as a right-sided or a two-sided test. Small values of L_n or the test statistics given in eqns (4.45) and (4.46) indicate an increasing termination rate, whereas large values indicate a decreasing termination rate. Therefore, the left-sided test is sensitive to convolutions and the right-sided test is sensitive to mixtures. The two-sided test is sensitive to mixtures as well as convolutions.

Comments: The test based on L_n (or L_n') is known to be very sensitive to small recording errors. Therefore, it is probably not very useful when acts have a relatively short duration. Goosen and Metz (1980) propose a slightly modified version of the test to reduce this effect when the errors are only due to the digital recording of the begin and end moments of bouts:

$$L_n^{**} = 2\left(n\log\bar{k} - \sum_{i=1}^{n} b(k_i)\right), \tag{4.49}$$

where k_i is the observed length of the ith interval, expressed in the number of enclosed clock ticks. For instance, if the accuracy of the recording equipment is 0.1 seconds, and the ith interval is 5.2 s, k_i is equal to 52. Furthermore:

$$\bar{k} = \frac{1}{n} \sum_{i=1}^{n} k_i \tag{4.50}$$

and

$$b(0) = -2 \quad \text{and}$$
$$b(k) = \log k + c(k), \quad k = 1, 2,... \tag{4.51}$$

with

$$c(k) = \tfrac{1}{2}\{(k+1)^2\log(1+k^{-1}) + (k-1)^2\log(1-k^{-1}) - 3\}. \tag{4.52}$$

$c(k)$ can mostly be neglected when k is larger than one. This modification is especially relevant when there are a few observations of apparent length zero.

Note that it is not known how sensitive other tests on exponentiality are for small recording errors. Adjustments such as in eqn (4.49) are only available for this particular test.

4.7   TESTS FOR RANDOMLY CENSORED OBSERVATIONS

To test whether the transition rates towards certain behavioural categories are constant, rather than whether the termination rate is constant, tests adjusted for random censoring are needed (see section 4.4). This is of special interest in the analysis of social interactions, where it may be that deviations from exponentiality are caused by only one of the interacting individuals. In this case, only the ethogram of one of the animals has to be reconsidered, whereas that of the other(s) can stay the same. See Box 4.14 for an illustration.

Box 4.14 Example of random censoring due to the behaviour of another individual

Suppose that we want to model the dyadic interaction of male rats, given in Box 1.4, by means of a CTMC. In this case, all combined behavioural categories must have exponentially distributed bout lengths. Suppose, however, that the bout length distribution of the combination *Walk+Immobile* is a mixture of exponentials. This indicates that the behavioural category should be split up further. There are several possibilities:
1. Rat 1 has a constant tendency to stop *Walking* while the other is *Immobile*, whereas rat 2's tendency to stop *Immobile* changes during the bout. This would mean that the act *Immobile* should be reconsidered.
2. Rat 2 has a constant tendency to stop *Immobile* and rat 1's tendency to stop *Walking* while the other is *Immobile* changes. We would have to take a closer look at *Walk*.
3. Neither rat 1's tendency to stop *Walking* nor rat 2's tendency to stop *Immobile* is constant. Both acts should be split up further.

Table 1 Record of dyadic interactions of male rats

Behavioural category	Begin	Duration	(*Walk+Immobile*) ended by	Censored/ uncensored
Crawl over + Immobile	0	10		
Rear + Immobile	10	5		
Walk + Immobile	15	5	rat 2	censored
Walk + Sniff	20	2		
Sniff + Sniff	22	11		
Sniff + Walk	33	2		
Rear + Walk	35	4		
Walk + Walk	39	5		
Groom Opp* + Walk	44	1		
Groom Opp* + Immobile	45	16		
Walk + Immobile	61	2	rat 2	censored
Walk + Walk	63	2		
Walk + Immobile	65	10	rat 1	uncensored
Groom Opp* + Immobile	75	14		

* *Opp = Opponent.*

To find out which is the case, we must study the tendencies of rat 1 and rat 2 to stop this combination of acts separately, by means of log-survivor plots and/or formal tests. When the tendency of rat 1 is considered, the bouts terminated by rat 2 are censored, and vice versa for the tendency of rat 2. This illustrates an important point: which bouts are considered as censored depend on what is investigated. An example of a record of such an interaction is given in Table 1 and it is indicated which bouts are censored and which are not when rat 1's tendency to stop *Walking* while rat 2 is *Immobile* is studied.

4.7.1 A visual scanning method: the Kaplan-Meier estimate

As stated in subsection 4.6.1 most graphical procedures for testing exponentiality are based on the cumulative distribution function (see Box 1.2):

$$F(x) = Pr\{X \leq x\}, \tag{4.53}$$

or on the survivor function:

$$\bar{F}(x) = 1 - F(x). \tag{4.54}$$

When the distribution of X is exponential, a plot of the logarithm of the estimated survivor function against x gives approximately a straight line through the origin. A graphical test can be based on this property. However, if there are censored observations, the survivor function cannot be estimated in the straightforward manner indicated in eqn (4.11), since this may lead to serious deviations of a straight line even if the data have an exponential distribution. See Fig. 4.9 for an example. It shows three estimated survivor functions, $\bar{F}_n^a$, $\bar{F}_n^i$ and $\bar{F}_n^o$, for a censored sample of 100 exponentially distributed observations. $\bar{F}_n^a$ is the correct estimate, which is adjusted for censors, $\bar{F}_n^i$ is the estimated survivor function when the censored observations are included as uncensored and $\bar{F}_n^o$ is the estimate when the censors are simply omitted. In this example 30% of the observations were censored with an exponential (Fig. 4.9a), a uniform (Fig. 4.9b) and a degenerate (Fig. 4.9c) censor mechanism. No matter what kind of censor mechanism is used, the adjusted estimate indeed indicates that the data are exponentially distributed. With censoring times from an exponential distribution (e.g. due to social interactions, see Example 3, section 4.4) or a uniform distribution (e.g. due to random entrance, see Example 2, section 4.4), log-survivors

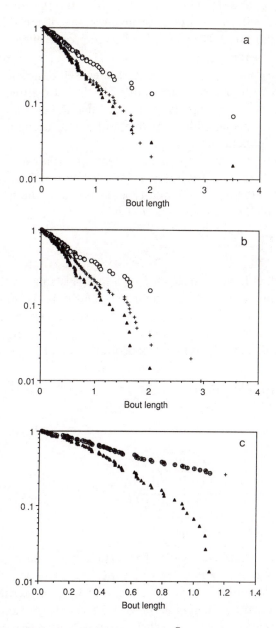

Figure 4.9 Log-survivor plots estimated using $\bar{F}_n^{\,a}$ (∘, correct estimate, i.e. adjusted for censors), $\bar{F}_n^{\,i}$ (+, censors included as if they were uncensored observations) and $\bar{F}_n^{\,o}$ (▲, censors omitted) for a set of 100 exponentially distributed variables.
a. exponential censoring mechanism
b. uniform censoring mechanism
c. degenerate censoring mechanism.

estimated by $\bar{F}_n^o$ and $\bar{F}_n^i$ may still resemble straight lines, as does $\bar{F}_n^a$. However, there will be a difference in the slope of the three lines. As a consequence, the mean bout length (the reciprocal of the slope) will be underestimated when $\bar{F}_n^o$ or $\bar{F}_n^i$ is used. If censoring times are from a degenerate distribution (e.g. in latency studies, see Example 1, section 4.4) and the censored observations are omitted in the estimation of the survivor function, it will often wrongly be concluded that the data are non-exponential. Hence, an adjusted estimation method has to be applied in order to avoid incorrect conclusions.

Several types of censoring mechanisms are relevant in behavioural research. See section 4.4 for an overview. We will only consider random censoring as it is the most common type. In this case the maximum likelihood estimate of the survivor function is

$$\bar{F}_n(x) = \prod_{\{j \,|\, y_j < x\}} \frac{n_j - d_j}{n_j} , \tag{4.55}$$

where the product is taken over all j such that y_j is smaller than x, n_j is the number of bouts larger than or equal to y_j and d_j is the number of bouts ended at y_j (Kaplan and Meier, 1958). As in subsection 4.6.1, a graphical procedure can be based on plotting log $\bar{F}_n(x)$ against x. See Box 4.15 for a worked example. A $(1 - \alpha)$ confidence intervals of log $\bar{F}(x)$ can be estimated by

$$\log\bar{F}_n(x)\exp[z_{1-\frac{1}{2}\alpha}s(x)] < \log\bar{F}(x) < \log\bar{F}_n(x)\exp[z_{\frac{1}{2}\alpha}s(x)],$$

$$\tag{4.56}$$

where

$$(s(x))^2 = \frac{\displaystyle\sum_{\{j \,|\, y_j < x\}} \frac{d_j}{n_j(n_j - d_j)}}{\left\{\displaystyle\sum_{\{j \,|\, y_j < x\}} \log\left(\frac{n_j - d_j}{n_j}\right)\right\}^2} \tag{4.57}$$

(Kalbfleisch and Prentice, 1980). The values $z_{1-\frac{1}{2}\alpha}$ and $z_{\frac{1}{2}\alpha}$ can be determined from Table A1. For large n the increments of log $\bar{F}_n(x)$ have approximately independent errors. The small-sample properties of $\bar{F}_n(x)$ for $n \leq 30$ were studied by Chen et al. (1982). Moeschberger and Klein (1985) discuss adjustments of $\bar{F}_n(x)$ when there are relatively many censoring times larger than the largest termination time (see also Wellner, 1985).

Box 4.15 The Kaplan-Meier estimator of the survivor function

1. The data are given in Table 1. Column 1 contains the bout lengths and in column 2 it is indicated whether the observation was uncensored ($\delta_i = 1$) or censored ($\delta_i = 0$).

2. First, order the distinct bout lengths according to increasing value (as in column 1 of Table 1).

Table 1 Observed bout lengths and estimated survivor functions of uncensored and censored bouts

| | | Survivor functions | |
y_i	δ_i	Uncensored bouts	Censored bouts
0.003	0	1	0.990
0.028	0		0.980
0.062	0		0.970
0.085	0		0.960
0.087	0		0.950
0.093	0		0.940
0.113	1	0.989	
0.136	0		0.930
0.138	0		0.920
0.150	0		0.910
0.173	0		0.900
0.174	0		0.890
0.177	0		0.879
0.201	1	0.978	
0.202	1	0.967	
0.270	0		0.869
0.282	0		0.859
0.321	1	0.955	
0.344	0		0.848
0.387	1	0.943	
0.395	1	0.931	
0.441	0		0.837
0.469	0		0.827
0.470	0		0.816
0.471	0		0.805

Table 1 Continued

y_i	δ_i	Survivor functions	
		Uncensored bouts	Censored bouts
0.529	1	0.919	
0.550	0		0.794
0.557	0		0.783
0.575	1	0.906	
0.581	0		0.772
0.624	0		0.761
0.630	1	0.893	
0.636	0		0.750
0.644	0		0.739
0.647	0		0.728
0.649	1	0.879	
0.650	0		0.716
0.668	0		0.705
0.817	1	0.865	
0.819	0		0.694
0.887	0		0.682
0.900	1	0.850	
0.912	0		0.670
0.914	1	0.836	
0.935	0		0.658
0.953	0		0.646
0.960	0		0.634
1.032	0		0.622
1.055	1	0.820	
1.071	1	0.803	
1.086	0		0.610
1.090	1	0.787	
1.115	0		0.597
1.156	1	0.770	
1.201	1	0.754	
1.241	0		0.584
1.250	1	0.736	
1.254	1	0.719	
1.270	0		0.570
1.280	1	0.702	
1.282	0		0.556
1.309	0		0.542
1.347	0		0.527
1.384	1	0.683	

Table 1 Continued

1.53	0		0.513
1.539	1	0.663	
1.545	0		0.498
1.553	0		0.482
1.565	0		0.467
1.623	0		0.452
1.672	1	0.641	
1.691	0		0.437
1.797	0		0.421
1.801	0		0.406
1.936	0		0.390
1.952	0		0.374
1.989	0		0.359
1.991	0		0.343
2.018	1	0.612	
2.136	1	0.583	
2.146	1	0.554	
2.298	0		0.325
2.467	0		0.307
2.481	1	0.521	
2.663	0		0.288
2.694	0		0.269
2.704	0		0.249
2.763	0		0.230
2.908	0		0.211
3.183	1	0.474	
3.243	1	0.426	
3.337	1	0.379	
3.592	1	0.332	
3.686	1	0.284	
3.858	0		0.176
4.053	0		0.141
4.212	1	0.213	
4.522	1	0.142	
5.699	1	0.071	
5.772	1	0	

3. For t smaller than the smallest uncensored bout length, the estimated survivor function is equal to one. For the rest, it changes only at the times corresponding to durations of uncensored bouts, so it is calculated only at these points. To do this, first calculate $(n_j - d_j)/n_j$ at each point (see eqn (4.55)). In the example, there are no equal uncensored bout lengths, so $d_j = 1$ in each case. The total number of observations is 100. The smallest

uncensored bout length is 0.113. Before then, six bouts have been censored. Thus, there are 94 bouts that are larger than or equal to 0.113. Accordingly, the estimated survivor function at this time is $(94 - 1)/94 = 0.989$. The next smallest uncensored bout length is 0.2011. There are 13 bouts that are smaller than this value, so $n_j = 87$ and $(n_j - d_j)/n_j = 0.988$. The estimated survivor function is $0.989 \times 0.988 = 0.978$. This calculation continues at subsequent points. The result is given in column 3 of Table 1.

4. Similarly, the survivor function of the censor times can be estimated. In that case, the role of the uncensored and censored bouts is reversed (see the remarks made in Box 4.14). The result is given in column 4 of Table 1.

5. As before, a logarithmic plot of the estimated survivor function can be used to detect deviations from exponentiality. Both log-survivor plots are given in Fig. 1. Apparently, the censor time distribution is more or less exponential, whereas the log-survivor of the uncensored bouts looks concave, which indicates an increasing termination rate.

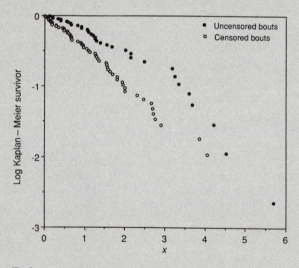

Figure 1 Estimated log-survivors of uncensored bouts and of censored bouts in Table 1.

4.7.2 Unspecified alternative hypotheses

In this subsection we do not discuss all the generalizations of the tests described in subsection 4.6.2, since not all the adjusted tests are as easy to apply as their counterparts for uncensored data. For instance, Habib and Thomas (1986) give a generalization of the chi-squared test of goodness of fit (see subsection 4.6.2) for randomly censored data with unknown censoring distribution. This test involves much computational effort and it is not known whether it gives much improvement over the other available tests. Therefore it is not included here. For a generalization of the Shapiro-Wilk test we refer to Samanta and Schwarz (1988).

In the three test procedures that we describe it is assumed that the censoring time distribution is exponential. In practice it is usually not known beforehand whether this assumption is justified. In that case it is advisable to check this assumption by means of visual inspection of the Kaplan-Meier estimate of the censor distribution. Note that the roles of censored and uncensored observations are reversed when this is done (see e.g. Box 4.15).

The adjusted Kolmogorov-Smirnov test

In subsection 4.6.2 we treated the Kolmogorov-Smirnov test for exponentiality when the mean is unknown, as proposed by Durbin (1975). The test can easily be adjusted when the censoring times are exponentially distributed.

Data: One sample of n bout lengths. The ith observation is described by the pair (x_i, δ_i), where x_i denotes the bout length and δ_i indicates whether this observation is censored ($\delta_i = 0$) or not ($\delta_i = 1$), $i = 1, ..., n$, see Table 1 in Box 4.15 for an example.

Assumptions:

A1: $Y_1, ..., Y_n$ are mutually independent and exponentially distributed with common and unknown parameter λ. (The Y_i represent the hypothetical bout lengths if there were no censoring. See section 4.4.)

A2: $C_1, ..., C_n$ are mutually independent and exponentially distributed with common and unknown parameter μ. (The C_i are the censoring times.)

A3: Under the null hypothesis, H_0, X_i is distributed as the minimum of Y_i and C_i.

A4: Under the alternative hypothesis, H_1, there are unspecified deviations of A1, A2 or A3.

Procedure: The unknown parameter λ is estimated by the maximum likelihood estimator

$$\lambda^c = \frac{n_f}{\sum_{i=1}^{n} x_i} , \qquad\qquad (4.58)$$

where n_f denotes the number of uncensored bout lengths (in survival analysis it is called the number of failures). See Box 4.16.

Box 4.16 Maximum likelihood estimation of λ in the presence of random censoring

When there is random censoring, the likelihood of the observations is

$$\prod_{k=1}^{n_f} f(s_k) \cdot \prod_{j=1}^{n_c} \bar{F}(c_j), \tag{1}$$

where $s_1,...,\, s_{nf}$ are the uncensored and $c_1,...,\, c_{nc}$ the censored observations. Thus, when $f(x)$ is the exponential density, the log-likelihood equals

$$n_f \log \lambda - \lambda \sum_{k=1}^{n_f} s_k - \lambda \sum_{j=1}^{n_c} c_j. \tag{2}$$

Thus, equating the derivative to λ to zero gives

$$\frac{n_f}{\lambda^c} - \sum_{k=1}^{n_f} s_k - \sum_{j=1}^{n_c} c_j = 0. \tag{3}$$

Solving this equation gives

$$\lambda^c = \frac{n_f}{\sum_{k=1}^{n_f} s_k + \sum_{j=1}^{n_c} c_j} = \frac{n_f}{\sum_{i=1}^{n} x_i}, \tag{4}$$

where $x_1,...,\, x_n$ denote the whole set of censored and uncensored observations.

In the case of an exponentially distributed variable the survivor function $\bar{F}(x) = Pr\{X > x\}$ is equal to $\exp[-\lambda x]$. The empirical survivor function can be estimated by the Kaplan-Meier estimate $\bar{F}_n(x)$, see subsection 4.7.1, eqn (4.55). The two-sided Kolmogorov-Smirnov test statistic is then defined by

$$D_n^c = \max \{D_n^{c+}, D_n^{c-}\}, \text{ where}$$

$$D_n^{c+} = \max_{y_1...y_k} \{\exp[-\lambda^c y_i] - \bar{F}_n(y_i)\} \tag{4.59}$$

$$D_n^{c-} = \max_{y_1...y_k} \{\bar{F}_n(y_{i-1}) - \exp[-\lambda^c y_i]\}$$

and $y_1, ..., y_k$ are the ordered *distinct uncensored* bout lengths. D_n^c can be considered as the maximum distance between the empirical survivor function and the theoretical survivor function for $\lambda = \lambda^c$. Burke (1982) has shown that when the distribution of the censoring times is exponential, $\sqrt{n}\, D_n^c$ has the same asymptotic distribution under H_0 as when there is no censoring, provided that the proportion of uncensored observations, $n_f\,/\,n$, becomes a positive constant which is smaller than (or equal to) one as n

tends to infinity. For small sample sizes one can also apply the same critical values. Hence, reject H_0 if $\sqrt{n}\, D_n^c > c_\alpha(n)$, where $c_\alpha(n)$ is obtained from Table A17, for sample sizes $n = 2\,(1)\,10\,(2)\,30\,(5)\,50\,(10)\,100$. See Box 4.17 for a detailed example.

Box 4.17 The adjusted Kolmogorov-Smirnov test

Table 1 Calculation of the adjusted Kolmogorov-Smirnov test statistic

y_i	$\bar{F}_n(y_i)$	$\bar{F}_n(y_{i-1})$	$\exp[-\lambda^c y_i]$	D^+	D^-
0.113	0.989	1	0.970	−0.019	0.030
0.201	0.978	0.989	0.948	−0.030	0.041
0.202	0.967	0.978	0.947	−0.020	0.031
0.321	0.955	0.967	0.918	−0.037	0.049
0.387	0.943	0.955	0.902	−0.041	0.053
0.395	0.931	0.943	0.900	−0.031	0.043
0.529	0.919	0.931	0.868	−0.051	0.063
0.575	0.906	0.919	0.858	−0.048	0.061
0.630	0.893	0.906	0.845	−0.048	0.061
0.649	0.879	0.893	0.841	−0.038	0.052
0.817	0.865	0.879	0.804	−0.061	0.075
0.900	0.850	0.865	0.786	−0.064	0.079
0.914	0.836	0.850	0.784	−0.052	0.066
1.055	0.820	0.836	0.755	−0.065	0.081
1.071	0.803	0.820	0.752	−0.051	0.068
1.090	0.787	0.803	0.748	−0.039	0.055
1.156	0.770	0.787	0.734	−0.036	0.053
1.201	0.754	0.770	0.726	−0.028	0.044
1.250	0.736	0.754	0.716	−0.020	0.038
1.254	0.719	0.736	0.716	−0.003	0.020
1.280	0.702	0.719	0.711	0.009	0.008
1.384	0.683	0.702	0.691	0.008	0.011
1.539	0.663	0.683	0.663	0	0.020
1.672	0.641	0.663	0.640	−0.001	0.023
2.018	0.612	0.641	0.584	−0.028	0.057
2.136	0.583	0.612	0.566	−0.017	0.046
2.146	0.554	0.583	0.564	0.010	0.019
2.481	0.521	0.554	0.516	−0.005	0.038
3.183	0.474	0.521	0.428	−0.046	0.093*
3.243	0.426	0.474	0.421	−0.005	0.053
3.337	0.379	0.426	0.411	0.032	0.015
3.592	0.332	0.379	0.384	0.052	−0.005
3.686	0.284	0.332	0.374	0.090	−0.042
4.212	0.213	0.284	0.325	0.112	−0.041
4.522	0.142	0.213	0.299	0.157	−0.086
5.699	0.071	0.142	0.219	0.148	−0.077
5.772	0	0.071	0.214	0.214*	−0.143

1. The data are as given in Box 4.15. We perform a Kolmogorov-Smirnov test on the distribution of uncensored bouts. The ordered (distinct) uncensored bout lengths are given in column 1 of Table 1.

2. The estimated Kaplan-Meier survivor, $\bar{F}_n(y_{(i)})$, is as given in Box 4.15. The values are listed in column 2 of Table 1. Column 4 gives the values of $\bar{F}_n(y_{(i-1)})$, with $\bar{F}_n(y_{(0)}) = 1$. The estimated survivor function under H_0 is equal to $\exp[-\lambda^c y_i]$. Application of eqn (4.58) gives $\lambda^c = 0.267$. The resulting survivor function is given in column 5 of Table 1.

3. The values in column 3 are subtracted from those in column 5 (results are given in column 6). The maximum difference is $D_n^+ = 0.214$. The values in column 5 are subtracted from those in column 4 (column 7). The maximum, D_n^{c-}, equals 0.093. It follows that $D_n = 0.214$.

4. The sample size is 100 (see Box 4.15), so the test statistic is 2.14. It follows from Table A17 that the null hypothesis is rejected at a significance level of 0.05. Apparently, the distribution of the bout lengths differs significantly from exponentiality.

Large-sample approximation: The approximation reported in Pearson and Hartley (1972) can be used. This is based on the statistic T_n^c, which is defined by

$$T_n^c = (D_n^c - 0.2/n)(\sqrt{n} + 0.26 + 0.5/\sqrt{n}).\tag{4.60}$$

Reject H_0 if $T_n^c > c^*_\alpha$, where the c^*_α are listed in Table A18 for the usual values of α.

Comments: Tables of the critical values of the one-sided analogue of this test can be found in Durbin (1975). The test is described by Burke (1982).

The adjusted Cramér-von Mises test

Data and assumptions: See the adjusted Kolmogorov-Smirnov test.

Procedure: The unknown parameter λ is estimated by the maximum likelihood estimator λ^c. The ordered uncensored bout lengths are denoted by $x_{(1)}$, $x_{(2)},..., x_{(nf)}$. Under the null hypothesis, the survivor function is estimated by

$$\bar{F}(x;\lambda^c)=\exp[-\lambda^c x].\tag{4.61}$$

The Cramér-von Mises test statistic W_c^2 is defined by

$$W_c^2 = n \sum_{i=1}^{n_f} \{ F(x_{(i)}; \lambda^c) - F_n(x_{(i)}) \}^2 \{ F_n(x_{(i)}) - F_n(x_{(i-1)}) \}, \qquad (4.62)$$

where n is the total number of bouts and $F_n(x_{(0)})$ is equal to zero. Hence, just as in the case of an uncensored sample, the test statistic is essentially equal to the sum of the squared differences between the empirical distribution function and the (estimated) exponential distribution function. Reject H_0 if W_c^2 exceeds the critical value $W^2(\alpha)$. See Box 4.18 for a worked example.

Burke (1982) has shown that when the distribution of the censoring times is exponential, the adjusted test statistic W_c^2 has the same asymptotic distribution under H_0 as when there is no censoring, provided that the proportion of uncensored observations converges to a positive constant (smaller than or equal to one) as n tends to infinity. For small sample sizes one can also apply the same critical values. Hence, reject H_0 if W_c^2 exceeds the critical value $W^2(\alpha)$ listed in Table A19.

Box 4.18 The adjusted Cramér-von Mises test

1. The data are identical to those of Box 4.15. In Box 4.17 we give the estimated survivor function and the differences between the empirical and the exponential survivor function. Since the distribution function is one minus the survivor function, eqn (4.62) is equivalent to:

$$W_c^2 = n \sum_{i=1}^{n_f} \left\{ \bar{F}(x_{(i)}; \lambda^c) - \bar{F}_n(x_{(i)}) \right\}^2 \left\{ \bar{F}_n(x_{(i)}) - \bar{F}_n(x_{(i+1)}) \right\}, \qquad (1)$$

with $\bar{F}_n(x_{(n_f+1)})$ equal to zero.

2. Columns 1 and 2 of Table 1 contain the data and estimated survivor function (as in Box 4.17). Column 3 shows the change in value of the estimated survivor function at each step and column 4 the squared differences between the empirical survivor function and the exponential (obtained by taking the square of the values given in column 4 of Table 1, Box 4.17).

3. Multiplying the values in column 3 with those in column 4 and summing the results gives 0.005535. The Cramér-von Mises test statistic is 0.5535, since $n = 100$. To improve the asymptotic approximation of the distribution

of the test statistic, we perform the transformation given in eqn (4.63), which leads to $(W_c^2)' = 0.5544$. It follows from Table A19 that the null hypothesis is rejected at a significance level of 0.05.

Table 1 Calculation of the adjusted Cramér-von Mises test statistic

s_i	$\bar{F}_n(s_i)$	$\bar{F}_n(s_i)-\bar{F}_n(s_{i+1})$	$(\bar{F}_n(s_i) - \exp[-\lambda^c s_i])^2$
0.113	0.989	0.011	0.000361
0.201	0.978	0.011	0.0009
0.202	0.967	0.012	0.00004
0.321	0.955	0.012	0.001369
0.387	0.943	0.012	0.001681
0.395	0.931	0.012	0.0000961
0.529	0.919	0.013	0.0002601
0.575	0.906	0.013	0.0002304
0.630	0.893	0.014	0.0002304
0.649	0.879	0.014	0.0001444
0.817	0.865	0.015	0.003721
0.900	0.850	0.014	0.004096
0.914	0.836	0.016	0.002704
1.055	0.820	0.017	0.004225
1.071	0.803	0.016	0.002601
1.090	0.787	0.017	0.001521
1.156	0.770	0.016	0.001296
1.201	0.754	0.018	0.000784
1.250	0.736	0.017	0.0004
1.254	0.719	0.017	0.000009
1.280	0.702	0.019	0.000081
1.384	0.683	0.020	0.000064
1.539	0.663	0.022	0.000
1.672	0.641	0.029	0.000001
2.018	0.612	0.029	0.000784
2.136	0.583	0.029	0.000289
2.146	0.554	0.033	0.0001
2.481	0.521	0.047	0.000025
3.183	0.474	0.012	0.002116
3.243	0.426	0.083	0.000025
3.337	0.379	0.047	0.001024
3.592	0.332	0.048	0.002704
3.686	0.284	0.071	0.0081
4.212	0.213	0.071	0.012544
4.522	0.142	0.071	0.024649
5.699	0.071	0.071	0.021904
5.772	0	0	0.045796

Large-sample approximation: Stephens (1976) gives asymptotic critical values. See Table A19. The transformation

$$(W_c^2)' = W_c^2(1 + 0.16/n) \qquad (4.63)$$

improves the accuracy considerably. As we have shown in subsection 4.6.2 the approximation is quite accurate even for very small sample sizes, provided that the level of significance is not too small.

Chen's correlation goodness-of-fit test

Chen (1984) also gives an overall test for exponentiality in which it is assumed that the distribution of the censoring times is exponential with unknown parameter.

Data and assumptions: See the Kolmogorov-Smirnov test adjusted for censors.

Procedure: The unknown parameter λ is estimated by the maximum likelihood estimator λ^c (eqn (4.58)). Next, k strictly increasing values between zero and one (denoted by $p_1,..., p_k$) are chosen. The quantiles at these values are subsequently estimated. For an arbitrary random variable X with cumulative distribution function F, the pth quantile is defined by

$$Q(p) = \min \{ x : F(x) < p \}, \qquad (4.64)$$

see Fig. 4.10.

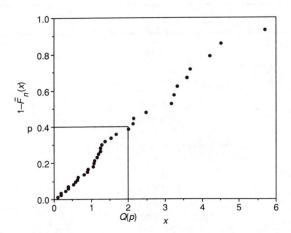

Figure 4.10 Determination of the pth quantile, $Q(p)$, from a plot of the estimated cumulative distribution function.

The quantiles can be estimated in two ways by using different estimates of the cumulative distribution function:

1. Under the null hypothesis of exponentiality, the quantiles are estimated by substitution of λ^c for λ in $F(x) = 1 - \exp[-\lambda x]$, the cumulative distribution function under the null hypothesis. This gives the following estimate for the p_j th quantile:

$$q_j = -(1/\lambda^c)\{\log(1 - p_j)\}. \tag{4.65}$$

2. Using the Kaplan-Meier estimate of the survivor function (Box 4.15) the distribution function can be estimated by $F_n(x) = 1 - \bar{F}_n(x)$. This gives estimates Q_j of the quantiles. This method is illustrated in Box 4.19.

Box 4.19 Chen's test

1. Here, we again use the data given in Box 4.15. However, since the proportion of censored observations is 0.63, Chen's test cannot be used to test whether the distribution of the uncensored bouts deviates significantly from exponentiality (Table A18 only gives the critical values for ε less than or equal to 0.50). Instead we use this test to find whether the distribution of the censored bouts in Box 4.15 deviates from exponentiality. To do this, the roles of censored and uncensored observations are reversed. The data are given in column 1 of Table 1.

Table 1 Estimation of empirical quantiles

s_i	$\bar{F}_n(s_i)$	$1-\bar{F}_n(s_i)$
0.003	0.990	0.010
0.028	0.980	0.020
0.062	0.970	0.030
0.085	0.960	0.040
0.087	0.950	0.050
0.093	0.940	0.060
0.136	0.930	0.070
0.138	0.920	0.080
0.150	0.910	0.090
0.173	0.900	0.100
0.174	0.890	0.110
0.177	0.879	0.121
0.270	0.869	0.131
0.282	0.859	0.141
0.344	0.848	0.152

Table 1 Continued

0.441	0.837	0.163
0.469	0.816	0.184
0.470	0.805	0.195
0.550	0.794	0.206 *
0.557	0.783	0.217
0.581	0.772	0.228
0.624	0.761	0.239
0.636	0.750	0.250
0.644	0.739	0.261
0.647	0.728	0.272
0.650	0.716	0.284
0.668	0.705	0.295
0.819	0.694	0.306
0.887	0.682	0.318
0.912	0.670	0.330
0.935	0.658	0.342
0.953	0.646	0.354 *
0.960	0.634	0.366
1.032	0.622	0.378
1.086	0.610	0.390
1.115	0.597	0.403
1.241	0.584	0.416
1.270	0.570	0.430
1.282	0.556	0.444
1.309	0.542	0.458 *
1.347	0.527	0.473
1.530	0.513	0.487
1.545	0.498	0.502
1.553	0.482	0.518
1.565	0.467	0.533
1.623	0.452	0.548
1.691	0.437	0.563 *
1.797	0.421	0.579
1.801	0.406	0.594
1.936	0.390	0.610
1.952	0.374	0.626
1.989	0.359	0.641
1.991	0.343	0.657 *
2.298	0.325	0.675
2.467	0.307	0.693
2.663	0.288	0.712
2.694	0.269	0.731
2.704	0.249	0.751
2.763	0.230	0.770
2.908	0.211	0.789
3.858	0.176	0.824 *
4.053	0.141	0.859

Tests for exponentiality

2. Calculate $1 - \bar{F}_n(s_i)$ (see column 4 of Table 1) and determine the quantiles at which $p = 0.20, 0.35, 0.45, 0.55, 0.65, 0.8$. To do this, determine the points at which $1 - \bar{F}_n(s_i)$ becomes larger than or equal to these p-values. These points are indicated by an asterisk in Table 1. The corresponding values of s_i are the quantiles. In this way we find the Q_j-values, given in Table 2.

Table 2 Calculation of Chen's test statistic for the data in Table 1

j	p_j	Q_j	$\log(1 - p_j)$	q_j	$Q_j q_j$
1	0.20	0.550	−0.223	0.511	0.281
2	0.35	0.953	−0.431	0.988	0.942
3	0.45	1.309	−0.598	1.372	1.796
4	0.55	1.691	−0.799	1.833	3.100
5	0.65	1.991	−1.050	2.408	4.794
6	0.8	3.858	−1.609	3.690	14.236

3. For the uncensored observations in Box 4.15, λ^c is 0.267. When we reverse the roles of the censored and uncensored observations, it follows from eqn (4.58) that λ^c becomes $0.267 \times 38/62 = 0.436$. The quantiles under the null hypothesis can be calculated by means of eqn (4.65). These quantiles are given in column 3 of Table 2.

4. The sum of the squared values in column 3 of Table 1 is 24.598, the sum of the squared values of column 5 is 25.894, and the sum of $Q_j q_j$ (column 6) is 25.149. Thus, the test statistic, given by eqn (4.66), is equal to

$$R_n^2 = \frac{(25.149)^2}{24.598 \times 25.894} = 0.993, \tag{1}$$

so $n(1 - R_n^2)$ is 0.701. The number of bouts is 62, so ε is equal to $1 - 0.62 = 0.38$ (since $n = 100$). It follows from Table A23 that under the null hypothesis, the p-value of the test statistic is larger than 0.05. Thus, the null hypothesis is not rejected.

The test statistic is the correlation between the two series of estimated quantiles:

$$R_n^2 = \frac{\left(\sum_{j=1}^{k} Q_j q_j\right)^2}{S_n^2 S_0^2}, \tag{4.66}$$

where

$$S_n^2 = \sum_{j=1}^{n} Q_j^2 \qquad (4.67)$$

and

$$S_0^2 = \sum_{j=1}^{n} q_j^2. \qquad (4.68)$$

The null hypothesis is rejected for large values of $n(1 - R_n^2)$. See Box 4.19 for a worked example. There are no critical values available for small sample sizes.

Large-sample approximation: In Table A23 the asymptotic critical values of $n(1 - R_n^2)$ are given for $k = 6$ and p_j-values 0.20, 0.35, 0.45, 0.55, 0.65, 0.80. These were calculated by Chen (1984), who also gives a method for calculating critical values when different p-values are chosen. In most cases, however, the values given in Table A23 will suffice. The critical values also depend on the proportion of censors, $\varepsilon = 1 - n_f /n$ (where n_f is the number of failures and n the sample size). The critical values are given for $\varepsilon = 0$ (0.05) 0.50. From simulation results it appears that the asymptotic approximation is already good for $n = 30$ (Chen, 1984).

Properties: Chen (1984) found that the test based on R_n^2 is robust for other censoring distributions, such as the gamma distribution with shape parameter $k = 2$ (see subsection 1.3.1, eqn (1.17)).

4.7.3 Broad classes of alternative hypotheses: adjusted 'total time on test' statistic

When there is random censoring, a generalization of the test based on V_n (see subsection 4.6.3) can be used.

Data: One sample of n bout lengths. The ith observation is described by the pair (x_i, δ_i), where x_i denotes the bout length and δ_i indicates whether this observation is censored ($\delta_i = 0$) or not ($\delta_i = 1$), $i = 1,..., n$.

Assumptions:
A1: $Y_1,..., Y_n$ are mutually independent and exponentially distributed with common and unknown parameter, the termination rate λ. (The Y_i represent the theoretical uncensored bout lengths.)

A2: The censoring times $C_1,..., C_n$ are mutually independent and identically distributed.

A3: Under the null hypothesis, H_0, X_i is distributed as the minimum of Y_i and C_i.

A4: Under the alternative hypothesis, H_1, the termination rate $\lambda(x)$ is a monotonically increasing or decreasing function of x.

Procedure: The 'cumulative total time on test' statistic V_c is defined by

$$V_c = \frac{\sum\limits_{j=1}^{n_f-1} \sum\limits_{i=1}^{j} \{n_{(i)}(s_{(i)} - s_{(i-1)})\}}{\sum\limits_{k=1}^{n_f} \{n_{(k)}(s_{(k)} - s_{(k-1)})\}}, \tag{4.69}$$

where (as before) n_f is the number of uncensored bout lengths, $s_{(0)}(=0) \le s_{(1)} \le ... \le s_{(n_f)}$ are the ordered uncensored bout lengths and $n_{(i)}$ is the number of the x_i that are larger than or equal to $s_{(i)}$.

Large values of V_c indicate an increasing termination rate, small values a decreasing one. Aalen and Hoem (1978) showed that when the number of censored observations is fixed, V_c has the same distribution as V_n, defined in subsection 4.6.3, eqn (4.34), with n replaced by n_f. Hence, the same critical values as in the uncensored case can be applied. The upper critical values for $n_f = 3$ (1) 25 are listed in Table A21. The lower critical values can be calculated from these (see subsection 4.6.3).

Large-sample approximation: Gill (1986) proved that when the number of censored observations is random, V_c still has the same asymptotic distribution as V_n (subsection 4.6.3). For large n, the distribution of

$$Z = \left(\frac{V_c}{N_f - 1} - \frac{1}{2}\right)\sqrt{12(n_f - 1)} \tag{4.70}$$

tends to a standard normal distribution. H_0 is rejected if Z exceeds the critical value z_α listed in Table A1. See Box 4.20 for an example.

Box 4.20 Barlow's test adjusted for censors

1. The data are as given in Box 4.15.

2. First, put the survival times in increasing order and determine the number of failures, $n(i)$, that occurred at or after each survival time (see

Table 1, column 3). Next, calculate the difference between each survival time and the preceding one, $s(i) - s(i-1)$ (Table 1, column 2).

Table 1 Calculation of V_c

$s_{(i)}$	$s_{(i)} - s_{(i-1)}$	$n_{(i)}$	$n_{(i)}(s_{(i)} - s_{(i-1)})$	Cumulative
0.1131	0.1131	94	10.63	10.63
0.2011	0.08802	87	7.658	18.29
0.2014	2.618×10^{-4}	86	0.02252	18.31
0.3207	0.1193	83	9.898	28.21
0.3869	0.06626	81	5.367	33.58
0.3951	8.157×10^{-3}	80	0.6525	34.23
0.5291	0.134	75	10.05	44.28
0.5746	0.04553	72	3.278	47.56
0.6302	0.05555	69	3.833	51.39
0.649	0.01879	65	1.222	52.61
0.8167	0.1677	62	10.4	63.01
0.9005	0.08379	59	4.944	67.96
0.9136	0.01318	57	0.7512	68.71
1.055	0.1412	52	7.344	76.05
1.071	0.01577	51	0.8045	76.86
1.09	0.01905	49	0.9336	77.79
1.156	0.06679	47	3.139	80.93
1.201	0.04447	46	2.046	82.97
1.25	0.04931	44	2.17	85.14
1.254	3.919×10^{-3}	43	0.1685	85.31
1.28	0.02589	41	1.061	86.37
1.384	0.1035	37	3.831	90.21
1.539	0.1556	35	5.448	95.65
1.672	0.1328	30	3.984	99.64
2.018	0.3461	22	7.615	107.3
2.136	0.1178	21	2.474	109.7
2.146	0.01034	20	0.2069	109.9
2.481	0.335	17	5.694	115.6
3.183	0.7017	11	7.719	123.3
3.243	0.05965	10	0.5965	123.9
3.337	0.0941	9	0.8469	124.8
3.592	0.2554	8	2.043	126.8
3.686	0.09413	7	0.6589	127.5
4.212	0.5261	4	2.104	129.6
4.522	0.3095	3	0.9286	130.5
5.699	1.177	2	2.355	132.9
5.772	0.07287	1	0.07287	–
Totals		133	2938	

3. The denominator of V_c (eqn (4.69)) consists of the summed products of $n(i)$ and $s(i) - s(i-1)$. In this case, it is equal to 133 (i.e. the summation of column 4 of Table 1). To determine the numerator, first calculate the cumulative sums of $n(i)\{s(i) - s(i-1)\}$ for $i = 1$ to $n_f - 1$. These are given in column 5. Subsequently, the total of these cumulative sums is calculated, which, in the example, gives 2938 (bottom of Table 1).

4. The test statistic is equal to $2938/133 = 22.09$. Since $n_f = 37$, we use the large-sample approximation given in eqn (4.70), which leads to $Z = 2.02$. From Table A1 it can be seen that the null hypothesis is rejected. Since Z is positive, the test indicates a departure in the direction of an increasing termination rate, i.e. a convolution. This is confirmed by the log-survivor plot given in Fig. 1 of Box 4.15.

Properties: Chen (1984) compared the power of the test based on V_c with that based on R_n^2 (eqn (4.66)), when 25% of the data is censored according to an exponential distribution. His results indicate that V_c performs better than R_n^2 against increasing as well as decreasing termination rate alternatives.

4.7.4 Specific alternative hypotheses

The likelihood ratio test against Weibull alternatives

The test statistic, t, given in subsection 4.6.4 can also be used in the case of random censoring.

Data: See subsection 4.7.3.

Assumptions: A1 to A3 as in subsection 4.7.3, and, moreover:
A4:   Under the alternative hypothesis, H_1, the termination rate is
$\lambda(x) = \lambda\rho(\lambda x)^{\rho-1}$, with λ and ρ unspecified.

Procedure: The test statistic is

$$t = \frac{n_f + \sum \log s_i - n_f \dfrac{\sum x_i \log x_i}{\sum x_i}}{I_{\rho\rho} - \dfrac{I_{\rho\lambda}^2}{I_{\lambda\lambda}}} , \qquad (4.71)$$

where the elements of the information matrix I are equal to

$$I_{pp} = n_f + \sum (\lambda^c x_i)\{\log(\lambda^c x_i)\}^2,$$

$$I_{p\lambda} = \sum x_i \log(\lambda^c x_i),$$

$$I_{\lambda\lambda} = \frac{n_f}{(\lambda^c)^2},$$

(4.72)

and λ^c is the MLE of λ under the null hypothesis, i.e. $n_f/\bar{x}$. As before, $s_1,..., s_{nf}$ denote the uncensored bout lengths. Positive values of t indicate an increasing termination rate, negative values a decreasing one. There are no critical values available for the small-sample case.

Large-sample approximation: As n tends to infinity the test statistic t has a standard normal distribution. See Table A1 for the critical values. In the right-sided case reject H_0 in favour of an increasing termination rate if $t > z_\alpha$ and in the left-sided case reject H_0 in favour of a decreasing termination rate if $t < z_{1-\alpha}$. In the two-sided case, reject H_0 if $t < z_{1-\frac{1}{2}\alpha}$ or if $t > z_{\frac{1}{2}\alpha}$.

Comments: The test is described by Cox and Oakes (1984).

4.8   FURTHER REMARKS

4.8.1 More about the performance of tests

Simulation results on power are available for the generalized two-sided Kolmogorov-Smirnov test (eqn (4.25), see Kimber, 1985), Barlow's 'cumulative total time on test' statistic V_c (or equivalently V_n, eqns (4.69) and (4.34), Chen, 1984) and Chen's correlation test based on R_n^2 (eqn (4.66), Chen, 1984). A comparison of these results indicates that the two-sided test based on V_c (or V_n when there are no censors) performs best against increasing as well as decreasing termination rates. There are no such results on the other tests. However, the test based on L_n or L_n' (eqns (4.44) and (4.45)) has good theoretical power properties against gamma alternatives (Shorack, 1972). It is expected to have good power against other convolutions of exponentials and against mixtures of exponentials as well.

Although most tests perform well with alternatives with monotone increasing or decreasing termination rates, they usually do not give significant results when the termination rate initially increases and later decreases, as in a log-normal bout length distribution (see e.g. Kalbfleisch and Prentice, 1980, Chapter 2). At least, this was found for the two-sided Kolmogorov-Smirnov test (Kimber, 1985), for Chen's correlation test and for Barlow's 'cumulative total time on test' statistic (Chen, 1984). It is to be expected that the other tests are not very powerful against such alternatives either. A termination rate of that form can occur when there are time-lags as well as lumped states. It is recommended that the log-survivor plots are inspected for such deviations.

Except for the tests based on L_n or L_n' (eqns (4.44) and (4.45)), it is not known how sensitive tests for exponentiality are to small recording errors. These tests are the only ones for which an adjustment (eqn (4.49)) is available.

Most tests do not perform well when a relatively large proportion of the data at large x is censored. In these cases we usually have to rely on graphical methods, with adjustments as described by Moeschberger and Klein (1985). When most of the larger bout lengths are censored, the survivor function can only be estimated accurately for small x and the power of tests for exponentiality is small.

4.8.2 Strategy for testing

Which type of test to use

We strongly recommend the use of graphical inspection procedures as well as formal tests. Graphical inspection gives an impression of the type of deviations, which may be used to select a test. Furthermore, these procedures indicate whether there are problematic deviations, such as combinations of time-lags and a mixture of exponentials, for which most tests are not sensitive. Graphical procedures can also indicate the direction of deviations when the selected test statistic does not. For instance, the value of the Cramér-von Mises test statistic (eqns (4.29) and (4.62)) does not indicate whether there is a deviation in the direction of an increasing or a decreasing termination rate.

In the initial stage it is best to use tests that are sensitive to increasing as well as decreasing termination rate alternatives, such as tests against unspecified alternatives or against monotone increasing or decreasing termination rates. When it is clear, either from previous (visual) inspection or from additional knowledge about the behaviour under study (e.g. about physiological processes), which type of departures can be expected, tests against specific alternatives may be used.

When many records are tested for exponentiality, the values of the test statistics can be combined to study the general direction of departures. For instance, Dienske and Metz (1977) used Moran's L_n statistic (eqn (4.44)) to this end (see Chapter 6).

Which actions to take

Whether or not there are time-lags can be checked by inspection of the log-survivor plot (see Fig. 4.6). If the time-lags have a relatively small variance, the average lag duration can be estimated from the log-survivor plot and subtracted from the bout lengths, after which values smaller than or equal to zero are left out and the sample is tested for exponentiality again. If there are no deviations after this 'correction', the behavioural category under consideration can be represented by a state of a semi-Markov chain (provided that there are no deviations from the sequential dependency properties; see Chapter 5).

As mentioned before, combinations of time-lags and mixtures of exponentials can also occur. When a graphical procedure indicates that this is the case, the average lag duration can be estimated from the plot and a procedure similar to the one described above can be used to test whether the resulting distribution is indeed a mixture of exponentials. Alternatively, the Shapiro-Wilk test (subsection 4.6.2) can be used to test for the presence of time-lags as well as other deviations from exponentiality.

Departures in the direction of decreasing termination rates indicate mixtures. Increasing termination rates can indicate a convolution of exponentials or the occurrence of time-lags. However, significant deviations in these directions may also imply that the termination rate indeed changes during a bout. Usually it is best to examine first whether acts with bout length distributions that deviate significantly from exponentiality can be considered as lumped Markov states. This implies that the behavioural category under consideration is split up on observable grounds and subsequently the bout lengths of the resulting subcategories are again tested for exponentiality.

When splitting up according to observable differences is not possible or does not lead to exponential bout length distributions, either a semi-Markov model or a function of a Markov model may be used. If there are no deviations from first-order sequential dependency (Chapter 5), acts with non-exponential bout length distributions can be modelled as states in a semi-Markov chain, with termination rates that change during a bout. When tests on first-order sequential dependency do give significant results, a function of a Markov chain is the appropriate model and methods as given by Putters *et al.* (1984) should be used (Chapter 8).

Interactive behaviour

When interactions between several individuals are studied, one can first test whether the bout lengths of combined behavioural categories (see Box 1.4) are exponentially distributed, using the methods of section 4.6. If there are deviations from exponentiality, a comparison of different test results can indicate which individual causes the deviations. For instance, when deviations occur in the bout length distribution of a combination of acts A by individual 1 and C by individual 2, the test results on bout lengths of all combinations of acts that contain A or C can be compared. If the bout length distributions of combinations with A deviate significantly from exponentiality, whereas most combinations with C do not, act A by individual 1 should be split up. However, usually this will not give a clear picture and tests adjusted for censors are needed to determine whether A, C or both should be split up (see also Box 4.14).

5 TESTS OF SEQUENTIAL DEPENDENCY PROPERTIES

5.1 INTRODUCTORY REMARKS

In a CTMC there is, at most, a first-order dependency in the sequence of states. Consequently, when it is applied to a behavioural record where states correspond to acts, it is assumed that

$$p_{AB} = Pr\{\text{act at } t + \Delta t \text{ is B} \mid$$

$$\text{act at } t \text{ is A and a transition occurs in } (t, t + \Delta t)\} \tag{5.1}$$

is independent of the sequence of preceding acts (section 1.2). This implies that given the sequence of acts the bout durations are mutually independent. In this chapter we discuss tests against possible additional dependencies that are relevant in ethological applications. In most of these tests it is not assumed that the bout length distributions are necessarily exponential. Since the sequential dependency properties of a semi-Markov chain are equal to those of a Markov chain (with one exception, discussed below, see eqn (5.2)), such tests may be regarded as tests of the semi-Markov property. Usually it is safer not to assume that bout lengths are exponentially distributed if this assumption is based only on the results of goodness-of-fit tests of exponentiality. In these cases the results of testing for sequential dependency properties depend on the outcomes of previously used tests, which is in general not desirable. If it is fairly certain that bout lengths are exponential, however, a test that uses this information will have more power. Where possible we therefore also give tests (or references) that can be used in such situations.

The tests discussed in section 5.3 and the following (sub)sections are carried out conditionally on the observed number of bouts of an act and/or the observed numbers of bouts of an act that are preceded or succeeded by certain other acts. Similar considerations as in subsection 1.5.1 apply here.

Deviations from Markovian sequential dependency properties can be caused by short-term dependencies as well as dependencies on a longer time scale. Among the first types of departures are deviations from first-order dependency in the sequence of acts (tests against such deviations are given in section 5.2), dependencies between bout lengths and preceding and/or following acts (section 5.3), and correlations between subsequent bouts (section 5.4). Long-term statistical dependencies can be due to time inhomogeneities. Methods for detecting such inhomogeneities are treated in Chapter 3. Another plausible source of long-term statistical dependencies are feedback mechanisms operating over longer periods. For instance, an animal may have a global recollection of what has happened during a certain period and this may affect its future behaviour. Some of the tests given in section 5.4 are mainly sensitive to dependencies of this kind. However, often the statistical structure of the behaviour resulting from such dependencies is indistinguishable from that due to time inhomogeneity. There may also be feedback mechanisms over shorter periods, for instance if an animal has to perform a certain act for an approximately fixed, and relatively short, time before it can proceed to some other activity. Such a mechanism was found, for example, by De Jonge and Ketel (1981) in the mating behaviour of voles, where the male needs an approximately constant amount of stimulation before ejaculation. Methods for detecting such dependencies are given in subsection 5.4.3.

Long-term statistical dependencies can sometimes be dealt with by distinguishing two or more different periods and incorporating parameter changes in the Markov model. See Chapter 3 for an extensive treatment of abrupt or gradual parameter changes. Tests for certain types of long-term dependencies are further discussed in section 5.4.

The interpretation of apparent non-Markovian sequential dependencies needs further consideration. Records should be thoroughly checked for time inhomogeneity before sequential properties are tested (see Fig. 1.5). However, it should be borne in mind that, even then, apparent deviations from the first-order Markov property can still be due to undetected inhomogeneity (see Fig. 5.1). For instance, suppose that there are two subperiods during an observation period. In the first period the transition rate from B

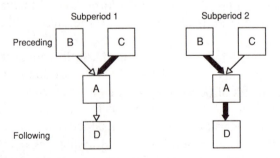

Figure 5.1 Example of how correlations between preceding acts and transition rates can be caused by time inhomogeneity. The widths of the arrows represent the relative magnitudes of the transition rates. For further explanation see text.

to A is low and the transition rate from C to A is high. In the second period the transition rate from B to A is high and that from C to A is low. Furthermore, the transition rate from A to D is low in the first period and high in the second period. As a result A is often preceded by C and rarely followed by D in period 1. In period 2, A is usually preceded by B and followed by D. If the periods are not distinguished, it will seem as though the transition rate from A to D is higher when B precedes A than when C precedes A, whereas in a CTMC these transition rates should not depend on preceding states.

Apparent correlations between bout lengths and preceding acts may result in a similar way. For example, when bouts of act A are on average short in period 1 and long in period 2, it seems as though bouts of A preceded by C are shorter than those preceded by B. Correlations between bout lengths and following acts or between subsequent bout lengths or bouts and gaps may also be due to time inhomogeneity. For instance, Dienske and Metz (1977) initially found that long bouts of the body contact state *Nipple* (where the rhesus monkey infant keeps its mother's nipple in its mouth, see subsection 4.2.1 and Box 4.1) were followed more often by vocalization by the infant than short bouts. Later, this apparent correlation was found to be due to the occurrence of 'drowsy phases', with long *Nipple* bouts, besides 'fully awake phases', with short *Nipple* bouts. Vocalizations nearly exclusively occur in drowsy phases (Dienske *et al.*, 1980).

In the following sections it is assumed that the records are either homogeneous or have been divided into homogeneous periods, i.e. deviations from Markov sequential dependency properties are assumed to have other causes.

A cause of departures from a (semi-)Markov sequential dependency structure that is of special ethological interest is that a behavioural category

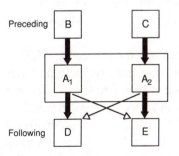

Figure 5.2 Example of how correlations between preceding acts and transition rates can be caused by lumping categories. The widths of the arrows represent the relative magnitudes of the transition rates. For further explanation see text.

consists of subcategories with different transition rates. In Fig. 5.2 it is indicated how the inadvertent lumping of categories may lead to departures from first-order Markov sequential dependency properties.

In this example act A consists of acts A_1 and A_2. A_1 is often preceded by B and followed by D, whereas A_2 is usually preceded by C and followed by E. If A_1 and A_2 are considered as one act A, it seems as though bouts of A that are preceded by B are often followed by D and bouts of A that are preceded by C are often followed by E. Apparent correlations between the bout length of A and preceding acts result when the mean bout lengths of A_1 and A_2 are different. Situations such as the one described in Fig. 5.2 occur relatively often in ethological experiments, for instance in the behaviour of male WE/zob rats (Haccou *et al.*, 1985). Rats of this strain frequently look up and move their heads slowly to and fro. This is called *Scanning*. Initially, it was recorded as one behaviourial category. There were apparent deviations from first-order sequential dependency in the records, since, for instance, when *Grooming* is followed by *Scanning* the probability that subsequently the rat starts grooming again is much higher than the probability that the rat will groom when *Scanning* is preceded by another act like *Walking*. Thus, it appeared that in a majority of the cases *Scanning* bouts are in fact interruptions of the current acts, i.e. during *Scanning* the tendency to go on with the current act is very high. (This situation is similar to that given in Fig. 5.2, with D and E replaced by B and C, respectively.) By subdividing the scanning bouts according to the preceding act, departures of first-order dependency could have been removed. Thus, one option is to consider all the different types of scanning as separate states in the models. This, however, would have led to a large number of behavioural categories. Therefore, the scanning interruptions were considered as part of the act during which they occurred. Further analysis showed that the remaining categories could also be considered as Markov states. Eventually, this procedure led to a smaller set of behavioural categories that can be represented by Markov states: the latter solution is the more economical.

Note that lumping does not always lead to Markov states or semi-Markov states. Especially when there are observable differences between the subcategories, it is sometimes better to do the reverse and to split act A up. In this context it is important to distinguish acts like short head movements or pauses, which can be considered as interruptions, from other types of acts. For instance, Goosen and Metz (1980) lumped together a large group of different acts, which were recorded as 'blanks'. Of course, this resulted in deviations from Markov sequential dependency properties, that cannot be handled in the way described above.

As mentioned, the sequential dependency properties of a CTMC and a semi-Markov chain are the same, with one exception. Whereas in a CTMC bout lengths do not depend on following acts, such dependencies may occur in a semi-Markov chain. For instance, suppose that after act A an animal can do either B or C. Furthermore, suppose that B can follow A immediately with constant probability per time unit, whereas the speed of a transition to C is subject to mechanical constraints. Hence the transition rate x time units after the beginning of act A, $\alpha_{AC}(x)$, is not constant. Very short bouts of A are never followed by C. Since the termination rate of act A, $\lambda_A(x)$, is the sum of the transition rates to B and C,

$$\lambda_A(x) = \alpha_{AB} + \alpha_{AC}(x), \tag{5.2}$$

this rate is not constant as well. Consequently, this behavioural process should be described by a semi-Markov chain rather than a Markov chain and apparent dependencies between bout lengths and succeeding acts should be expected. In these cases it is interesting when such dependencies are absent, since this suggests that an animal terminates an act before it decides what will be its next act.

In the past, ethologists have often examined sequences of acts by means of tests based on higher-order discrete Markov chains (Box 5.1), ignoring the information of bout durations. The procedure used was to perform a sequence of tests of kth-order dependency against $(k-1)$st-order dependency, for $k = m, m-1$, etc. When, for a certain value of k, say l, a test gave a significant result, it was concluded that the sequence was lth-order dependent, and the process was modelled by means of an lth-order discrete Markov chain (e.g. Altmann, 1965). However, the tests used are also sensitive to, for instance, time inhomogeneity or inadvertently lumped categories. Therefore, a significant result by itself is not enough 'evidence' for higher-order dependency. Furthermore, higher-order Markov chains are difficult to interpret as models for ethological data, since they do not have a continuous time analogue and it is unlikely that sequential dependencies only manifest themselves in the so-called 'embedded' discrete process. The lumping of Markov states usually leads to sequential dependencies of an infinite order. For these reasons, it is preferable to search for possible causes of deviations from first-order sequential dependency within the framework of the continuous time models considered here. In practice this strategy has proved to be successful: deviations from first-order dependency often disappear when acts are split up or lumped (as in the examples described previously).

Box 5.1 First- and higher-order discrete Markov chains

The discrete Markov chains we consider are processes with a finite (or at least countable) state space. The states are usually labelled by non-negative integers or by A, B, C,.... . The probability p_{AB} that the next state is B, given that the process is in state A, defined by

$$p_{AB} = Pr\{ \text{ the next state is B} \,|\, \text{process is in A} \} \qquad (1)$$

(see also section 1.2), is assumed to be independent of time and the number of states already visited. The Markov chain is called 'first-order dependent' if the probability p_{AB} is independent of the states visited before the process reached state A, i.e.

$$p_{AB} = Pr\{ B\,|\,A \} = Pr\{ B\,|\,Z_1, Z_2,..., Z_n, A \} \qquad (2)$$

for all possible states $Z_1, Z_2,..., Z_n$. The process is called second-order dependent if the transition probability p_{AB} does depend on the state, say X, which preceded A, but not on the other states before A is reached, i.e.

$$p_{XAB} = Pr\{ B\,|\,X, A \} = Pr\{ B\,|\,Z_1, Z_2,..., Z_n, X, A \} \qquad (3)$$

for all possible states $Z_1, Z_2,..., Z_n$ which might be visited. The definitions of third- or higher-order processes are analogous. A process with transition probabilities independent of time is called stationary.

Equation (2) is the 'lack of memory' property of a first-order Markov process. It implies that the probability of any particular future behaviour of the process when its present state is known is not influenced by additional knowledge concerning its past behaviour. Consequently, a first-order discrete time Markov chain arises from the continuous counterpart by omitting the durations leading to a sequence of states that are visited successively.

The main property of the matrix with all transition probabilities p_{AB}, defined by eqn (2), is

$$p_{AA} = 0 \quad \text{and} \quad \sum_{B \neq A} p_{AB} = 1, \qquad (4)$$

for all states A and B.

The theory of discrete time Markov chains is extensively treated in Feller (1968). The most important computational procedures, illustrated with many examples, are described in Kemeny and Snell (1976). A thorough treatment of the analysis of the structure of Markov chains is given by Chung (1967).

5.2   Sequences of acts

As was pointed out in section 5.1, the sequence of states in a CTMC is a first-order discrete Markov chain. In the context of this book a state corresponds to an act in the ethogram. Deviations from first-order dependency in a sequence of acts can be detected by applying a test for first- against second-order dependency. We discuss two such tests, which are expected also to be sensitive against deviations in the direction of a function of a

Markov chain. However, to test really against a function of a Markov chain, the alternative hypothesis should be a Markov process of infinite order. Such a process can be approximated by a kth-order Markov chain, provided that k is chosen sufficiently large as to assure an accurate description of the data. One might consequently question whether the restriction of the alternative hypothesis to second-order dependency is justified; strictly speaking it is not. One should apply a sequence of tests on a hierarchy of nested subhypotheses, where jth-order dependency is tested against $(j - 1)$th-order, for $j = k, k - 1, ..., 2$.

There are two major objections to this approach. Firstly, as discussed in the previous section, higher-order Markov chains are difficult to interpret as models for ethological data, and secondly, it leads to time-consuming and very tedious calculations. Moreover, in practice there are usually not enough data to meet the requirements for testing second- against third-order, let alone third- against fourth-order dependency.

In our experience the methodological approach, described in section 1.4 and visualized in Fig. 1.5, reduces the risk of loss of power (and/or an undesired increase of the level of significance) because the process is actually of order k ($k > 2$), when a test is applied for first- against second-order dependency. The detection of deviations from exponentiality in the early stage of the analyses, for instance, frequently leads to adjustments of the ethogram, which often results in the disappearance of higher-order sequential dependency, and consequently to a more parsimonious and ethologically well interpretable description of the records.

5.2.1 Chi-squared test for first- against second-order dependency in sequences of acts

Let (X, A, Y) be a triplet of acts. For each act A it must be tested whether the preceding act X influences the chance of transitions to the following act Y. The chi-squared test, based on the chi-squared test for goodness of fit, is described by e.g. Anderson and Goodman (1963).

Data: One sequence of acts $a_1, a_2, ..., a_{n-1}, a_n$.

Assumptions:
A1: Under H_0, $p_{XAY} = p_{AY}$, for all acts A, X and Y, where p_{AY} is defined by eqn (5.1) and $p_{XAY} = Pr\{$the present act A is succeeded by Y given that it is preceded by X$\}$.
A2: Under H_1, $p_{XAY} \neq p_{AY}$ for at least one combination of acts.

Procedure: We start with a test of whether the probability that a certain act A is succeeded by an act Y depends on the act which preceded A. Let N_A denote the total number of acts A in the sequence $a_2,..., a_{n-1}$, i.e. N_A is the number of the A, if the first and the last act are omitted. For every act X and act Y, N_{XA} denotes the number of times a transition from X to A has been observed, N_{AY} denotes the number of transitions from A to Y and N_{XAY} denotes the number of triplets X, A, Y, i.e. transitions from X via A to Y.

The (unconditional) transition probability from A to Y is estimated by

$$\hat{p}_{AY} = \frac{N_{AY}}{N_A}. \tag{5.3}$$

The chi-squared test statistic is

$$C_A = \sum_X \sum_Y \frac{(N_{XAY} - N_{XA}\hat{p}_{AY})^2}{N_{XA}\hat{p}_{AY}}, \tag{5.4}$$

where the summation is taken over all acts X and Y. Note that $X \neq A$ and $Y \neq A$ (see Box 5.2 for an example). This restricted H_0 is rejected if C_A exceeds the critical value at the chosen level of significance. For the small sample-case, no critical values are available.

Box 5.2 Chi-squared test for sequential dependency

1. We illustrate the procedure with the data from Table 1 in Box 1.1: a record of solitary rat behaviour (see also Fig. 1 in Box 1.1).

2. The transition matrix for this record is:

To: From:	Care	Walk	Sit	Rear	Shake
Care	0	0	9	0	4
Walk	2	0	18	3	0
Sit	10	19	0	7	6
Rear	0	0	10	0	0
Shake	1	5	5	0	0

3. For each possible middle act the matrices with the observed number of triplets are given below. When the transition frequencies are non-zero, the expected numbers are given in brackets.

a. Number of triplets of the form (A, *Care*, B):

A	Care	Walk	Sit	Rear	Shake
			B		
Care	0	0	0	0	0
Walk	0	0	1 (1.38)	0	1 (0.62)
Sit	0	0	7 (6.92)	0	3 (3.08)
Rear	0	0	0	0	0
Shake	0	0	1 (0.69)	0	0 (0.31)

The (conditional) expected number of triplets, for instance $E\ N_{Walk,Care,Sit}$ is calculated as follows (by applying eqns (5.3) and (5.4):

$$E\ N_{Walk,Care,Sit} = N_{Walk,Care}\ \hat{p}_{Care,Sit} = N_{Walk,Care} N_{Care,Sit}\ /\ N_{Care} = 2{\times}9/13 = 1.38.$$

b. Number of triplets of the form (A, *Walk*, B):

A	Care	Walk	Sit	Rear	Shake
			B		
Care	0	0	0	0	0
Walk	0	0	0	0	0
Sit	2 (1.65)	0	14 (14.87)	3 (2.48)	0
Rear	0	0	0	0	0
Shake	0 (0.43)	0	4 (3.91)	0 (0.65)	0

c. Number of triplets of the form (A, *Sit*, B):

	Care	Walk	Sit	Rear	Shake
			A		
Care	3 (2.14)	6 (4.07)	0	0 (1.50)	0 (1.29)
Walk	3 (4.29)	6 (8.14)	0	5 (3.00)	4 (2.57)
Sit	0	0	0	0	0
Rear	3 (2.38)	4 (4.52)	0	1 (1.67)	2 (1.43)
Shake	1 (1.19)	3 (2.26)	0	1 (0.83)	0 (0.71)

d. Number of triplets of the form (A, *Rear*, B):

A	Care	Walk	Sit	Rear	Shake
			B		
Care	0	0	0	0	0
Walk	0	0	3 (3.00)	0	0
Sit	0	0	7 (7.00)	0	0
Rear	0	0	0	0	0
Shake	0	0	0	0	0

e. Number of triplets of the form (A, *Shake*, B):

	B				
A	*Care*	*Walk*	*Sit*	*Rear*	*Shake*
Care	0 (0.36)	3 (1.82)	1 (1.82)	0	0
Walk	0	0	0	0	0
Sit	1 (0.54)	1 (2.73)	4 (2.73)	0	0
Rear	0	0	0	0	0
Shake	0	0	0	0	0

4. According to eqn (5.4), the statistic C_{Care} is equal to

$$C_{Care} = \frac{(1 - 1.38)^2}{1.38} + \frac{(1 - 0.62)^2}{0.62} + \frac{(7 - 6.92)^2}{6.92}$$

$$+ \ldots + \frac{(0 - 0.31)^2}{0.31} = 0.793.$$

5. The other statistics are respectively equal to
$C_{Walk} = 1.323$, $C_{Sit} = 8.555$, $C_{Rear} = 0$, $C_{Shake} = 3.567$.

6. The overall test statistic
$C = 0.793 + 1.323 + 8.555 + 0 + 3.567 = 14.238$.

7. Although the sample size is not large enough to apply the chi-squared approximation with a sufficient degree of accuracy, we calculate the number of degrees of freedom according to eqn (5.6) as an illustration. The parameters k and l in the present example are as follows:

			A		
	Care	*Walk*	*Sit*	*Rear*	*Shake*
k_A	2	3	1	3	3
l_A	3	2	1	4	2

Hence, the overall number of degrees of freedom $k = 2 \times 1 + 1 \times 2 + 3 \times 3 + 1 \times 0 + 1 \times 2 = 15$ according to eqn (5.6) and it can be concluded that there is no evidence for sequential dependency in the recorded rat behaviour, see Table A2. However, as said before, it must be kept in mind that this result has to be treated with some reservation due to the limited sample size.

Large-sample approximation: When the N are large, i.e. for large observation times, C_A has approximately a chi-squared distribution. Let m be the number of different acts; then the number of degrees of freedom is equal to $(m - 2)^2$. Reject H_0 if $C_A \geq \chi^2_k(\alpha)$, where $k = (m - 2)^2$. Moreover, since

C_A and C_B are stochastically independent when A ≠ B, an overall test for sequential dependency can be based on

$$C = \sum_A C_A,$$
(5.5)

where the summation is taken over all acts A. When the length of the sequence of acts tends to infinity, the statistic C follows a chi-squared distribution with $k = m(m - 2)^2$ degrees of freedom.

In most experiments, however, the large-sample approximation is not accurate enough and cannot be applied in this way, since some transitions are impossible (for instance, due to mechanical or physical constraints) or are extremely rare. See Box 5.2 for an example of an observed transition matrix.

For the sake of simplicity we first consider C_A. The large-sample approximation can be improved by a reduction of the number of degrees of freedom from the chi-squared statistic. Let k_A denote the number of transitions towards A and l_A the number of transitions from A which cannot occur. Then the adjusted number of degrees of freedom for C_A is $(m - k_A - 1)(m - l_A - 1)$. Since transitions from an act to itself cannot occur, $k_A \geq 1$ and $l_A \geq 1$. If these are the only impossible transitions, there is no reduction in the number of degrees of freedom.

The number of degrees of freedom for the overall chi-squared test C (eqn (5.5)) is

$$k = \sum_A (m - k_A - 1)(m - l_A - 1).$$
(5.6)

Besides transitions which are impossible due to mechanical, physical or other reasons, there are transitions that can occur in principle, but do not occur in practice, i.e. certain acts almost never succeed each other. This can be detected by scanning a large number of records of different individuals. When certain transitions do not occur in any of the records it can be assumed that the chance of such transitions is negligible and the number of degrees of freedom can be adjusted accordingly. There may also be incidental zeros, transitions that can occur but do not occur in the behavioural record under consideration. Some authors (e.g. Kullback *et al.*, 1962) argue that the number of degrees of freedom should be reduced accordingly. Others (e.g. Baker *et al.*, 1985) claim that this should not be done. However, W.M.A. Bressers (pers. comm.) has shown by a large number of simulation runs that the approximation with a chi-squared distribution with the 'full' number of degrees of freedom is very poor when relatively many transition frequencies are equal to zero. Her main results are as follows.

Replacing k_A and l_A in the definition of k, eqn (5.6), by k_A^* and l_A^* respectively (the observed number of zero transition frequencies towards and from A, respectively) gives a satisfactory improvement in the approximation if the number of triplets is larger than $4k$. The approximation is better the less the transition probabilities differ.

A more refined approximation appears to depend on these (unknown) transition probabilities and cannot be applied in practice, since these probabilities (and consequently the correction terms) as well as the test statistic have to be based on the same data.

Comments: The test statistic in eqn (5.4) is equal to the sum of the squared differences between the observed and expected frequencies under H_0, divided by the expected frequency. If the test gives a significant result, the individual components of the test statistic can be used to get indications about which of the transition(s) cause(s) deviations. The test is discussed by Anderson and Goodman (1963) and Kullback *et al.* (1962).

5.2.2 Likelihood ratio test for first- against second-order dependency in sequences of acts

A well-known alternative for the chi-squared test for testing first- against second-order dependency in sequences of acts is the likelihood ratio test, which is also described by Anderson and Goodman (1963).

Data and assumptions: See the chi-squared test treated in subsection 5.2.1.

Procedure: Let N_A denote the total number of acts A in the sequence $a_2, ..., a_{n-1}$ (i.e. N_A is the number of the A, if the first and the last act are omitted). For every act X and act Y, N_{XA} denotes the number of times a transition from X to A has been observed and N_{XAY} denotes the number of triplets X, A, Y, i.e. transitions from X via A to Y.

The (unconditional) transition probability from A to Y is estimated by

$$\hat{p}_{AY} = \frac{N_{AY}}{N_A} \tag{5.7}$$

and the probability of the occurrence of the triplet (X, A, Y) by

$$\hat{p}_{XAY} = \frac{N_{XAY}}{N_{XA}}. \tag{5.8}$$

When only act A is considered the likelihood ratio test statistic is

$$L_A = 2\left\{\sum_X \sum_Y N_{XAY}(\log \hat{p}_{XAY} - \log \hat{p}_{AY})\right\}. \tag{5.9}$$

The statistic for the overall test is

$$L = \sum_A L_A. \tag{5.10}$$

Reject the restricted or the overall H_0 if L_A or L exceeds the corresponding critical values for the chosen level of significance. No critical values are available for the small-sample case.

Large-sample approximation: Exactly the same chi-squared approximation applies as for the chi-squared test described in subsection 5.2.1. See Box 5.3 for a worked example.

Box 5.3   Likelihood ratio test for sequential dependency

1. We illustrate the procedure with the data from Table 1 in Box 1.1: a record of solitary rat behaviour (see also Fig. 1 in Box 1.1). Hence, see Box 5.2 for the transition matrix and the matrices with the numbers of triplets.

2. The test statistic L_A, defined by eqn (5.9), can be rewritten for ease of computation by substituting eqns (5.7) and (5.8) in (5.9):

$$L_A = 2\left(\sum_X \sum_Y N_{XAY}\log N_{XAY} - \sum_X N_{XA}\log N_{XA}\right.$$

$$\left. - \sum_Y N_{AY}\log N_{AY} + N_A\log N_A\right). \tag{1}$$

3. The numbers of transitions and triplets can be found in Box 5.2. Substitution for *Care* in eqn (1) gives

$$
\begin{aligned}
L_{Care} &= 2\times(1\log1 + 1\log1 + 7\log7 + 3\log3 + 1\log1 - 2\log2 - 10\log10 \\
&\quad - 1\log1 - 9\log9 - 4\log4 + 13\log13) \\
&= 2 \times (0 + 0 + 13.62137 + 3.29584 + 0 - 1.38629 - 23.02585 \\
&\quad - 0 - 19.77502 - 5.54518 + 33.34434) = 2 \times 0.5294 \\
&= 1.0584.
\end{aligned}
$$

4. The other statistics are respectively equal to
$$L_{Walk} = 2.184, \; L_{Sit} = 11.9471, \; L_{Rear} = 0, \; L_{Shake} = 4.637$$

5. The overall test statistic
$$L = 1.0584 + 5.4941 + 11.9471 + 0 + 5.6555 = 19.827.$$

6. The number of degrees of freedom is equal to that in Box 5.2 (15) and it can be concluded that there is some evidence for sequential dependency in the recorded rat behaviour, see Table A2: the critical value at a level of significance of 0.05 is 25.0; thus this test almost results in the rejection of H_0. However, as stated in Box 5.2, it must be kept in mind that this outcome has to be treated with some reservation due to the small sample size.

Comments: Simulations indicate that for this test statistic the convergence to the asymptotic chi-squared distribution is slower than for the statistic of eqn (5.4) (Bressers, pers. comm.). If the test gives a significant result, the contributions of the different transitions can be studied by comparing $\log \hat{p}_{XAY} - \log \hat{p}_{AY}$ for different triplets XAY. These statistics are, however, very sensitive to small values of $\hat{p}_{AY}$. The test is described by Anderson and Goodman (1963) and Kullback *et al.* (1962).

5.3   RELATIONS BETWEEN BOUT LENGTHS AND PRECEDING AND/OR FOLLOWING ACTS

5.3.1 Bout lengths and preceding acts

A non-parametric test based on the Kruskal-Wallis test

When a behavioural category, say A, can be considered as a (semi-)Markov state, the preceding act should have no effect on the bout length of A. For instance, let $x_1,..., x_n$ be the bouts of A that are preceded by act B and $y_1,..., y_n$ the bouts preceded by act C. Then $x_1,..., x_n$ and $y_1,..., y_n$ should have the same distribution. More generally, when there are k different acts that can precede act A, the corresponding k samples of bouts should all have the same distribution.

The independence of bout lengths and preceding acts can be tested without making assumptions about the bout length distribution, by means of a non-parametric k-sample test such as the Kruskal-Wallis test.

Data: The original record consists of a sequence of N^* acts with corresponding bout lengths:

Act	Bout length
a_1	x_1
a_2	x_2
.	.
.	.
.	.
a_{N^*}	x_{N^*}

Consider the subset of bouts of act A which are preceded by one of the k acts denoted by $B_1,..., B_k$. (Hence, if the first observed act is A, that bout must be omitted.) Sort the bout lengths according to the preceding act and renumber the bout lengths as follows:

Preceding act	Bout lengths of act A preceded by
B_1	$y_{1,1}, y_{1,2}, ..., y_{1,n_1}$
B_2	$y_{2,1}, y_{2,2}, ..., y_{2,n_2}$
.	.
.	.
.	.
B_k	$y_{k,1}, y_{k,2}, ..., y_{k,n_k}$

Assumptions:

A1: The bout lengths $Y_{i,j}$ are mutually independent and continuously distributed. The distributions of the $Y_{i,j}$ are identical, except possibly the means.

A2: The mean bout lengths are $E\,Y_{i,j} = \mu + \tau_i$, where τ_i denotes the effect of the preceding act on the mean bout length, $i = 1,..., k$, and $\Sigma\,\tau_i = 0$.

A3: Under H_0 the mean bout length is independent of the preceding act, i.e. $\tau_1 = ... = \tau_k = 0$.

A4: Under H_1 there is at least one non-zero τ_i.

Procedure: Rank all $N = \Sigma\,n_i$ observations in increasing order. In the case of ties (equal bout lengths) take the average rank of the group of tied observations. Let r_{ij} denote the rank of $y_{i,j}$, the jth bout length of A which is preceded by act B_i. Calculate the sum R_i of the ranks of each sample i, $i = 1,..., k$. The Kruskal-Wallis test statistic, K_k, for testing the equality of k mean rank sums is defined as

$$K_k = \left(\frac{12}{N(N+1)} \sum_{i=1}^{k} \frac{R_i^2}{n_i} \right) - 3(N + 1) \tag{5.11}$$

in the absence of ties, and as

$$K_k' = \frac{K_k}{1 - \dfrac{\displaystyle\sum_{l=1}^{g} (t_l^3 - t_l)}{N^3 - N}} \tag{5.12}$$

otherwise, where g is the number of tied groups and t_l is the size of the lth tied group (see e.g. Siegel and Castellan, 1988). Reject H_0 if the test statistic exceeds the critical value corresponding to the chosen level of significance listed in Table A24 (only available for $k = 3$).

Box 5.4   Testing relations between mean bout length and preceding acts:
application of the Kruskal-Wallis test

1. We consider the effect of the preceding act on the duration of *Sit* bouts in solitary rat behaviour (see Table 1 and Fig. 1 in Box 1.1). In Table 1 the bouts are arranged according to these acts with the corresponding ranks.

Table 1 *Sit* bouts arranged according to the preceding act

	Care		Walk		Rear		Shake	
j	y_{1j}	r_{1j}	y_{2j}	r_{2j}	y_{3j}	r_{3j}	y_{4j}	r_{4j}
1	36.20	35	10.51	14	1.74	1	18.14	28
2	22.10	31	20.25	30	22.97	32	15.44	24
3	67.96	41	10.79	15	53.13	37	5.46	11
4	64.47	40	3.18	5	55.11	39	4.64	10
5	3.31	6	2.67	3	12.71	20	14.20	22
6	90.28	42	28.09	33	2.90	4		
7	11.18	16	11.76	17	5.64	12		
8	53.81	38	1.90	2	51.27	36		
9	12.48	19	3.47	7	19.51	29		
10			17.05	25	32.01	34		
11			9.93	13				
12			17.58	26				
13			15.06	23				
14			13.08	21				
15			4.41	9				
16			17.71	27				
17			4.31	8				
18			12.06	18				
	$R_1 = 268$		$R_2 = 296$		$R_3 = 244$		$R_4 = 95$	
	$n_1 = 9$		$n_2 = 18$		$n_3 = 10$		$n_4 = 5$	

2. The rank sums per sample are substituted in the test statistic defined in eqn (5.11):

$$K_4 = \{\frac{12}{42\times43}(\frac{268^2}{9} + \frac{296^2}{18} + \frac{244^2}{10} + \frac{95^2}{5})\} - 3\times43$$

$$= \frac{12}{42\times43}(7980.44 + 4867.55 + 5953.6 + 1805) - 129$$

$$= \frac{12}{42\times43}20606.6 - 129 = 136.921 - 129 = 7.921.$$

3. Under the null hypothesis that the preceding acts have no effect on the mean duration of Sit bouts, K_4 is chi-squared distributed with three degrees of freedom. Hence, the result is just significant: the critical value at a 5% level is 7.81 (see Table A2), whereas $K_4 = 7.921$. Accordingly, H_0 is rejected and there is an indication that the preceding act affects the duration of the *Sit* bouts.

Large-sample approximation: For large n_i ($i = 1,..., k$), K_k (or K_k') has approximately a chi-squared distribution with $k - 1$ degrees of freedom. See Table A2 for the critical values and Box 5.4 for an example.

Multiple comparison: If the test based on (5.11) (or (5.12)) gives a significant result, the differences between the k samples can be further examined by means of the multiple comparison method described by Hollander and Wolfe (1973). Let

$$R_{i.} = \frac{R_i}{n_i} \tag{5.13}$$

be the average rank in the ith sample, $i = 1,..., k$. The difference between the ith and the jth sample is significant when

$$(R_{i.} - R_{j.})^2 > c_\alpha \frac{N(N + 1)}{12}\left(\frac{1}{n_i} + \frac{1}{n_j}\right), \tag{5.14}$$

where c_α is the critical value of the Kruskal-Wallis test at significance level α. This procedure is conservative. For large sample sizes, c_α in eqn (5.14) can be approximated by the upper critical value of a standard normal distribution at significance level $\alpha/\{k(k - 1)\}$ (Dunn, 1964).

The results of a multiple comparison method can indicate how a category A could be split up. For instance, when bout lengths of A preceded by B and C are similar and those preceded by D and E are similar but different from those preceded by B or C, it can be investigated whether there are observable differences between the two types of bouts of A.

When there are prior clues about differences between bouts of A in connection with different groups of acts instead of single acts, the procedure can also be applied to samples of bouts preceded by groups of acts. In that case B_i denotes a group of acts.

The *k*-sample test for exponentially distributed bout lengths

If it can be assumed that the bout lengths of each of the samples are exponentially distributed the *k*-sample likelihood ratio test for homogeneity can be applied. This test has more power when the distribution is indeed exponential.

Data: See under Kruskal-Wallis test at the beginning of this subsection for a description of the initial sorting and renumbering of the data:

Preceding act	Bout lengths of act A preceded by
B_1	$y_{1,1}, y_{1,2},..., y_{1,n_1}$
B_2	$y_{2,1}, y_{2,2}..............., y_{2,n_2}$
.	.
.	.
.	.
B_k	$y_{k,1}, y_{k,2}.........., y_{k,n_k}$

Assumptions:

A1:   The bout lengths $Y_{i,j}$ are mutually independent and exponentially distributed.

A2:   The mean bout lengths are $E\,Y_{i,j} = \lambda_i$, $i = 1,..., k$.

A3:   Under H_0 the mean bout length is independent of the preceding act, i.e. $\lambda_1 = ... = \lambda_k = \lambda$, where λ is unknown.

A4:   Under H_1 there is at least one inequality.

Procedure: Calculate the likelihood ratio test statistic Λ, defined by

$$\Lambda = 2\{\sum_{i=1}^{k}(n_i\log n_i - n_i\log S_i) - N\log N + N\log(\sum_{i=1}^{k}S_i)\}, \qquad (5.15)$$

where

$$S_i = \sum_{j=1}^{n_i} y_{i,j} \quad \text{and} \quad N = \sum_{i=1}^{k} n_i. \qquad (5.16)$$

Reject H_0, the hypothesis of equal mean bout lengths, if Λ is larger than the critical value corresponding to the level of significance one has chosen. No small-sample critical values are available.

Large-sample approximation: It can be proved that Λ is asymptotically chi-squared distributed with $k - 1$ degrees of freedom if the n_i tend to infinity. Hence, in practice H_0 must be rejected if $\Lambda > \chi^2_{k-1}(\alpha)$. See Table A2 for the critical values.

5.3.2 Bout lengths and following acts

Tests for correlations between bout lengths and succeeding acts can also be based on the test statistic K_k (or K_k') defined in subsection 5.3.1. In a CTMC the following state has no influence on the residence-time distribution of the current state. However, as noted in section 5.1, in a semi-Markov chain it may have an effect. Thus, if it is already suspected that a process should be described by a semi-Markov chain rather than a CTMC, a significant correlation between bout lengths and succeeding acts does not give additional clues. However, if there is no such dependency and the process is a semi-Markov chain, the mechanism by which bouts are terminated is apparently independent of the following bout. This leads to a special form of the semi-Markov model, namely one with proportional transition rates (see section 7.5).

5.3.3 Testing relations between preceding acts, bout lengths and following acts

The (semi-)Markov property implies that the way in which an act is started should not affect the way in which it is terminated. Thus, the transition rate from A to B should not be different when A is preceded by B, C or D, etc.

To compare the transition rates from act A to B for different preceding acts, the lengths of bouts of A that are succeeded by B are considered as termination times and those not followed by B are treated as censoring times (see section 4.4), as is illustrated in Fig. 5.3. In the first part of this subsection we treat a non-parametric test for the equality of transition rates under censoring, i.e. the most general case. Under the assumptions of a CTMC bout lengths are exponentially distributed and the simpler likelihood ratio test described in the second part of subsection 5.3.1 can be applied. Furthermore we describe in this subsection a graphical method for comparing transition rates, based on log-survivor plots (see subsection 4.6.1).

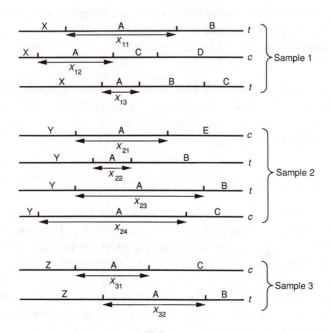

Figure 5.3 Study of the effect of the preceding act on the transition rates from act A to B. Bouts of A are arranged according to these acts. If A is succeeded by B, x_{ij} is considered as a termination time (t). In the other cases it is treated as a censoring time (c).

A non-parametric test: the log-rank test

Data: The original record consists of N^* acts with corresponding bout duration, as indicated in subsection 5.3.1. We assume that besides act A the ethogram contains k other acts. Consider all possible triplets of three succeeding acts, where the second act is A. Arrange the triplets in k samples according to the first act in the triplet. The jth bout length of A in the ith sample is denoted by x_{ij}, $i = 1,..., k$, $j = 1,..., n_i$, where n_i is the size of the ith sample. To each bout length x_{ij} a number δ_{ij} is attributed which is one if A is followed by B (an uncensored observation in the present context) and zero if A is followed by one of the other acts (hence it is considered to be a censored observation). See Fig. 5.3 for an illustration. This manipulation of the data eventually results in the following data set:

Sample 1: $(x_{1,1}, \delta_{1,1}), (x_{1,2}, \delta_{1,2}), ..., (x_{1,n_1}, \delta_{1,n_1})$
Sample 2: $(x_{2,1}, \delta_{2,1}), (x_{2,2}, \delta_{2,2}), ..., (x_{2,n_2}, \delta_{2,n_2})$

.   .
.   .
.   .

Sample k: $(x_{k,1}, \delta_{k,1}), (x_{k,2}, \delta_{k,2}), ..., (x_{k,n_k}, \delta_{k,n_k})$

Assumptions:
A1:   Define $X_{ij} = \min(X_{ij}^*, C_{ij})$, where the X_{ij}^* are the hypothetical bout durations and C_{ij} the random censoring times. The X_{ij}^* are mutually independent with $E\, X_{ij}^* = \mu_i$, $i = 1,..., k$. The C_{ij} are mutually independent and identically distributed censoring times.
A2:   Under H_0, $\mu_1 = ... = \mu_k = \mu$, where μ is unknown.
A3:   Under H_1 there is at least one inequality.

Procedure: Let $y_{(1)} < ... < y_{(N)}$ be the N ordered termination times in the combined sample, formed by pooling the k samples of bout lengths. Furthermore, let d_{ij} be the number of terminations at time $y_{(j)}$ in the ith sample and let n_{ij} be the number of bouts in the ith sample that are larger than or equal to $y_{(j)}$. Let d_j and n_j denote the corresponding numbers in the combined sample. The test statistic is

$$S = v'V^{-1}v \tag{5.17}$$

where the vector v is defined by

$$v' = \sum_{j=1}^{N} v_j', \tag{5.18}$$

with

$$v_j' = (d_{1j} - w_{1j}, ..., d_{k-1,j} - w_{k-1,j})' \tag{5.19}$$

and

$$w_{ij} = \frac{n_{ij} d_j}{n_j}, \ i = 1,..., k-1; \ j = 1,..., N. \tag{5.20}$$

Furthermore,

$$V = \sum_{j=1}^{N} V_j, \tag{5.21}$$

where V_j is a $(k - 1) \times (k - 1)$ matrix with elements

$$(V_j)_{ii} = \frac{n_{ij}(n_j - n_{ij})d_j(n_j - d_j)}{n_j^2(n_j - 1)}, \quad i = 1,..., k - 1 \qquad (5.22)$$

and for $i \neq l$

$$(V_j)_{il} = (V_j)_{li} = \frac{n_{ij}n_{lj}d_j(n_j - d_j)}{n_j^2(n_j - 1)}, \qquad (5.23)$$

for $i = 1,..., k - 1$ and $l = 1,..., k - 1$. Note that if $n_j = 1$ (and hence $d_j = 1$) the quotient

$$\frac{n_j - d_j}{n_j - 1} = 1 \qquad (5.24)$$

in eqn (5.23).

Reject H_0, the hypothesis of equal termination rates, irrespective of the preceding and the following act, for large values of S (eqn (5.17)). No critical values are available for the small-sample case.

Large-sample approximation: Under the hypothesis that the termination rates are equal, S has asymptotically (when the total number of bouts, N^*, tends to infinity and N/N^* goes to a positive constant smaller than one) a chi-squared distribution with $k - 1$ degrees of freedom. Reject H_0 if S is larger than $\chi^2_{k-1}(\alpha)$. See Table A2 for the critical values and Box 5.5 for a detailed example.

Box 5.5   The log-rank test for testing the relation between preceding acts and the transition rate

1. We consider the effect of the preceding act on the transition rate of the act *Sit* to *Walk* in rat behaviour (see Table 1 and Fig. 1 in Box 1.1). The smallest Sit bout (1.74 s) started 140.92 s from the beginning of the record, is preceded by *Rear* and followed by *Walk*; hence it is an uncensored observation. The next smallest bout lasted 1.90 s (started on 286.30 s) and is followed by *Shake*; thus it is a censor, etc.

2. The ordered *Sit* bouts, arranged according to the preceding act, are listed in Table 1, where $\delta = 1$ denotes an uncensored bout length, i.e. a transition to Walk, and $\delta = 0$ a censored bout length, i.e. a transition to one of the other acts.

Table 1 Ordered *Sit* bouts arranged according to the preceding act (uncensored, i.e. *Sit* followed by *Walk*: δ = 1; and censored, i.e. transition to another act: δ = 0)

j	Care	δ	Walk	δ	Rear	δ	Shake	δ
1					1.74	1		
2			1.90	0				
3			2.67	0				
4					2.90	0		
5			3.18	1				
6	3.31	1						
7			3.47	0				
8			4.31	0				
9			4.41	0				
10							4.64	0
11							5.46	1
12					5.64	0		
13			9.93	0				
14			10.51	1				
15			10.79	1				
16	11.18	1						
17			11.76	0				
18			12.06	0				
19	12.48	1						
20					12.71	0		
21			13.08	0				
22							14.20	1
23			15.06	0				
24							15.44	0
25			17.05	0				
26			17.58	1				
27			17.71	0				
28							18.14	1
29					19.51	0		
30			20.25	1				
31	22.10	1						
32					22.97	0		
33			28.09	1				
34					32.01	0		
35	36.20	1						
36					51.27	1		
37					53.13	1		
38	53.81	1						

Table 1 Continued

j	Care	δ	Walk	δ	Rear	δ	Shake	δ
39					55.11	1		
40	64.47	0						
41	67.96	0						
42	90.28	0						

3. We only give a few indications of the calculation of the statistic S, defined by eqn (5.17). For instance, let $j = 15$: the *Sit* bout lasted 10.79 s, is preceded and followed by a *Walk* bout (hence, it is not censored). The number of *Sit* bouts larger than or equal to 10.79 s, preceded by *Care*, $n_{1,15}$, is equal to 8, the number of terminations of such bouts, $d_{1,15}$, is zero. The total number of *Sit* bouts larger than or equal to 10.79 s, n_{15}, is 28, (see Table 2). The expected number of terminations, $w_{1,15}$, given in eqn (5.20), is equal to $8 \times 1/28$, etc., see also Table 2.

Table 2 Survey of results of the calculations

j	$d_{j,15}$	$n_{j,15}$	$w_{1,15}$
1	0	8	8/28
2	1	10	10/28
3	0	7	7/28
4	0	3	3/28
	$d_j = 1$	$n_j = 28$	

4. According to Table 2 and eqn (5.19), the vector v_j is equal to $(-8/28, 18/28, -7/28)$.

The elements V_{11} and V_{12} of the covariance matrix V_j, eqns (5.22) and (5.23), are equal to, respectively
$8 \times (28 - 8) \times 1 \times (28 - 1)/(28^2 \times (28 - 1))$ and
$8 \times 10 \times 1 \times (28 - 1)/(28^2 \times (28 - 1))$, etc.
All other computations are performed analogously.

Literature: The log-rank test is described, for instance, by Kalbfleisch and Prentice (1980). Fleming and Harrington (1991) discuss the properties of the log-rank test and several alternative tests at a more advanced level.

The exponential case: a graphical procedure

Data: See the log-rank test.

Assumptions:

A1, A2: See the log-rank test.

A3:   The $X_{ij}*$ defined in assumption A1 are exponentially distributed with parameter λ_i.

A4:   Under H_0 the mean bout length is independent of the preceding act, i.e. $\lambda_1 = \ldots = \lambda_k = \lambda$, where λ is unknown.

A5:   Under H_1 there is at least one inequality.

Procedure: Metz *et al.* (1983) proposed a graphical inspection procedure based on log-survivor plots (see subsection 4.6.1). Assumptions A1, A2 and A3 imply that under H_0 the behavioural process can be adequately described by a CTMC. The distinction between uncensored and censored bout lengths therefore disappears: all data are treated as uncensored observations (except in the case of an external censor mechanism, see section 4.4 for a few examples).

An easy graphical test can be based on the CTMC properties. Let

$$\bar{F}_{ZB}(x) = Pr\{X > x \text{ and next act is B} \mid \text{current} \tag{5.25}$$
$$\text{act is A and preceding act was Z}\}.$$

Then

$$\bar{F}_{ZB}(x) = p_{AB}\exp[-\lambda x], \tag{5.26}$$

and thus

$$\log \bar{F}_{ZB}(x) = \log p_{AB} - \lambda x, \tag{5.27}$$

where p_{AB} is the transition probability from A to B (see eqn (1.5)). Thus, under H_0 plots of estimates of $\log \bar{F}_{ZB}(x)$ against x for different acts Z should look like identical straight lines, and for different B the plots should be parallel. $\bar{F}_{ZB}(x)$ is estimated as usual (subsection 4.6.1) by the number of bouts of A that are larger than x, preceded by B and succeeded by C, divided by the total number of bouts of A that are preceded by B (Box 5.6). The procedure can be used for initial inspection of the data. Furthermore, if a formal test, such as the log-rank test described in the previous subsection, gives a significant result, this graphical procedure can give indications about the causes of deviations.

Box 5.6   Log-survivor plots for testing relations between preceding act, bout length and following acts

1. We consider the effect of the preceding act on the duration of *Sit* bouts followed by a *Walk* bout in rat behaviour (see Table 1 and Fig. 1 in Box 1.1). The first *Sit* bout followed by a *Walk* bout lasted 10.51 s and is preceded by a *Walk* bout, the second had a duration of 20.25 s and is also preceded by a *Walk* bout, etc. In Table 1 all *Sit* bouts followed by a *Walk* bout are arranged according to the preceding acts.

Table 1 Lengths of *Sit* bouts followed by a *Walk* bout arranged according to the preceding act

Care	Walk	Rear	Shake
36.20	10.51	1.74	18.14
22.10	20.25	53.13	5.46
3.31	10.79	55.11	14.20
11.18	3.18	51.27	
53.81	28.09		
12.48	17.58		

2. Figure 1 shows the four log-survivor functions (see Box 1.2). Under the assumptions of a CTMC they are supposed to be identical. Obviously there are deviations, perhaps due to the considerable variability resulting from the small sample sizes. Nevertheless, Fig. 1 indicates that there is non-Markov sequential dependency.

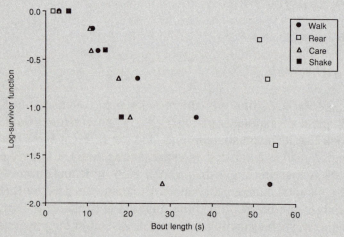

Figure 1 Four empirical log-survivor functions based on the data listed in Table 1.

The exponential case: the likelihood ratio test

Data and assumptions: See the above-mentioned assumptions for the graphical test.

Procedure: Calculate the likelihood ratio statistic Λ, defined by eqns (5.15) and (5.16), where n_i denotes the number of uncensored bouts in the ith sample, and hence $n_i = \Sigma \delta_{ij}$, $i = 1,..., k$. Note that under the assumptions of a CTMC bouts are to be considered as uncensored, except in the case when an external censor mechanism is operating.

Reject H_0, the hypothesis of equal mean bout lengths, if Λ is larger than the critical value corresponding to the level of significance one has chosen. No small-sample critical values are available.

Large-sample approximation: It can be proved that Λ is asymptotically chi-squared distributed with $k - 1$ degrees of freedom if the n_i tend to infinity and the fractions of censored observations converges to Δ_i ($\varepsilon < \Delta_i < 1 - \varepsilon$, $\varepsilon > 0$), $i = 1,..., k$. Hence, H_0 must be rejected if $\Lambda > \chi^2_{k-1}(\alpha)$. See Table A2 for the critical values.

5.4 RELATIONS BETWEEN SUBSEQUENT BOUTS AND/OR GAPS

In this section we consider tests of independence between subsequent bout lengths of one and the same act as well as between different acts. This independence, the so-called 'renewal property' (subsection 1.3.1), is one of the main features of a (semi-)Markov chain.

Testing for renewal properties is difficult, owing to the large number of assumptions which have to be made under H_0 and the unspecified character of the alternative hypothesis. Therefore we treat only a few aspects of this property, which are especially important in the present context. If no departures from these aspects are found, it can be assumed that departures from other properties are only slight or absent.

Aspects of the renewal property can be studied at either a short or a long time scale. Tests for dependencies over a relatively short period can be based on a non-parametric autocorrelation coefficient if the bout lengths of one act are considered or in the case of two different acts on a non-parametric (cross-)correlation coefficient. Tests for dependencies over a longer time scale can be based on correlation coefficients with lag j, where j is sufficiently large, or on power spectra. In subsections 5.4.1 and 5.4.2 we treat the autocorrelation and the cross-correlation test and we indicate how to apply these tests when investigating several aspects of the renewal property of a (semi-)Markov model. References for tests against dependencies on a large time scale are given in subsection 5.4.3. One test based on the occurrence of a specific act within a group of acts is worked out in detail.

5.4.1 Subsequent bouts of one act: the non-parametric auto-correlation coefficient

The so-called serial or auto correlation coefficient indicates whether the length of a bout of a certain act affects one or more subsequent bouts of the same act (say A, see Fig. 5.4).

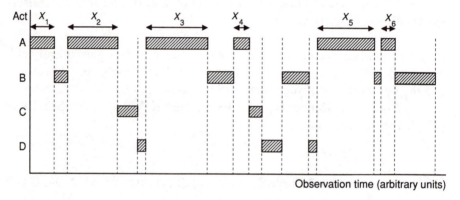

Observation time (arbitrary units)

Figure 5.4 Analysis of autocorrelation between bout lengths in a behavioural record: subsequent bouts of act A of length x_i.

Data: A sequence of n bout lengths of a specific act: $x_1,..., x_n$.

Assumptions:
A1: $X_1,..., X_n$ are identically distributed random variables.
A2:   Under H_0 the X_i are mutually independent.
A3:   Under H_1 there are dependencies.

Procedure: To each x_i a rank r_i is assigned. Take average ranks in the case of tied observations. A non-parametric test for serial correlation with lag 1, i.e. between two adjacent bouts, is based on R_1 defined by

$$R_1 = \sum_{i=1}^{n-1} r_{i+1} r_i. \tag{5.28}$$

Reject H_0 if $|R_1|$ exceeds the critical value. Hallin and Mélard (1988) give small sample critical values. They compare the properties of a few tests.

Large-sample approximation: Under the hypothesis of no correlation, $(R_1 - \mu)/\sigma$ has asymptotically (as n tends to infinity) a standard normal distribution (Wald and Wolfowitz, 1943), where

$$\mu = \frac{(n - 1)(n + 1)(3n + 2)}{12} \tag{5.29}$$

and

$$\sigma^2 = \frac{5n^6 + 16n^5 - 14n^4 - 80n^3 - 35n^2 + 64n + 44}{720(n - 1)}. \tag{5.30}$$

Reject H_0 if $(R_1 - \mu)/\sigma < z_{1-\alpha}$ or if $(R_1 - \mu)/\sigma > z_\alpha$ in the one-sided cases, and if $|(R_1 - \mu)/\sigma| > z_{\frac{1}{2}\alpha}$ in the two-sided case. See Table A1 for the critical values, and Box 5.7 for an example. Since the rate of convergence is not known, we cannot tell at which sample sizes the asymptotic approximation is sufficiently accurate.

Box 5.7   Correlation in mean bout length between subsequent bouts of one act: the non-parametric autocorrelation coefficient

1. We illustrate the use of the non-parametric autocorrelation coefficient for testing correlation between subsequent bout lengths of a specific act with the data of solitary rat behaviour in Table 1 from Box 1.1. For this act we choose *Walk*. The n (= 24) *Walk* bouts are listed in the second column of Table 1.

2. Next the corresponding ranks are assigned to the *Walk* bouts and the test statistic R_1 (eqn (5.28)) is calculated. See Table 1 for details.

3. The expected value μ of R_1, given by eqn (5.29), is
$23 \times 25 \times (72 + 2)/12 = 3545.83$
and the variance σ^2, (eqn (5.30)), is
$(955514880 + 127401984 - 4644864 - 1105920 - 20160 + 1536 + 44)/$
$720 \times 23 = 1077147500/16560 = 65045.14$.
Hence, the standard deviation, σ, is equal to 255.04 and the test statistic $z = (3590 - 3545.83)/255.04 = 0.173$, i.e. the approximate right-sided *p*-value is equal to 0.4314 (= 1 - 0.5686) (see Table A1). It can be concluded that no serial or autocorrelation between subsequent *Walk* bouts is present in this record.

Table 1 Calculation of the non-parametric autocorrelation coefficient between subsequent *Walk* bouts

i	x_i	r_i	r_{i+1}	$r_i r_{i+1}$
1	0.59	6	9	54
2	0.89	9	22	198
3	6.19	22	23	506
4	6.25	23	3	69
5	0.30	3	10	30
6	0.95	10	20	200
7	5.56	20	11	220
8	1.02	11	18	198
9	3.22	18	14	252
10	2.19	14	24	336
11	8.03	24	1	24
12	0.15	1	13	13
13	1.42	13	12	156
14	1.18	12	7	84
15	0.72	7	2	14
16	0.27	2	4	8
17	0.33	4	5	20
18	0.50	5	21	105
19	6.09	21	8	168
20	0.78	8	17	136
21	2.78	17	15	255
22	2.20	15	16	240
23	2.51	16	19	304
24	4.74	19	–	–

$$\Sigma \, r_i r_{i+1} = 3590$$

Comments: A non-parametric test is preferable to the test based on the serial (product-moment) correlation coefficient (which is asymptotically most powerful when the X_i are normally distributed), since bout length distributions are in general very skew, and the parametric test appears to be very sensitive to departures from the normality assumption.

When the bout length distribution is (close to) exponential the serial correlation R_1 can be calculated by means of 'exponential scores'. The ranks r_i in (5.28) are then replaced by the corresponding expected order statistics from the unit exponential distribution. For further details see Cox and Lewis (1978). According to Cox and Lewis, the test based on (5.28) emphasizes correlations between short intervals, whereas the exponential score test emphasizes correlations between long intervals.

5.4.2 Subsequent bouts of different acts: Spearman's correlation coefficient

To test the independence between the lengths of two adjacent bouts of different acts (see Fig. 5.5) it is best to use a non-parametric cross-correlation coefficient. However, in most applications the gaps between the occurrence of two adjacent bouts of the type considered are so large that one of the more well-known coefficients, e.g. Spearman's correlation coefficient, can be used.

Data: n pairs of lengths of adjacent bouts $(x_1,y_1),...,(x_n,y_n)$, where the bout lengths x_i are of act A and the y_i of act B.

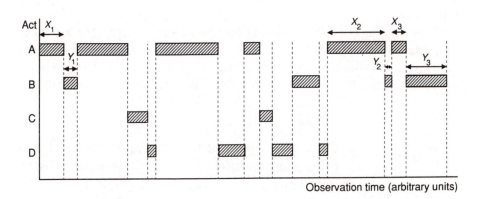

Figure 5.5 Analysis of correlation between the bout lengths of two different acts: adjacent bouts of act A and act B of length x_i, respectively y_i.

Assumptions:
A1:   The pairs (X_i,Y_i) are identically and independently distributed.
A2:   Under H_0, X_i and Y_i are stochastically independent for $i = 1,...,n$.
A3:   Under H_1, X_i and Y_i are dependent.

Procedure: Ranks r_i are assigned to the x_i, $i = 1,..., n$, in the same way as described in subsection 5.3.1. Similarly, ranks s_i are assigned to the y_i ($i = 1,..., n$). Let $d_i = r_i - s_i$, $i = 1,..., n$. If there are tied observations, assign average ranks. Spearman's rank correlation is defined as

$$r_S = 1 - \frac{6 \sum_{i=1}^{n} d_i^2}{n^3 - n} . \tag{5.31}$$

Siegel and Castellan (1988) give the following correction for the case when a large proportion of the observations is tied (otherwise eqn (5.31) may be used for computation):

$$r_S = \frac{(n^3 - n) - 6 \sum_{i=1}^{n} d_i^2 - \dfrac{T_x + T_y}{2}}{\sqrt{(n^3 - n)^2 - (T_x + T_y)(n^3 - n) + T_x T_y}} \tag{5.32}$$

where T_x is defined by

$$T_x = \sum_{i=1}^{g} (t_i^3 - t_i) \tag{5.33}$$

with a number of g tied groups of size t_i, $i = 1,..., g$. T_y is defined in the same way.

Critical values, c_α, for $n = 4$ (1) 50 and several values of α are listed in Table A25. Reject H_0 if r_S exceeds the critical value in the right-sided case (or if it is smaller than $-c_\alpha$ in the left-sided case) and if $|r_s| > c_{\frac{1}{2}\alpha}$ in the two-sided test. In Box 5.8 we give an example of an application.

Box 5.8   Spearman's non-parametric correlation coefficient for testing correlation between bout lengths of different acts

1. We illustrate the use of Spearman's correlation coefficient for testing correlation between bout lengths of different acts with the data of solitary rat behaviour in Table 1 from Box 1.1. For the first act we choose *Walk* and for the second *Sit*. The 18 pairs of bout lengths of consecutive acts *Walk*, *Sit* are listed in Table 1. The first pair concerns the second and the third bout: *Walk* 0.59 s and *Sit* 10.51 s; the second pair is: *Walk* 0.89 s and *Sit* 20.25, and so on.

Table 1 Calculation of Spearman's rank correlation coefficient

i	Walk x_i	r_i	Sit y_i	s_i	d_i	d_i^2
1	0.59	5	10.51	8	−3	9
2	0.89	6	20.25	17	−11	121
3	6.25	18	10.79	9	9	81
4	0.30	2	3.18	3	−1	1
5	0.95	7	2.67	2	5	25
6	5.56	16	28.09	18	−2	4
7	1.02	8	11.76	10	−2	4
8	3.22	15	1.90	1	14	196
9	2.19	11	3.47	4	7	49
10	1.42	10	17.05	14	−4	16
11	1.18	9	9.93	7	2	4
12	0.27	1	17.58	15	−14	196
13	0.33	3	15.06	13	−10	100
14	0.50	4	13.08	12	−8	64
15	6.09	17	4.41	6	11	121
16	2.78	14	17.71	16	−2	4
17	2.20	12	4.31	5	7	49
18	2.51	13	12.06	11	2	4

$$\Sigma \, d_i^2 = 1048$$

2. Next the corresponding ranks are assigned for both the *Walk* and the *Sit* bouts and the differences and the squared differences are calculated.

3. Substitution of the data in eqn (5.31) yields

$$r_s = 1 - \frac{6 \times 1048}{18^3 - 18} = 1 - \frac{6288}{5814} = 1 - 1.08153 = -0.08153. \tag{1}$$

4. Hence, according to Table A25 the two-sided H_0 is not rejected: for $n = 18$, $c_{0.975} = -0.472$ and $c_{0.025} = +0.472$. The conclusion is that no correlation between *Walk* and *Sit* bouts can be shown.

5. It follows from Table A25 that the p-value of r_s lies between 0.25 and 0.75.

6. Student's statistic for the large-sample approximation, eqn (5.34), is $t = -0.08153 \, \{16/(1 - 0.006647)\}^{\frac{1}{2}} = -0.08153 \times 4.0134 = -0.3272$. The number of degrees of freedom is equal to $18 - 2 = 16$ and the result is also not significant (see Table A3).

Large-sample approximation: For large n the distribution of r_S can be approximated by Student's distribution with $n - 2$ degrees of freedom:

$$t = r_S \sqrt{\frac{n - 2}{1 - r_S^2}} \, . \tag{5.34}$$

The critical values of this transformed statistic are listed in Table A3. The approximation already appears to be satisfactory if $n \geq 10$.

5.4.3 Dependencies on a large time scale

Dependencies on a large time scale can be studied in several ways. In this subsection we treat some of them.

Serial correlation

The definition of the serial (or auto)correlation coefficient given in subsection 5.4.1 (eqn (5.28)) can be generalized to R_j, the lag j non-parametric correlation coefficient, defined as

$$R_j = \sum_{i=1}^{n-j} r_{i+j} r_i, \, j = 1, 2,\ldots \, . \tag{5.35}$$

R_j is sensitive to dependencies between pairs of bout lengths of an act, say A, separated by exactly $j - 1$ bouts of that act (see Fig. 5.6). The time scale at which the dependencies are examined depends on j. Note that the correlation can be based on $n - j$ pairs of bout lengths. To our knowledge there are as yet no results available concerning the distribution of R_j, for $j > 1$, and hence no tests can be based on it.

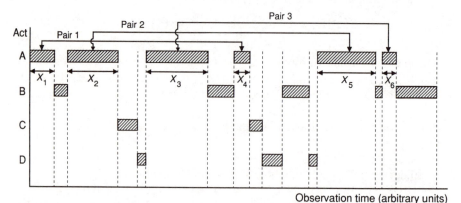

Figure 5.6 Analysis of autocorrelation between bout lengths in a record: pairs of bouts of act A with lag 3, with length x_i, respectively x_{i+3}.

Another type of application of the lag 1 correlation coefficient R_1 (eqn (5.28)) is the following: in a (semi-)Markov chain the points of entry in, or exit from, a certain state are renewal points. Therefore, the times between such points (bout lengths in the broad sense, see Fig. 5.7) should be independent. This can be tested with R_1. A positive correlation can indicate time inhomogeneity. When there is a negative correlation, there are probably deviations in the direction of a function of a Markov chain.

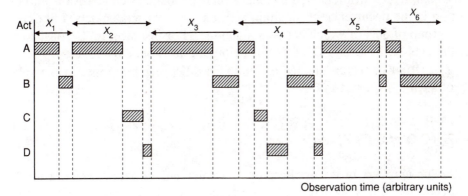

Figure 5.7 Analysis of autocorrelation between renewal points: the x_i's are the observed times between two subsequent start times of act A.

Cross-correlation

The concept of 'bout' or 'bout length' can be used in different senses as pointed out in Box 1.1. It can denote the duration of an uninterrupted period in which an animal grooms, eats or runs. However, it can also stand for the interval between two acts: a gap; for instance, the interval between just having stopped grooming and starting grooming again (Fig. 5.8).

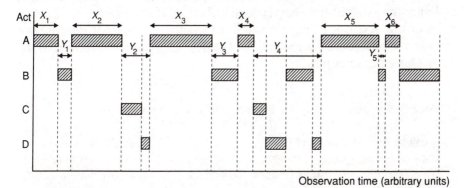

Figure 5.8 Analysis of the correlation between bout lengths of an act and the gaps in between: the x_i are the lengths of bouts of act A, the y_i the lengths of the gaps.

The analysis of dependencies between adjacent bout durations applies to all these more general interpretations of the concept 'bout'. Tests can be based on Spearman's correlation coefficient discussed in subsection 5.4.2.

Occurrence within a group of acts

A completely different type of long-term dependencies concerns the variation in the occurrence of a certain act in a group of related acts. Let G be a group of acts of which act A is a member. For instance, let G consist of the acts A, B, C and D. Acts not belonging to G will be denoted by Z_1, Z_2,.... Runs of acts belonging to G are placed between brackets. An example of such a sequence is

(A B $\underline{A}$ D) Z_3 Z_2 (D C A B $\underline{A}$ C D C B $\underline{A}$) Z_3 Z_2 Z_4 Z_3 (B C B D C B D) Z_1 (C D A B C) Z_1 (A) Z_4 Z_1 .

The numbers of occurrences of A within runs of acts belonging to G, given that A has occurred once, are respectively 2, 3, 1 and 1. We denote these numbers by x_1, x_2, x_3, x_4.

Under the assumptions of a (semi-)Markov process the probability of a transition from A to one of the other acts in G is a constant, say p, and the probability of one of the acts not belonging to G is $1 - p$. Hence, the probability distribution of X, the number of the A, provided that the group already contains one, is a so-called geometric distribution:

$$Pr\{X = x\} = p^x(1 - p), \quad x = 1, 2, 3,... . \tag{5.36}$$

Departures from a constant p result in two different classes of alternatives. If p varies in the course of the experiment, the variance of X will be larger than that of the geometric distribution. If the number of the A is more or less fixed, it leads to a smaller variance. In Box 5.9 we describe an example of mechanisms which lead to deviations in the direction of the latter type of alternatives.

Box 5.9 An example: mating behaviour of voles

It can be proved that in a CTMC the distribution of the total time spent in state A during a bout of the group G, given that A has occurred at least once, is an exponential distribution. Deviations towards a distribution with an increasing termination rate indicate that animals perform A for a more or less fixed total time, before G is left. This was found by De Jonge and

Ketel (1981) in the mating behaviour of continental voles (*Microtis arvalis*). They considered bursts of mountings that are ended by ejaculation and separated by gaps. Within each mounting burst, the behaviour is adequately described by a two-state Markov model, where the states represent mounting and pauses between mounting. However, the total mounting duration within a burst is approximately normally distributed instead of exponential. This led to the conclusion that the male needs an approximately constant amount of stimulation before ejaculation.

Such a mechanism leads to a deviation in the direction of a smaller variance of X than expected under the assumption of a geometric distribution. This can be detected by applying the left-sided version of the test based on T, defined by eqn (5.41).

Data: A sequence $x_1, x_2,..., x_n$ of positive integers.

Assumptions:

A1: Under, H_0 $X_1, X_2,..., X_n$ are independent and have a geometric distribution with the same unknown parameter p.

A2: Under H_1 there are unspecified departures from H_0.

Procedure: The test statistic T_g is defined by

$$T_g = \sum_{i=1}^{n} x_i^2.$$ (5.37)

The test is a conditional test, given that

$$\sum_{i=1}^{n} x_i = S.$$ (5.38)

Large values of T_g indicate a mixture of geometric distributions, i.e. a varying parameter p. Small values indicate that the number of occurrences of A is less variable than expected on the basis of a geometric distribution.

Reject H_0 of a constant probability p if $T_g < c_{1-\alpha}$, if $T_g > c_\alpha$, or if $T_g < c_{1-\frac{1}{2}\alpha}$ or $T_g > c_{\frac{1}{2}\alpha}$ depending on the nature of the alternative hypothesis one wants to test.

See Table A26 for the right-sided small-sample case ($n = 3$, $s = 13$ (1) 48; $n = 4$ $s = 10$ (1) 39; $n = 5$, $s = 10$ (1) 35; $n = 6$, $s = 10$ (1) 31; $n = 7$, $s = 11$ (1) 27; $n = 8$, $s = 12$ (1) 28; $n = 9$, $s = 13$ (1) 29 and $n = 10$, $s = 13$ (1) 30). Other critical values are not available for the small-sample case. An example of an application is given in Box 5.10.

Box 5.10 Testing for constant occurrence in a group of acts

1. The data are taken from Table 1 in Box 1.1: a record of solitary rat behaviour. The group G of acts consists of *Walk*, *Sit* and *Shake*. The first run of acts belonging to G is: *Shake, Walk, Sit, Walk, Sit, Walk*; the second run is *Sit, Walk, Sit, Walk, Sit, Walk, Sit*, etc.

2. Suppose we are interested in the occurrence of *Sit* bouts within runs of acts belonging to G. The number of *Sit* bouts in the n (= 23) succeeding runs are respectively:
2, 4, 3, 1, 3, 1, 1, 1, 1, 1, 1, 2, 2, 2, 1, 2, 2, 3, 2, 3, 1, 2, 1.

3. The total number of *Sit* bouts, s (eqn (5.38)), within runs of acts belonging to G is equal to 42. The test statistic
$T_g = 2^2 + 4^2 + ... + 2^2 + 1^2 = 94$.
The expected value, given by eqn (5.39), is
$42 \times (84 - 23 + 1)/(23 + 1) = 108.5$.
The variance, given by eqn (5.40), is for $n = 23$ and $s = 42$:
$25672.9846 - 3178.56 + 301 - 22650.25 = 145.1746$.
Hence, the standard deviation is equal to 12.05 and T, defined by eqn (5.41), is equal to $(94 - 108.5)/12.05 = -1.203$. Accordingly, the approximate left-sided p-value equals 0.1145 (= 1 - 0.8855), see Table A1.

4. Based on the record of Table 1, Box 1.1, it can be concluded that the occurrence of *Sit* bouts within runs of G is constant.

Large-sample approximation: For large n, T_g has approximately a normal distribution. On condition that $S = s$ the expectation and variance of T_g are

$$E(T_g \mid S = s) = \frac{s(2s - n + 1)}{n + 1} \, , \tag{5.39}$$

$$Var(T_g \mid S = s) = \frac{(4n + 20)s(s + 1)(s + 2)(s + 3)}{(n + 1)(n + 2)(n + 3)}$$

$$- \frac{24s(s + 1)(s + 2)}{(n + 1)(n + 2)} + \frac{4s(s + 1)}{n + 1}$$

$$- \frac{4s^2(s + 1)^2}{(n + 1)^2} \, . \tag{5.40}$$

Reject H_0 if $T < z_{1-\alpha}$, if $T > z_\alpha$, or if $|T| > z_{\frac{1}{2}\alpha}$, depending on whether the test is left-sided, right-sided or two-sided, where

$$T = \frac{T_g - E(T_g | S=s)}{\sqrt{Var(T_g | S=s)}} .$$

(5.41)

(When s/n tends to a positive constant γ for large n, the conditional expectation and variance of T_g tend to, respectively, $s(2\gamma - 1)$ and $4s\gamma(\gamma - 1)^2$.)

Properties: It can be proved that the test based on T_g is locally most powerful against a mixture of geometric distributions (see Meelis, 1974). It is to be expected that an analogous result holds for the left-sided test, against a more or less constant number of additional acts A. When the two-sided test is applied, significant large values of T_g indicate time inhomogeneity (or possibly deviations in the direction of a function of a Markov chain) and significant small values a 'constant' number of acts A per run of G.

Time series analysis

To analyse dependencies on a longer time scale, tests based on power spectra can be used, see for instance Cox and Lewis (1978), or Chatfield (1980) for an introduction. Priestley (1981) treats this topic on a more advanced level and Kendall and Ord (1990) give an up-to-date survey of time series analysis.

Power spectra tests, however, require very large data sets, involve a considerable computational effort and their outcome is not easily interpreted in a behavioural context. These methods therefore fall beyond the scope of this book.

The methods treated in this chapter are simpler to apply and provide better keys for an understanding of behaviour.

5.5 FURTHER REMARKS

As mentioned in Chapter 1, it is best to use tests for sequential dependency properties after tests for exponentiality. If the latter tests indicate deviations in the direction of mixtures (see eqn (4.1)), but the considered behavioural category cannot be split up according to observable criteria, the methods of section 5.3 can give further indications. The other methods given in this chapter are also very likely to give significant results in these cases. If they do not, the process can presumably be described by a function of a Markov chain that is also a semi-Markov chain. This simplifies further analysis considerably. Even if there are no apparent deviations from exponential bout length distributions in the direction of mixtures, there may still be inadvertently lumped categories or long-term dependencies. Therefore tests for sequential dependency properties should still be used as a final check whether a (semi-)Markov model fits the data.

The methods of section 5.4 can be used as initial tests against long-term dependencies. When these tests indicate deviations from the semi-Markov property, the methods of sections 5.2 and 5.3 can be applied to see whether such dependencies can be explained in terms of short-term dependencies. If so, lumping and/or splitting up behavioural categories may remove deviations. If not, models incorporating long-term dependencies should be used. Sometimes it is possible to use a (semi-)Markov model on a relatively short time scale and to formulate a model that incorporates the dependencies on a longer time scale. Examples are inhomogeneous Markov models, where long-term dependencies are taken into account by means of parameter changes (see Chapter 8).

The test concerning the number of occurrences of an act within a run of acts belonging to a certain group, described in subsection 5.4.3, is sensitive to a special type of long-term dependencies. Such dependencies (described in Box 5.9) can occur when an animal is able to 'keep track' of the total time that it has performed a certain act, for instance by means of physiological processes. Such mechanisms are likely to occur in e.g. mating behaviour or foraging behaviour. If the test indicates that animals indeed perform the considered act for an approximately fixed time within runs of a group of acts, these runs should be analysed further. If possible, a model for the behaviour within such runs of bouts should be formulated, and on a longer time scale, a model for the alternation between runs of 'bouts' of the group of acts in question and other behavioural categories.

When there is prior knowledge about the mechanisms underlying long-term dependencies, e.g. based on functional considerations, such information can be used to formulate a model. An example can be found in Haccou *et al.* (1991), who studied the effects of intra-patch experiences on patch leaving decisions of parasitic wasps by means of a semi-Markov model, using prior knowledge based on optimality models.

If there are no apparent long-term dependencies, or if these have been taken care of previously, dependencies on a shorter time scale can be investigated further. It is recommended to use tests for first- against second-order dependency, e.g. C_A (see eqn (5.4)) for the initial detection of short-term dependencies. Subsequently, the methods of section 5.3 can be used for closer inspection. After the splitting and/or lumping of acts the tests given in section 5.2 and/or those given in section 5.4 can be used again, for a last global check.

6 SIMULTANEOUS TESTS

6.1 INTRODUCTORY REMARKS

In analysing large data sets, for instance when testing for goodness of fit of a (semi-)Markov model, one can follow two different strategies. The first is to test every assumption or subhypothesis separately. Afterwards all results are combined by means of so-called omnibus or combination tests. This strategy can be considered as a 'bottom-up' approach. The second strategy is to apply an 'overall test' and to split up significant results, to find out which of the (sub)hypotheses have to be rejected, by means of so-called multiple comparison methods. This strategy is a 'top-down' approach. The two strategies are complementary to each other. Which one is to be preferred depends on the special characteristics of the problem under study. Combination tests and multiple comparison methods are called 'simultaneous tests' in the literature. We shall illustrate and discuss the procedures for the case of Markov modelling.

There are, in general, several points at which simultaneous tests are needed: in the explorative as well as in the testing stage. In the explorative stage the 'bottom-up' approach is normally the appropriate way of analysis: usually each hypothesis is tested in several samples, since modelling is in general based on more than one behavioural record. A combination test can indicate whether the total picture of individual test results from a set of samples gives reason to reject a null hypothesis at a previously chosen level of significance. Since usually several different aspects of (semi-)Markovity are tested, the chance of finding at least one significant test result (when the null hypothesis of Markovity is true) is much larger than the average significance level used for each separate test.

For instance, when two independent tests are performed with a significance level of 0.05, the chance of finding at least one significant result is equal to $1 - (1 - 0.05)^2 = 0.0975$. For three independent tests it is equal to $1 - (1 - 0.05)^3 = 0.142625$, and so on. Hence, application of many statistical tests usually leads to one or more significant deviations of the null hypothesis. Therefore, a combination procedure is needed if a restricted overall level of significance (probability of rejection of at least one null hypothesis) is desired.

The problem of a large overall level of significance can become even more serious if not all test results are independent, for instance when several Markov properties are tested: the dependency may increase the total significance level considerably.

In some situations, especially in the early explorative phase of research, an increase in the level of significance is not of major importance. In this phase, it is perhaps much more important to detect deviations than to avoid wrongly rejected null hypotheses. The latter type of error generally leads to a model which is too complicated, for example with a larger number of behavioural categories than necessary. Such 'over-parametrization' is very likely to be detected at later stages of the research, when parameter values are tested and confidence intervals are determined. At these stages, we may get indications of whether certain categories can be combined. Thus, whether or not a combination procedure is desired depends on the specific objectives one has in mind. (Note, however, that at each stage in the analysis the simplest possible model should be chosen.)

Combination procedures are discussed in section 6.2. In principle, one can combine dependent as well as independent test results. If the tests are performed on different data sets, independence of the tests is assured. To deal with this restricted type of combination problem, several well-known procedures are given, which can readily be applied. Furthermore, we treat one procedure for combining dependent test results.

After the explorative stage of research, in the testing stage, the top-down approach is more appropriate, for instance for testing the goodness of fit of a complete model with new data, or for testing the significance of treatment effects. In both cases an overall test with a preset significance level is needed. If the overall hypothesis is rejected, we normally want to know which one(s) of the subhypotheses must be rejected. Multiple comparison methods offer the possibility to determine this. In a few cases these methods are directly connected to, and derived from, the overall test. Well-known examples are the Scheffé method and the Tukey method, derived from the F-test in the analysis of variance. When available, we treat these specific multiple comparison methods after the description of the test procedure. In most cases such specific methods are either not yet derived, or too complicated for practical use, and one has to apply general methods. Which methods can be used depends on whether the test statistics are stochastically independent or not. This is determined, for instance, by which tests for Markov properties are used. The tests for sequential dependency given in section 5.2 are independent of tests for exponentiality, since they only concern the sequence of states. The test results on relations between bout lengths and preceding and/or following acts described in section 5.3, however, also concern the residence times, and their results are therefore correlated with the results of tests for exponentiality.

In section 6.3 several generally applicable procedures are proposed to disentangle the properties of a complicated model, by means of multiple comparison methods for both independent and dependent tests. Special emphasis is given to the testing of Markov properties.

As before, testing proceeds conditionally on the observed numbers of bouts (see section 1.5 for a more extensive discussion).

6.2 PROCEDURES FOR COMBINING TEST RESULTS

In this section we consider how to combine the results of independent tests when the n null hypotheses, $H_{0,i}$, say $\theta_i = \theta_{0,i}$, are tested in n different samples. Thus, suppose that the n test statistics, denoted by $Q_1,..., Q_n$, are stochastically independent. Furthermore, it is assumed that the hypotheses are of the same type and that it is required to detect whether there are deviations in the same direction in the different samples, i.e. one wants to combine the results from one-sided tests which are either all right sided or all left sided. Hence, for all i, the alternative hypotheses, $H_{1,i}$, are either all of the form $\theta_i > \theta_{0,i}$ or all of the form $\theta_i < \theta_{0,i}$. This has consequences if we want to combine results from two-sided tests. First of all it must be clear what is meant by combining two-sided tests. A combination of procedures which test $\theta_i = \theta_{0,i}$ against $\theta_i \neq \theta_{0,i}$ does not make sense, because if this leads to a significant result the deviations in the samples may be in different directions and insight would not be increased. A significant result of such an overall test is not unambiguously interpretable and is thus meaningless. Furthermore, possible significant left- and right-sided deviations may cancel each other out.

A combination procedure of two-sided tests can only be applied if it can be assumed that if any deviation of the overall null hypothesis H_0 (the intersection of the n hypotheses $H_{0,i}$) is present, it has the same direction for all i: $H_{1,i}$ is of the form $\theta_i > \theta_{0,i}$ or of the form $\theta_i < \theta_{0,i}$. For example, suppose that one investigates the influence of the administration of a drug on the bout duration of a certain act. It is not known if the drug will either increase or decrease the mean bout duration. It is reasonable, however, to expect that the effect will be the same for all experimental animals. In that case it is justified to apply two one-sided combination procedures successively, each with a level of significance $\frac{1}{2}\alpha$. See Oosterhoff (1969, section 1.1) for a more extensive discussion.

In subsection 6.2.1 we describe unweighted combination procedures and in the next subsection we treat the weighted procedures. Weighted procedures are especially relevant if the sample sizes differ strongly or if some observations are less accurate than others. In subsection 6.2.3 we drop the assumption of independence of the n test results and give a method for combining dependent statistics.

6.2.1 Unweighted combination of independent tests

If the test statistics are equivalent an unweighted combination procedure, where each test result is treated in the same way, is to be preferred. For the sake of simplicity we assume that the tests are right sided (for the left-sided test the procedures are analogous and can be adjusted in a straightforward manner).

Data: n test statistics, $q_1,..., q_n$.

Assumptions:
A1: The test statistics $Q_1,..., Q_n$ are mutually independent.
A2: Q_i is a continuously distributed statistic for testing the null hypothesis $H_{0,i}$, $\theta_i = \theta_{0,i}$, versus the alternative hypothesis $H_{1,i}$, $\theta_i > \theta_{0,i}$. $H_{0,i}$ is rejected in favour of $H_{1,i}$ if $Q_i > q_i(\alpha)$, where $q_i(\alpha)$ is the upper critical value at a level of significance α: $Pr\{Q_i > q_i(\alpha)\} = \alpha$ under $H_{0,i}$.
A3: Under the combined null hypothesis H_0: $H_{0,i}$ is true for $i = 1,..., n$. H_0 is tested against H_1: $H_{0,i}$ is false for at least one i.

We distinguish two special cases – the Q_i are identical and normally distributed or they are chi-squared distributed – and we treat a few generally applicable combination procedures.

Normally distributed test statistics

Many test statistics are (approximately) standard normally distributed under the null hypothesis. When combining test results this fact can be used. The type of significant departure from standard normality of the set of statistics can indicate the nature of the deviations of the null hypotheses.

Graphical test: normal-probability paper
By transforming the vertical scale on a graph of the cumulative normal distribution function, Φ, it is possible to transform the curve to a straight line. There is special graph paper, called 'normal-probability paper', which is scaled in such a way that any Φ (i.e. for each mean μ and variance σ^2) is transformed to a straight line. A plot of data on this graph paper is an easy and yet powerful way to detect departures from normality and is even applicable for small sample sizes. The use of normal-probability paper is discussed in most introductory textbooks on statistics, e.g. Sokal and Rohlf (1981) and Dixon and Massey (1969).

Data and assumptions: See subsection 6.2.1. Moreover:

A4:   The Q_i are identically and standard normally distributed.

Procedure: The ordered data are denoted by $q_{(1)}, q_{(2)}, ..., q_{(n)}$. Plot these data on a graph, where the ith point is given by $q_{(i)}$ on the horizontal axis and by i/n on the vertical axis. See Box 6.1 for an example.

Box 6.1   Normal-probability paper

1. The data are: 2.9257, 1.6534, 1.4377, 3.8816, 3.3497, 3.3016, 2.4127, −0.50509, 0.39453, 0.30879. The sample size $n = 10$.

2. The ordered set is plotted in Fig. 1: the smallest value, −0.50509, has y-coordinate 0.10, etc.

3. Clearly, there are no major departures from normality.

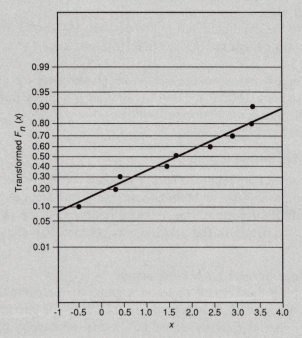

Figure 1 Plot of the data on normal-probability paper.

Properties: The advantage of using this visual scanning method is that departures from a straight line are very conspicuous. The test is, however, not a formal method. Furthermore, it is not sensitive to departures in the tails of the distribution, such as apparent outliers, etc.

The sum test

Oosterhoff (1969) describes a combination test based on the standardized sum of the different test statistics.

Data and assumptions: See subsection 6.2.1. Moreover:

A4:   The Q_i are standard normally distributed.

Procedure: Calculate the (standardized) sum of the n test statistics:

$$q = \frac{1}{\sqrt{n}} \sum_{i=1}^{n} q_i. \tag{6.1}$$

Since the sum of independent normal variables is normally distributed as well, H_0 is rejected if $q > z_\alpha$. See Table A1 and Box 6.2 for an example.

Box 6.2   The sum test

1. Assume that the test statistics are: 0.40349, 0.59619 and 1.20509.

2. $q = (0.40349 + 0.59619 + 1.20509)/\sqrt{3} = 1.273$.

3. The right-sided p-value is equal to 0.1015 (Table A1, linear interpolation between 1.27 and 1.28).

4. This result is not significant at a level of significance of $\alpha = 0.05$, H_0 is not rejected.

The Shapiro-Wilk test for departures from normality

Data and assumptions: See subsection 6.2.1. Moreover:

A4:   the Q_i are normally distributed.

Procedure: The test is essentially based on the regression of the ordered n test statistics on normal scores. We denote these ordered values by $q_{(1)}, \ldots, q_{(n)}$. Calculate the statistic W defined by

$$W = \frac{\left(\sum_{i=1}^{n} a_{i,n} q_{(i)}\right)^2}{\sum_{i=1}^{n} (q_i - \bar{q})^2}. \tag{6.2}$$

The coefficients $a_{i,n}$ are given in Table A27 for $n = 2,..., 50$. Note that $a_{n-i+1,n} = -a_{i,n}$. If two statistics have the same value, i.e. $q_{(i)} = q_{(i+1)}$, it is advised to multiply both by $\frac{1}{2}(a_{i,n}+a_{i+1,n})$. The critical values are given in Table A28. See Box 6.3 for an example.

Box 6.3 The Shapiro-Wilk test for departures from normality

1. Assume that the test statistics are: 0.40349, 0.59619 and 1.20509.

2. The corresponding weight coefficients are equal to 0.7071, 0 and −0.7071, see Table A27 (for $n = 3$) and use the fact that $a_{n-i+1,n} = -a_{i,n}$.

3. Substitution in eqn (6.2) yields:

$$W = \frac{(0.7071\times0.40349+0\times0.59619-0.7071\times1.20509)^2}{(0.40349-0.73492)^2+(0.59610-0.73492)^2+(1.20509-0.73492)^2}$$

$$= \frac{0.32126}{0.35015}.$$

4. The right-sided p-value lies in the interval 0.90–0.50 (see Table A28). Hence, H_0 is not rejected.

Properties: The test is one of the most powerful 'omnibus' procedures for testing for departures from normality. See Pearson *et al.* (1977) who compare the power of various tests of normality.

Comments: The test was proposed by Shapiro and Wilk (1965). Pearson and Hartley (1972) give an extensive description of it. Royston (1982) has extended the procedure to $n \leq 2000$ and gives a transformation of W which is normally distributed for large sample sizes.

Chi-squared-distributed test statistics

As with normally distributed test statistics, the sum of different chi-squared distributed statistics can be used for a combination test (Oosterhoff, 1969).

Data and assumptions: See subsection 6.2.1. Moreover:
A4:   The Q_i are chi-squared distributed with k_i degrees of freedom.

Procedure: Calculate the sum of the n test statistics:

$$q = \sum_{i=1}^{n} q_i. \qquad (6.3)$$

Since the sum of independent chi-squared variables is also chi-squared distributed, with a number of degrees of freedom equal to the sum of the degrees of freedom of the individual statistics, H_0 is rejected if $q > \chi^2_k(\alpha)$ with $k = \Sigma\, k_i$. See Table A2 and Box 6.4 for an example.

Box 6.4   Combining chi-squared test statistics

1. Suppose we have five chi-squared test statistics: 4.972 (5), 0.044 (1), 4.751 (3), 12.536 (6), 5.411 (2), where the number of degrees of freedom is indicated in brackets.

2. The sum, q, of the statistics is equal to 27.714 and the sum of the number of degrees of freedom is equal to 17.

3. At a level of significance of 0.05 H_0 is rejected: the corresponding p-value is 0.0487 (Table A2). Hence, although none of the individual test statistics deviates significantly from the null hypothesis of a chi-squared distribution, the combined test gives a significant result.

General case: the combination of independent one-sided p-values

The main tool for combining independent test results is based on the fact that, if the test statistics are (approximately) continuously distributed, the p-values should be uniformly (= homogeneously) distributed between 0 and 1 under the null hypothesis.

More precisely: let $p_1,...,\, p_n$ be the one-sided p-values of the different tests, i.e. for right-sided tests

$$p_i = Pr\{Q_i \geq q_i\} \qquad (6.4)$$

and for left-sided tests

$$p_i = Pr\{Q_i \leq q_i\}, \qquad (6.5)$$

where q_i is the observed value of test statistic Q_i. If the null hypothesis is true, the p_i are realizations of a uniformly [0,1] distributed random variable (see Box 6.5 for a proof), regardless of the cumulative distribution functions of the Q_i. (These should only be strictly increasing and continuous.)

Box 6.5   Proof that under H_0 p-values are uniformly distributed

1. Let the right-sided p-value be defined by

$$p = Pr\{Q > q\} = \bar{F}(q),\tag{1}$$

where $\bar{F}$ is the survivor function of Q.

2. Define the random variable P by

$$P = \bar{F}(Q).\tag{2}$$

3. For each p such that $0 \leq p \leq 1$:

$$Pr\{P > p\} = Pr\{\bar{F}(Q) > p\} = Pr\{Q > \bar{F}^{-1}(p)\}\tag{3}$$

$$= \bar{F}(\bar{F}^{-1}(p)) = p.$$

(Where $\bar{F}^{-1}$ denotes the inverse function of $\bar{F}$, which exists, since it is a strictly decreasing function.) Hence, $Pr\{P > p\} = p$, i.e. the right-sided p-values are uniformly distributed between 0 and 1.

4. The proof for the left-sided p-values is analogous (in this case the survivor function is replaced by the cumulative distribution function F).

We start the discussion with a graphical test and the chi-squared test for goodness of fit. Subsequently, we give three well-known formal combination methods and we end this subsection with two formal tests for the case when the test statistics to be combined have a discrete distribution.

Visual scanning of p-values and chi-squared test for goodness of fit
Whether or not the p-values have a uniform (or homogeneous) distribution (see Fig. 6.1) can be tested graphically, either by plotting the data on a line (which is especially useful in the small-sample case, see Fig. 6.2) or by plotting a frequency histogram (which is to be preferred if many p-values are available, see Fig. 6.3). Plots like Fig. 6.2 or 6.3 also give indications whether departures from the null hypothesis tend to be left sided or right sided, or two sided. (In these cases the departures are obviously two sided.)

In addition, a formal test, like the chi-squared test for goodness of fit, can be applied. See subsection 4.6.2 for a more extensive description in the special case of testing for exponentiality. In the present case, when testing for a uniform distribution, a few adjustments have to be made.

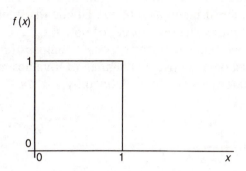

Figure 6.1 The uniform [0,1] distribution.

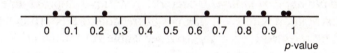

Figure 6.2 Visual scanning of (a small number of) *p*-values by plotting them on a line.

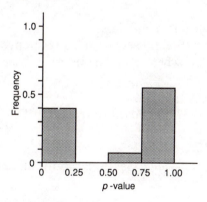

Figure 6.3 Frequency histogram of *p*-values.

Data: n one-sided *p*-values: $p_1,..., p_n$.

Assumptions:

A1:   Under H_0 the *p*-values, $P_1,..., P_n$, are mutually independent and uniformly [0,1] distributed.

A2:   Under H_1 there are unspecified deviations from A1.

Procedure: Choose k, the number of distinct classes. As a rule of thumb: let k be of the order of magnitude of $\sqrt{n}$. Divide the interval [0, 1] into k classes of equal width. The probability of obtaining a p-value in a certain class is $1/k$, and hence under H_0 the expected number of p-values per class is equal to n/k. Denote the observed number of p-values in the k classes by $O_1,..., O_k$. Calculate the statistic χ^2 defined by

$$\chi^2 = \sum_{i=1}^{k} \frac{(O_i - \frac{n}{k})^2}{\frac{n}{k}} \, , \tag{6.6}$$

Reject H_0 for large values of χ^2. No small-sample critical values are available.

Large-sample approximation: Under H_0, χ^2 follows approximately a chi-squared distribution with $k - 1$ degrees of freedom if n is sufficiently large. H_0 should be rejected if $\chi^2 \geq \chi^2_{k-1}(\alpha)$, see Table A2 for the critical values. The approximation is accurate enough for practical purposes if $n/k > 5$.

Tippett's combination test

Data and assumptions: See subsection 6.2.1. Moreover:
A4: $p_i = Pr\{Q_i > q_i\}$, where q_i is the observed value of the test statistic Q_i.

Procedure: Tippett (1931) proposed to reject H_0 if

$$t < 1 - (1 - \alpha)^{\frac{1}{n}}, \tag{6.7}$$

where t is the smallest p-value. See Box 6.6 for an example.

Box 6.6 Tippett's combination method

1. Assume that the one-sided p-values are equal to 0.315, 0.063, 0.562 and 0.259. Hence, the minimum value is $t = 0.063$.

2. At a level of significance $\alpha = 0.05$, the critical value for the combination test is equal to $1 - 0.95^{1/4} = 0.0127$. Thus, there are no indications of the presence of a departure from the null hypothesis.

Properties: Westberg (1985) compared the power of Tippett's test with those of Fisher's omnibus procedure. It turns out that neither of the two methods is generally more powerful than the other.

Comments: This combination method is described in Tippett (1931). Wilkinson (1951) derived a generalization and suggested the use of the *r*th smallest *p*-value. Under H_0 this statistic has a beta distribution with parameters *r* and $k - r + 1$.

Westberg (1985) proposed so-called 'adaptive methods' which are always better than the worst of Fisher's and Tippett's. These methods, however, require more computational effort.

Fisher's omnibus test

One of the oldest combination procedures, proposed by Fisher (1932), uses the fact that −2 times the natural logarithm of a uniformly distributed random variable has a chi-squared distribution with two degrees of freedom. Hence, under H_0 the sum of the *n* independent log-transformed one-sided *p*-values has a chi-squared distribution with $2n$ degrees of freedom.

Data: *n* one-sided *p*-values.

Assumptions: See subsection 6.2.1. Moreover:
A4: $p_i = Pr\{Q_i > q_i\}$, where q_i is the observed value of the test statistic Q_i.

Procedure: Calculate *f*, defined by

$$f = -2\sum_{i=1}^{n} \log p_i. \tag{6.8}$$

Reject H_0 if $f > \chi^2_{2n}(\alpha)$. See Table A2 for the critical values and Box 6.7 for an application.

Properties: See under Tippett's test.

Comments: The test was introduced by Fisher (1932). See also Oosterhoff (1969) for a historical survey, and Westberg (1985) for a comparison of its power properties with those of Tippett's test.

Box 6.7   Fisher's omnibus test

1. Assume that the one-sided p-values are equal to 0.315, 0.063, 0.562 and 0.259.

2. The statistic
$$f = -2 \times (\log 0.315 + \log 0.063 + \log 0.562 + \log 0.259)$$
$$= -2 \times (-1.1552 - 2.7646 - 0.5763 - 1.3509) = -2 \times (-5.847)$$
$$= 11.69.$$

3. The p-value of the omnibus test can be read from Table A2. The probability of finding an f equal to or larger than 11.69 is approximately equal to 0.166. (Calculated by interpolation between 10.2 and 13.4, Table A2, with eight degrees of freedom.) Accordingly, we conclude that there are no significant deviations from the null hypothesis.

Liptak's combination test

The test proposed by Liptak (1958) is based on so-called 'probits'. The probit, $\Phi^{-1}(p)$, is the value of a standard normal variable at which the cumulative distribution function, Φ, has value p. These values are listed in (or can be approximated by linear interpolation between the tabulated values from) Table A1 or they can be determined more accurately by means of extensive tables of the normal distribution, such as the one given by Zelen and Severo (1972). Pearson and Hartley (1954) give a table of probits. Note that in old tables the number 5.0 has been added to the probits to avoid negative numbers when using these data in 'probit analysis'. The standardized sum of the (thus transformed) right- or left-sided p-values has a standard normal distribution.

Data and assumptions: See subsection 6.2.1. Moreover:
A4: $p_i = Pr\{Q_i > q_i\}$, where q_i denotes the observed value of the test statistic Q_i.

Procedure: Calculate the statistic l:

$$l = \frac{1}{\sqrt{n}} \sum_{i=1}^{n} \Phi^{-1}(p_i). \tag{6.9}$$

Under H_0, l has a standard normal distribution. H_0 is rejected for large values of l. See Table A1 for the critical values and Box 6.8 for an example of the application of this test.

Box 6.8 Liptak's combination test

1. Assume that the one-sided p-values are equal to 0.315, 0.063, 0.562 and 0.259.

2. The corresponding 'probits', $\Phi^{-1}(0.315),..., \Phi^{-1}(0.259)$, are equal to, respectively $-0.482, -1.530, 0.156, -0.646$. See Table A1 and apply linear interpolation where necessary. For p-values smaller than 0.5, we use the fact that the standard normal distribution is symmetric around zero. Hence, for the cumulative distribution function, Φ, probits of p-values between 0 and 0.5 can be determined as follows:
$$\Phi^{-1}(p) = -\Phi^{-1}(1 - p), \text{ e.g. } \Phi^{-1}(0.315) = -\Phi^{-1}(0.685).$$
This value obviously lies somewhere between -0.480 and -0.490. Linear interpolation yields:
$$\Phi^{-1}(0.315) = -\{0.480 + (0.6850 - 0.6844)/(0.6879 - 0.6844)$$
$$\times (0.490 - 0.480)\} = -\{0.480 + 0.002\} = -0.482.$$

3. $l = -(-0.482 - 1.530 + 0.156 - 0.646)/\sqrt{4} = 2.502/2 = 1.251$. Hence, the p-value of the combination test is equal to 0.1054 and the result is not significant at a level of 0.05.

Comments: In the special case when the statistics $Q_1,..., Q_n$ have a standard normal distribution (6.9) is equal to

$$\frac{1}{\sqrt{n}}\sum_{i=1}^{n} q_i \qquad (6.10)$$

in the case of a right-sided test and to

$$-\frac{1}{\sqrt{n}}\sum_{i=1}^{n} q_i \qquad (6.11)$$

for a left-sided test, and Liptak's test reduces to the sum test (apart from the minus sign in eqn (6.11)).

Interval-valued or discrete p-values: the likelihood ratio and the chi-squared test for goodness of fit
Up to now we have assumed in this subsection that the test statistics $Q_1,..., Q_n$ are continuously distributed. Consequently, the one-sided p-values are uniformly distributed under H_0 (see Box 6.5). If the Q_i follow a discrete distribution this is no longer true. Another problem in practice is that p-values have to be determined, while in many cases tables consist merely of a few critical values, so that either a possibly inaccurate estimate or an

upper and lower bound can be provided, for instance: $0.10 \leq p \leq 0.25$. For such cases combination methods must be adjusted. We describe two combination procedures for the latter case which are also applicable if n, the number of tests to be combined, is small: the likelihood ratio test and the chi-squared test for goodness of fit.

We introduce the following notation. Let $0 = a_0 < a_1 < ... < a_{k-1} < a_k = 1$ be the class bounds of intervals of p-values. See for instance Table A2: $a_0 = 0$, $a_1 = 0.005,..., a_{13} = 0.995$, $a_{14} = 0.999$ and $a_{15} = 1$. Hence, application of the chi-squared table results in an interval-valued estimate of the p-value: $a_{i-1} < p \leq a_i$, where $i = 1$, or $2,....,$ or 15. Under H_0, the probability of a p-value between a_{i-1} and a_i is known and equal to the difference $a_i - a_{i-1}$. Let X_i denote the number of p-values between a_{i-1} and a_i, $i = 1,..., k$. Consequently, n, the total number of p-values we want to combine, is equal to the sum of the X_i. Since the tests are assumed to be stochastically independent, it follows that the vector $X = (X_1,..., X_k)$ is multinomially distributed. The combination tests are based on the latter property.

Data: k non-negative integers $x_1,..., x_k$.

Assumptions:

A1:   Under H_0, X follows a multinomial distribution with parameters n and $(a_i - a_{i-1})$, $i = 1,..., k$.

A2:   Under H_1, there is an unspecified departure from this distribution.

Procedure: Calculate the likelihood ratio statistic L:

$$L = 2\{ \sum_{i=1}^{k} x_i \log x_i - \sum_{i=1}^{k} x_i \log(a_i - a_{i-1}) - n \log n \}, \qquad (6.12)$$

or the chi-squared test statistic:

$$\chi^2 = \sum_{i=1}^{k} \frac{\{ x_i - n(a_i - a_{i-1}) \}^2}{n(a_i - a_{i-1})}. \qquad (6.13)$$

The null hypothesis of uniformly distributed p-values is rejected if L or χ^2 exceeds the corresponding critical value, which depends on $a_1,..., a_k$. Therefore, tables are not available for the small-sample case and critical values, c_α, must be calculated for each case. In Box 6.9 we give an example of such a computation.

Box 6.9 The exact likelihood ratio test for combining discrete or interval-valued p-values

1. We treat an example of the computation of the exact critical value (for the likelihood ratio test) for the case when we want to combine three test results, where $k = 4$ and the class bounds are 0.8, 0.9 and 0.95. Hence, for each test it is known if the p-value lies between 0 and 0.8, or between 0.8 and 0.9, and so on.

2. The total set of all possible test results, i.e. (x_1, x_2, x_3, x_4) values, are listed in Table 1. For instance, (3, 0, 0, 0) means that the three p-values lie in the interval [0, 0.8].

3. Next we calculate the probability for each test result:

$$Pr\{X_1 = x_1, ..., X_4 = x_4\} = \frac{n!}{x_1! x_2! x_3! x_4!} 0.8^{x_1} \, 0.1^{x_2} \, 0.05^{x_3} \, 0.05^{x_4}, \qquad (1)$$

according to the multinomial distribution (see the fifth column in Table 1).

Table 1 Calculation of critical values

x_1	x_2	x_3	x_4	$Pr\{X = x\}$		L
3	0	0	0	0.8^3	= 0.512	1.339
2	1	0	0	$3 \times 0.8^2 \times 0.1$	= 0.192	1.679
2	0	1	0	$3 \times 0.8^2 \times 0.05$	= 0.096	3.065
2	0	0	1	$3 \times 0.8^2 \times 0.05$	= 0.096	3.065
1	2	0	0	$3 \times 0.8 \times 0.1^2$	= 0.024	5.838
1	1	1	0	$6 \times 0.8 \times 0.1 \times 0.05$	= 0.024	4.451
1	1	0	1	$6 \times 0.8 \times 0.1 \times 0.05$	= 0.024	4.451
1	0	2	0	$3 \times 0.8 \times 0.05^2$	= 0.006	8.610
1	0	1	1	$6 \times 0.8 \times 0.05^2$	= 0.012	5.838
1	0	0	2	$3 \times 0.8 \times 0.05^2$	= 0.006	8.610
0	3	0	0	1×0.1^3	= 0.001	13.816
0	2	1	0	$3 \times 0.1^2 \times 0.05$	= 0.0015	11.383
0	2	0	1	$3 \times 0.1^2 \times 0.05$	= 0.0015	11.383
0	1	2	0	$3 \times 0.1 \times 0.05^2$	= 0.00075	12.769
0	1	1	1	$6 \times 0.1 \times 0.05^2$	= 0.0015	9.996
0	1	0	2	$3 \times 0.1 \times 0.05^2$	= 0.00075	12.769
0	0	3	0	1×0.05^3	= 0.000125	17.974
0	0	2	1	3×0.05^3	= 0.000375	14.155
0	0	1	2	3×0.05^3	= 0.000375	14.155
0	0	0	3	1×0.05^3	= 0.000125	17.974

4. The corresponding values of the likelihood ratio statistic are listed in the sixth column. For instance, if two p-values lie between 0 and 0.8 and one between 0.8 and 0.9 (i.e. $x = 2\ 1\ 0\ 0$), L, given by eqn (6.7), is equal to

$$2 \times \{(2\log2 + 1\log1) - (2\log0.8 + 1\log0.1) - 3\log3\}$$
$$= 2 \times \{(1.386 + 0) - (-0.446 - 2.303) - 3.296\} = 2 \times 0.839 = 1.678.$$

5. By ordering the L-values and the corresponding probabilities, it follows that $Pr\{L \geq 8.610\} = 0.021$ and $Pr\{L \geq 5.838\} = 0.057$. Accordingly, when testing at a level of approximately 0.05, we can use 5.838 as the critical values.

Large-sample approximation: Under H_0, L as well as χ^2 have asymptotically a chi-squared distribution with $k - 1$ degrees of freedom as n tends to infinity, i.e. $c_\alpha = \chi^2_{k-1}(\alpha)$. See Table A2 for the critical values. The approximation is sufficiently accurate in practice if the expected number of observations per class is larger than one, i.e. $n(a_i - a_{i-1}) > 1$, $i = 1,..., k$, and if, moreover, 80% of them are larger than or equal to five. This can be achieved by combining adjacent classes.

Comments: Only a few authors treat the problem of combining p-values from discrete variables, e.g. Lancaster (1949), Cochran (1954) and Kincaid (1962) who consider the combination of $k \times r$ tables. They suggest modifications of procedures designed for continuous data, or pooling data in order to improve the approximations. We strongly recommend, however, the application of an exact test in the small-sample case if one has the computer facilities. We especially dissuade to pool data, since that may obscure the fact that different data sets may deviate in a different way so that departures can cancel each other.

We have assumed implicitly that the class bounds (the a_i) coincide if different tables are used. If this is not the case, the procedure can be repeated for each table and the resulting chi-squared values can be combined, as illustrated in Box 6.4.

If the distribution of the Q_i is discrete and known under H_0, the procedure is analogous: n is the sample size and the differences $(a_i - a_{i-1})$ are to be interpreted as the probabilities $Pr\{Q = q\}$, where q is one of the k elements of the sample space.

Discussion

Marden (1985) compared the performance of the procedures listed above. His results indicate that the test based on f (eqn (6.8)) has favourable properties. Furthermore, he proved that the procedures given above are all 'admissible', i.e. there are no other procedures that are at least as good for all alternatives and better for some alternatives. (For a formal definition see Ferguson, 1967.) Further results on the performance of the procedures can be found in Liptak (1958) and Oosterhoff (1969).

6.2.2 Weighted combination of independent tests

In the procedures described in subsection 6.2.1, the results of different samples are treated equivalently. Sometimes there are reasons to treat test results differently; for instance, if sample sizes differ strongly, or if some of the observations are less accurate than others. In these cases it is better to use a weighted procedure, where the weight coefficients are a function of the sample size and/or the accuracy.

Weighted procedures other than those discussed in this subsection can be found in Oosterhoff (1969, section 1.2).

The sum test

Data and assumptions: See subsection 6.2.1. Moreover:

A4:   The Q_i are standard normally distributed, where Q_i is based on m_i observations, $i = 1,..., n$.

Procedure: Calculate the weighted sum of the n test statistics:

$$q = \sum_{i=1}^{n} w_i q_i, \tag{6.14}$$

where the weights are defined by

$$w_i = \sqrt{\frac{m_i}{\sum m_i}}. \tag{6.15}$$

The weights are chosen in such a way that the variance of Q is equal to one. Hence, under H_0, Q has a standard normal distribution. H_0 is rejected if $q > z_\alpha$. See Table A1 and Box 6.10 for an example.

Comments: To a certain extent the choice of the weight coefficients is arbitrary. An intuitive justification of the weights as defined in eqn (6.15) is, that they are a measure of the accuracy with which the possible departures of H_0, if any, are determined. To illustrate this: assume that Q_i is normally distributed under $H_{1,i}$ with expectation μ_i and variance 1 and that the parameter μ_i is estimated by $\bar{x}_i$, the mean of the ith sample. The variance of this estimator is proportional to $1/m_i$. Its accuracy can be defined as the inverse of the standard deviation (which determines the width of its confidence interval), i.e. as the square root of m_i.

With this choice of weights, the power of the different combined tests is to a certain extent equalized. The weights can also be chosen in direct relation to power properties of the tests. However, to do that properly, the direction in which good power is desired must be taken into account, since the tests may differ with respect to their sensitivity to alternatives in different directions.

Box 6.10 The weighted sum test

1. Assume that the test statistics are as in Box 6.2, i.e. 0.40349, 0.59619 and 1.20509, based on sample sizes 4, 19 and 85, respectively.

2. The weights are equal to $(\sqrt{4}, \sqrt{19}, \sqrt{85}) / \sqrt{(4+19+85)}$
   $= (2.0000, 4.3589, 9.2195) / 10.392 = 0.1924, 0.4194, 0.8872$.

3. $q = (0.1924 \times 0.40349 + 0.4194 \times 0.59619 + 0.8872 \times 1.20509) = 1.397$.

4. The right-sided p-value is equal to 0.0814 (Table A1, linear interpolation between 1.390 and 1.400).

5. This result is not significant at a level of significance of $\alpha = 0.05$. H_0 is not rejected, although the p-value is smaller than that of the unweighted sum test applied in Box 6.2.

Lancaster's generalization of Liptak's test

Lancaster (1961) proposed a procedure where positive weights w_i ($i = 1,..., n$) can be assigned to the p-values. It is a generalization of the test based on the statistic l, defined in eqn (6.9).

Data: n one-sided p-values.

Assumptions: See 6.2.1. Moreover:

A4: $p_i = Pr\{Q_i > q_i\}$, where q_i is the observed value of the test statistic Q_i.

Procedure: Calculate

$$l' = \frac{-\sum\limits_{i=1}^{n} w_i \Phi^{-1}(p_i)}{\sum\limits_{i=1}^{n} w_i^2} , \tag{6.16}$$

where $\Phi^{-1}(p)$ is the value of a standard normal variable at which the cumulative distribution function has value p. Under H_0, this statistic has a standard normal distribution. Reject H_0 if $q > z_\alpha$. See Table A1 and Box 6.11 for an example.

Box 6.11 Lancaster's generalization of Liptak's test

1. Assume that the one-sided p-values are equal to 0.315, 0.063, 0.562 and 0.259, based on sample sizes 3, 15, 4 and 105, respectively.

2. The corresponding 'probits', $\Phi^{-1}(0.315),...,$ $\Phi^{-1}(0.259)$, are equal to, respectively: -0.482, -1.530, 0.156, -0.646. See Box 6.8.

3. The weights are equal to
   $(\sqrt{3}, \sqrt{15}, \sqrt{4}, \sqrt{105})/\sqrt{(3 + 15 + 4 + 105)}$
   $= (1.7321, 3.8730, 2,0000, 10.2469)/11.2694$
   $= 0.1537, 0.3437, 0.1775, 0.9093.$

4. The statistic $l' =$
   $-(-0.1537 \times 0.482 - 0.3437 \times 1.530 + 0.1775 \times 0.156 - 0.9093 \times 0.646)$
   $= 1.1596$. Hence, the p-value of the combination test is equal to 0.123 and the result is not significant at a level of 0.05.

Comments: If $Q_1,..., Q_n$ are N(0,1) distributed this test reduces to the weighted sum test.

6.2.3 Combination of dependent tests: an improved Bonferroni procedure

If the test statistics are not independent, a combination procedure can be based on the Bonferroni inequality, which requires no assumptions about the distribution of $Q_1,..., Q_n$. It is a conservative procedure, however, and has little power if the Q_i are highly correlated. To increase the power,

Simes (1986) proposed a modified procedure, which was extended by Hochberg (1988).

Data: n test statistics, $q_1, ..., q_n$.

Assumptions:

A1: $Q_1, ..., Q_n$ are continuously distributed statistics for testing the null hypotheses $H_{0,1}, ..., H_{0,n}$, versus the alternative hypotheses $H_{1,1}, ..., H_{1,n}$.

Procedure: Let $p_{(1)} < ... < p_{(n)}$ be the ordered p-values of the tests, and denote the corresponding null hypotheses by $H_{(0,1)}, ..., H_{(0,n)}$. One starts with the largest p-value: $p_{(n)}$. If $p_{(n)} < \alpha$ the overall H_0 as well as all separate null hypotheses are rejected. If not, $H_{(0,n)}$ cannot be rejected and one goes on to compare $p_{(n-1)}$ with $\alpha/2$. If it is smaller, H_0 and all $H_{(0,i)}$, $i = n - 1, ..., 1$, are rejected. If not, $H_{(0,n-1)}$ cannot be rejected and one proceeds to compare $p_{(n-2)}$ with $\alpha/3$, etc.

Thus, the overall hypothesis H_0 is rejected if at least one of the n null hypotheses is rejected. See Box 6.12 for an application of this procedure.

Box 6.12 The improved Bonferroni procedure for combining dependent test results

1. We illustrate the procedure with the data from Box 6.1: 2.9257, 1.6534, 1.4377, 3.8816, 3.3497, 3.3016, 2.4127, −0.50509, 0.39453, 0.30879. The sample size $n = 10$, and under H_0 we assume that the Q_i are standard normally distributed. The corresponding p-values are (Table A1): 0.0018, 0.0495, 0.0755, 0.0000, 0.0004, 0.0005, 0.0079, 0.6932, 0.3465 and 0.3787. The level of significance is 0.05.

2. The largest p-value, $p_{(10)} = 0.6932 > 0.05$; hence $H_{(0,10)}$ is not rejected. The next largest value, $p_{(9)} = 0.3787 > 0.05/2 = 0.025$, is not significant; hence $H_{(0,9)}$ is not rejected. Continuing we find:

$p_{(8)} = 0.3464 > 0.05/3 = 0.0167$, not significant;
$p_{(7)} = 0.0755 > 0.05/4 = 0.0125$, not significant;
$p_{(6)} = 0.0495 > 0.05/5 = 0.01$, not significant;
$p_{(5)} = 0.0079 > 0.05/6 = 0.0083$, not significant;
$p_{(4)} = 0.0018 < 0.05/7 = 0.0071$, significant.

Hence $H_{(0,1)}$, $H_{(0,2)}$, $H_{(0,3)}$, $H_{(0,4)}$ (corresponding to, respectively, the fourth, the fifth, the sixth and the first observation), as well as the overall null hypothesis, are rejected.

Comments:

1. The Bonferroni inequality states that, when several tests are performed with significance levels $\alpha_1, ..., \alpha_n$, the overall significance level is at most equal to the sum of the αs (or, more precisely, the minimum of one and $\Sigma \alpha_i$). This inequality holds for arbitrary dependence structures. In the original Bonferroni test procedure the overall null hypothesis H_0 is rejected if any of the p_i are less than or equal to the corresponding level of significance α_i. Furthermore, the corresponding $H_{0,i}$ are rejected. It is obvious that smaller values of α_i should be chosen if n increases, since otherwise $\Sigma \alpha_i$ would become even larger than one. Obviously, such an overall test makes no sense. In general one chooses equal αs, i.e. $\alpha_1 = ... = \alpha_n = \alpha/n$.

2. The improved procedure is generally applicable since no assumptions are made about the form of the distributions of the test statistics. All that is required is that the p-values can be calculated with a certain degree of accuracy. Thus it can be applied with test statistics whose distributions can be explicitly calculated or are well tabulated, as is the case, for instance, for the normal distribution.

3. Hommel (1989) describes a slightly better procedure. The computation of the one described here, however, is much simpler.

4. Note also that this procedure can even be used (in the case of dependent as well as independent statistics) with different types of alternative hypotheses, provided it has a clear interpretation. For instance, a two-sided test of the equality of termination rates can be combined with a right-sided test on the absence of a minimum bout duration. A possible outcome of a two-sample testing problem could be that the apparent difference in mean bout duration is not due to differences in termination rate, but to the presence of a minimum bout duration in one of the two samples.

We have stated in the introduction of section 6.2 that it is generally preferable to restrict the combination of tests to hypotheses which are of the same nature. There are well-interpretable exceptions, however, and in those cases the modified Bonferroni procedure offers a way out.

6.3   MULTIPLE COMPARISON METHODS

6.3.1 Preliminary remarks

Multiple comparison methods are especially designed for detecting which of the subhypotheses must be rejected if an overall hypothesis has been rejected. Therefore, they are powerful procedures in the top-down approach discussed in section 6.1. In most practical cases such methods cannot be derived directly from the overall test, as is the case, for example, when the *F*-test is applied in the analysis of variance. In this section we treat methods which can be applied in the general case.

We consider the overall hypothesis, H_0, which can be subdivided into n related subhypotheses $\{H_{0,1}, H_{0,2},..., H_{0,n}\}$. The level of significance of the overall hypothesis is α. Besides the overall test of H_0, we also want to apply a procedure which allows conclusions to be drawn about the separate subhypotheses with a level of significance smaller than or equal to α. One might expect that procedures like the combination test statistics of section 6.2 would provide useful methods. Since most of them are based on linear combinations of *p*-values or observed values of test statistics for the sub-hypotheses, this might lead to relatively simple procedures. In general, however, this is not the case. It often leads to situations where the overall null hypothesis, H_0, is rejected, whereas no unambiguous indications are provided concerning the subhypotheses which are responsible for the significant departures from H_0.

Ideally the procedure should have the following property: if the overall hypothesis is not rejected, the same is concluded for all separate subhypotheses and conversely, if the overall hypothesis is rejected, at least one subhypothesis is also rejected. Only in a few special cases, however, is it possible to fulfil these requirements at the preset level of significance α.

In subsection 6.3.2 we treat a procedure for the case of independent test statistics and one for the case of dependent tests. The methods are applicable without making assumptions concerning the distributions and, in the case of dependent tests, the dependence structure.

In subsection 6.3.3 we describe how the test procedures given in 6.3.2 can be applied for simultaneous testing of whether a behavioural process is a CTMC. It is assumed that the records are homogeneous and that dependencies on a large time scale have previously been taken into account (see Chapter 5). The alternatives considered are that a process is a function of a Markov chain, or a semi-Markov chain. In order to avoid unnecessary complications, which obscure the application of the multiple comparison methods we illustrate, we shall not consider tests for exponentiality in the presence of censoring, such as was considered in section 4.7.

6.3.2 General procedures

Miller's multi-stage procedure for independent test statistics

Let $Q_1,..., Q_n$ be the test statistics for the subhypotheses $H_{0,1},..., H_{0,n}$ of the overall hypothesis H_0. If the Q_i are independent under H_0, a generalization of the maximum modulus test (Tukey, 1952) can be used. This multi-stage procedure, where the number of steps depends on the outcomes of the tests, was proposed by Miller (1981a). For simplicity, it is assumed that all the subhypotheses are tested at the same significance level α. The procedure can easily be adjusted for unequal significance levels.

Data: n test statistics, $q_1,..., q_n$.

Assumptions:
A1: The test statistics $Q_1,..., Q_n$ are mutually independent.
A2: Q_i is a continuously distributed statistic for testing the null hypothesis $H_{0,i}$ versus the alternative hypothesis $H_{1,i}$ ($i = 1,..., n$).

Procedure: Let $p_1,..., p_n$ be the p-values of the tests. For one-sided tests they are defined as in eqns (6.4) and (6.5): for right-sided tests $p_i = Pr\{Q_i \geq q_i\}$, for left-sided tests $p_i = Pr\{Q_i \leq q_i\}$ and for two-sided tests

$$p_i = 2 \times \min(Pr\{Q_i \geq q_i\}, Pr\{Q_i \leq q_i\}). \tag{6.17}$$

Define α_1 by $1 - (1 - \alpha)^{1/n}$ and reject the overall hypothesis H_0 if there is at least one p_i less than or equal to α_1. Since the tests are independent,

$$Pr\left\{\bigcup_{i=1}^{n}(p_i < \alpha_1)\right\} = 1 - Pr\left\{\bigcap_{i=1}^{n}(p_i > \alpha_1)\right\}$$
$$= 1 - (1 - \alpha_1)^n \tag{6.18}$$

and thus the significance level for testing H_0 is equal to α as a result of the definition of α_1.

If H_0 is not rejected, the procedure is stopped: obviously no significant deviations can be shown. However, if n_1 of the p_i are smaller than or equal to α_1 ($n_1>0$), the corresponding subhypotheses are rejected. Subsequently, the remaining m (= $n - n_1$) p-values are compared with $\alpha_2 = 1 - (1 - \alpha)^{1/m}$. The procedure is stopped if all p-values are larger than α_2. If not, those hypotheses for which the p-value is smaller than or equal to α_2 are rejected and the procedure is continued in a similar way until there are no significant results. An example is given in Box 6.13.

At each stage in the procedure the significance level at which the set of remaining hypotheses is tested is equal to α. Furthermore, it can be shown that the significance level at which each subhypothesis is tested is at most α for the whole procedure.

Box 6.13 Miller's multi-stage procedure for independent test statistics

1. We illustrate the procedure with the data from Box 6.1, see also Box 6.12. Hence, $n = 10$ and we choose $\alpha = 0.05$. Consequently,
$$\alpha_1 = 1 - (0.95)^{1/10} = 0.0051.$$
From Box 6.12 it can be seen that four p-values are smaller than α_1. Hence, the corresponding subhypotheses are rejected. The remaining p-values are compared with $\alpha_2 = 1 - (0.95)^{1/6} = 0.0085$. One p-value is smaller and accordingly the subhypothesis is rejected. The other p-values are compared with $\alpha_3 = 1 - 0.95^5 = 0.010$. As can be seen in Box 6.12 all remaining p-values are larger.

2. By using this procedure it turns out that the subhypotheses corresponding to observation numbers 1, 4, 5, 6 and 7 are rejected, and consequently the overall null hypothesis is rejected.

The improved Bonferroni procedure for dependent test statistics

If the test statistic for the overall test cannot be split up into independent statistics, the Bonferroni inequality can provide a useful multiple comparison procedure in the top-down approach.

Data: n test statistics, $q_1, ..., q_n$.

Assumptions:
A1: Q_i is a continuously distributed statistic for testing the null hypothesis $H_{0,i}$ versus the alternative hypothesis $H_{1,i}$ ($i = 1, ..., n$).

Procedure: Let $p_{(1)} < ... < p_{(n)}$ be the ordered p-values of the tests for the subhypotheses, and denote the corresponding null hypotheses by $H_{(0,1)}, ..., H_{(0,n)}$. The overall hypothesis H_0 is rejected if for at least one i

$$p_{(i)} < \frac{\alpha}{n - i + 1} . \tag{6.19}$$

Let j denote the maximum of all i for which the inequality (6.19) holds. The corresponding subhypotheses $H_{(0,1)}, ..., H_{(0,j)}$ are rejected. See Box 6.14 for an example.

Box 6.14 Improved Bonferroni procedure for multiple comparison

1. Assume that the ordered one-sided *p*-values of four dependent test statistics are equal to 0.0037, 0.0061, 0.047 and 0.153.

2. The 'adjusted' critical values, according to eqn (6.19), are, respectively, for $\alpha = 0.05$: 0.0125, 0.0163, 0.025 and 0.05. Hence, for $i = 1$ and 2 the corresponding null hypothesis is rejected. Besides, the overall hypothesis is rejected.

6.3.3 Multiple comparison procedures for testing Markov properties

Let H_M be the (overall) hypothesis that the process under consideration is a CTMC. In general, the test statistics will be dependent and the improved Bonferroni procedure will be the most appropriate multiple comparison method. It can be very useful to distinguish two or more families of related subhypotheses, so that the total set of statistics for testing the hypotheses belonging to one family are independent. For those sets of statistics the more powerful multi-stage procedure developed by Miller (see subsection 6.3.2) can be applied. Moreover, such an approach enlarges insight into the nature of the main departures of the assumptions of Markovity.

A natural division of the hypotheses constituting H_M is to take on the one hand the hypotheses concerning the bout length distributions, this subfamily of hypotheses being denoted by H_B, and on the other hand the hypotheses concerning sequential dependency properties, denoted by H_S. The maximum significance levels at which these subfamilies of hypotheses are tested are denoted respectively by α_B and α_S. It must be assured that the significance level at which H_M is tested does not exceed a chosen value α_M. The relation between α_M, α_B and α_S depends on which test statistics are used for testing H_S. If the tests only concern the sequence of states, they are independent of tests of H_B and therefore

$$\alpha_M = 1 - (1 - \alpha_B)(1 - \alpha_S) = \alpha_B + \alpha_S - \alpha_B\alpha_S. \tag{6.20}$$

If the tests also concern the bout lengths, as in section 5.3, their results presumably depend on the tests of H_B. In this case α_B and α_S should be chosen such that

$$\alpha_B + \alpha_S \leq \alpha_M, \tag{6.21}$$

to ensure that the significance level at which H_M is tested is smaller than or equal to α_M.

Whether Miller's multi-stage procedure can be applied depends on whether one wants to test many aspects of the Markov assumptions. In other words, if H_M is subdivided into several families of subhypotheses, it will in general be necessary to use the improved Bonferroni procedure. Miller's multi-stage procedure, which is only applicable if the test statistics are independent, can be used, for instance, in the following cases.

1. If we only want to test the hypothesis that the bout lengths of the several acts are exponentially distributed, H_M is restricted to H_B; then H_B is $\{H_{B(1)} \cap ... \cap H_{B(k)}\}$, where $H_{B(i)}$ denotes the hypothesis that the bout lengths of act i are exponential. To test $H_{B(i)}$, any of the test statistics given in section 4.6 can be used. For different acts the tests are independent. H_M ($= H_B$) can be tested with an overall level of significance $\alpha_B = \alpha_M$.

2. Suppose that we only consider the hypothesis that there is first-order dependence in the sequence of acts; then H_M is restricted to H_{S*}: $\{H_{S*}(1),..., H_{S*}(k)\}$, where $H_{S*}(i)$ denotes the null hypothesis that the transition probability from act A_i to a following act is not influenced by the preceding act, $i = 1,..., k$, where k is the number of acts under consideration. The chi-squared test for first-order dependency (subsection 5.2.1) or the corresponding likelihood ratio test (subsection 5.2.2) can be applied. In both cases, the test statistics ($C_{A(i)}$, respectively $L_{A(i)}$) are mutually independent. H_M ($= H_{S*}$) can be tested with an overall level of significance $\alpha_S = \alpha_M$.

3. A more realistic case is when H_M is defined by $H_B \cap H_{S*}$, where H_B and H_{S*} are as defined in cases 1 and 2. The two statistics for testing H_B and H_S are independent and thus α_B and α_S can be chosen according to eqn (6.20). Miller's procedure can be applied to tests for exponentiality of the bout length distributions of all the acts, as well as the first-order dependency of the sequence of acts, and to trace departures in the case of rejection of the overall hypothesis H_M.

As mentioned in section 5.2, a test of first- against second-order sequential dependency may be based on C, the sum of the statistics C_A (see eqn (5.5)), or on the statistic L, the sum of the L_A, defined by eqn (5.10). However, this can cause problems with the subsequent testing of which acts cause deviations. It may lead to rejection of the overall hypothesis, while none of the subhypotheses can be rejected. Therefore it is advisable to apply Miller's procedure straightforwardly to the k different $C_{A(i)}$ or $L_{A(i)}$ test statistics.

In practice it seldom occurs that the statistics are all mutually independent. Sometimes it is possible to group the hypotheses in a number of families of subhypotheses, where each family consists of hypotheses which can be tested independently. An important example is the case where $H_M = \{\cap H(i,j)\}$ denotes the hypothesis that the bout lengths of act number i, $i = 1,..., k$, do not depend on the preceding act j, $j = 1,..., k$, $i \neq j$. The test statistics are independent for different i. Accordingly, Miller's multi-stage procedure can be applied in so far as it concerns the family of hypotheses $H(j) = \{H(1,j) ,..., H(k,j)\}$. Hence, in order to test the overall hypothesis H_M in the most 'economic' way (with respect to significance levels) the subhypotheses grouped into distinct families $H(j)$ are each tested at a preset level of significance. The statistics for testing $H(1),..., H(k)$ are dependent. Next one can apply the improved Bonferroni procedure on this set of p-values.

6.3.4 Final remarks

Multiple comparison procedures are usually applied after data exploration. These procedures should provide more insight into possible departures of the assumptions. If significant deviations from H_M are found, the procedures give an indication of the nature of the deviation so that, if necessary, it can be investigated further by means of the more detailed procedures discussed in Chapters 4 and 5.

Before the procedures can be applied, choices concerning α_M, α_B and α_S must be made. The choice of α_M depends on whether one wants to detect any deviations or rather to avoid wrongly rejecting the null hypothesis. In the former case it is recommended to choose α_M somewhat higher than the usual 5%, e.g. 10% or 15%, in order to have more power at later stages of the procedure. Subsequently, α_B and α_S are chosen according to eqns (6.19) or (6.20). In this case there is not much difference between the two. For instance, if α_B and α_S are equal, they both should be 5.13% if we take α_M equal to 10% and the tests of H_B and H_S are independent (eqn (6.19)). If the tests are dependent and α_B and α_S are chosen to be equal, they must be 5% in order to make the overall significance level less than 10% (eqn (6.20)). The significance levels α_B and α_S may also differ. Within the framework of alternative models considered here, deviations from H_B imply that the process is a semi-Markov chain and deviations from H_S indicate that it may be considered as a function of a Markov chain. If the latter type of deviations is not detected it may have more serious consequences for later analysis. Therefore, it is sometimes better to choose α_S somewhat larger than α_B. (See Miller, 1981a, section 3.5 for a general discussion about multiple comparison procedures.)

7 ANALYSIS BASED ON A (SEMI-) MARKOV DESCRIPTION

7.1 INTRODUCTORY REMARKS

In previous chapters, the emphasis was on model formulation and valida-tion. This is useful in itself, since it will often increase insight into the temporal structure of the behaviour studied. In most cases, however, we want more than just to describe behaviour. For instance, we may want to study the effects of experimental treatments in order to gain insight into the mechanisms regulating behaviour. A validated model can be used as a firm basis for such analyses. A major advantage of an approach based on the formulation of Markov chain models and many of their generalizations is that, on the time scale(s) where such models can be applied, all the information contained in the behavioural records can be summarized concisely by means of the estimated transition rates.

Besides statistics based on (functions of) transition rates, there are of course more traditional statistics to describe a behavioural process, such as the proportion of time spent on a certain act, or frequencies of occur-rence of acts. These will not be considered here, since they are not charac-teristic of an analysis within the framework of a (semi-)Markov chain description. If such a description fits, analyses based on the statistics given in sections 7.2 and 7.4 are more efficient; since the large-sample distributio-nal properties of these statistics are optimal, they are easily calculated and, moreover, many of them are asymptotically independent of one another.

In this chapter we will describe the way in which many important etholo-gical research questions can be answered. We do not have the illusion, however, that it is possible to discuss all potential questions that can be encountered in practice. Therefore, we treat a number of frequently occur-ring meaningful research questions. Related questions can be handled analogously. In Chapter 8 we give a number of detailed examples of applications of the methods given here.

The methods for estimation and testing described in this chapter are based on likelihood theory (see Cox and Hinkley (1974) for an introduction and examples). The most important characteristics of maximum likelihood estimators (MLEs) are listed in Box 7.1. The likelihood ratio test was described in Box 2.10. Asymptotic properties for MLEs and likelihood ratio tests were originally derived for n independent, identically distributed random variables (where n is non-stochastic), as n goes to infinity. In (semi-)Markov chains, there are dependencies between the data. Moreover, the numbers of observed acts are random variables. However, Billingsley (1974) proved that under certain regularity conditions the key results remain true for CTMCs. His results were generalized for semi-Markov chains by Moore and Pyke (1968) and Gill (1980). Many large-sample results for (semi-)Markov processes are updated and well described by Fleming and Harrington (1991). In the cases considered here, the necessary regularity conditions are satisfied, provided that point events (subsection 1.3.1) are considered as a limiting case of events with a small finite duration (Metz, 1981). This is a reasonable assumption for behavioural processes, since it is unlikely that animals can perform acts infinitely quickly.

Box 7.1   Some basic results from likelihood theory

It is assumed that the likelihood function of the observations can be defined explicitly as a function of an unknown vector of parameters $\theta = (\theta_1,..., \theta_p)'$ and the vector of observations $x = (x_1,..., x_n)'$. The log-likelihood will be denoted by $L(x;\theta)$.

We denote the vector of partial derivatives of the log-likelihood function with respect to θ, the so-called score functions, by $U(x;\theta)$:

$$U(x;\theta)=(U_1(x;\theta),...,U_p(x;\theta))' = \left(\frac{\partial L}{\partial \theta_1}, ..., \frac{\partial L}{\partial \theta_p} \right)' . \tag{1}$$

The MLE of θ is that value for which $L(x;\theta)$ is maximal and can be found by equating (1) to zero and solving the resulting p equations.

The second partial derivatives with respect to θ

$$\frac{\partial^2 L(x;\theta)}{\partial \theta_i \partial \theta_j} , \tag{2}$$

are also functions of the observations as well as the unknown parameters. The ijth element of the so-called information matrix $I(\theta)$ is defined by

$$I_{ij}(\theta) = -E \frac{\partial^2 L(x;\theta)}{\partial \theta_i \partial \theta_j} . \tag{3}$$

It can be shown that this expression is identical to

$$I_{ij}(\theta) = E\{ \sum_{k=1}^{n} \frac{\partial L_k(x_k;\theta)}{\partial \theta_i} \frac{\partial L_k(x_k;\theta)}{\partial \theta_j} \} , \tag{4}$$

where $L_k(x_k;\theta)$ denotes the log-likelihood of the kth observation, which sometimes can be calculated more easily.

Tests of hypotheses about θ can be based on the score function (eqn (1)) since under certain regularity conditions $U(x;\theta)/\sqrt{n}$ has a multivariate normal distribution with expectation a vector of zeros and covariance matrix $I(\theta)/n$ as n goes to infinity (Cox and Hinkley, 1974). For instance, the hypothesis $\theta = \theta_0$ can be tested with the statistic

$$U(x;\theta_0)'I^{-1}(\theta_0)U(x;\theta_0), \tag{5}$$

which has approximately a chi-squared distribution with p degrees of freedom.

Alternatively, tests and confidence intervals for θ can be based on the asymptotic distribution of the MLE. Let $\hat{\theta}$ be the MLE of θ, then it can be proved that $(\hat{\theta} - \theta)\sqrt{n}$ is asymptotically multivariate normally distributed with expectation a vector of zeros and covariance matrix $nI^{-1}(\theta)$. This result implies that, for sufficiently large n, $(\hat{\theta}_j - \theta_j)\sqrt{n}$ is approximately normal with zero expectation and variance n times the jjth element of the inverse of the information matrix. Under the hypothesis $\theta = \theta_0$ the statistic

$$(\hat{\theta} - \theta_0)'I(\theta_0)(\hat{\theta} - \theta_0) \tag{6}$$

therefore has approximately a chi-squared distribution with p degrees of freedom.

It is not always possible to calculate the elements of the information matrix explicitly by the expressions given in eqns (3) and (4). In that case the MLE for θ can be substituted in either eqn (3) or (4). This implies that, for example, in the test statistics given in eqns (5) and (6), $I(\theta_0)$ is replaced by $I(\hat{\theta})$. It can be proved that this leads to asymptotically equivalent procedures.

The MLE $\hat{\theta}$ is strongly consistent, which means that for large n, $\hat{\theta}$ goes to θ almost surely. For large n, $\hat{\theta}$ has the smallest possible variance that an (unbiased) estimate of θ can have. It is therefore the asymptotically most accurate unbiased estimator of θ.

Example
Let $X_1,..., X_n$ be independent random variables and let X_i be exponentially distributed with parameter λ for $i \leq m$ and parameter $a\lambda$ for $i > m$. In this case θ is the two-dimensional vector $(\lambda,a)'$. The log-likelihood of the observations is equal to

$$L(x;\theta) = n\log\lambda + (n - m)\log a - \lambda\sum_{i=1}^{m}x_i - a\lambda\sum_{i=m+1}^{n}x_i. \tag{7}$$

The score functions are equal to

$$U_1(x;\theta) = \frac{\partial L}{\partial \lambda} = \frac{n}{\lambda} - \sum_{i=1}^{m} x_i - a \sum_{i=m+1}^{n} x_i \tag{8}$$

$$U_2(x;\theta) = \frac{\partial L}{\partial a} = \frac{n-m}{a} - \lambda \sum_{i=m+1}^{n} x_i. \tag{9}$$

Equating (8) and (9) to zero and solving the equations gives the MLEs of λ and a:

$$\hat{\lambda} = \frac{m}{\sum_{i=1}^{m} x_i} \tag{10}$$

$$\hat{a} = \frac{n-m}{m} \frac{\sum_{i=1}^{m} x_i}{\sum_{i=m+1}^{n} x_i}. \tag{11}$$

To calculate the elements of the information matrix, we take the second derivatives of the log-likelihood (eqn (7))

$$\frac{\partial^2 L}{\partial \lambda^2} = -\frac{n}{\lambda^2} \tag{12}$$

$$\frac{\partial^2 L}{\partial a^2} = -\frac{n-m}{a^2} \tag{13}$$

$$\frac{\partial^2 L}{\partial \lambda \partial a} = \frac{\partial^2 L}{\partial a \partial \lambda} = -\sum_{i=m+1}^{n} x_i. \tag{14}$$

Since the expected value of x_i is equal to $1/a\lambda$ for $i > m$, it follows from eqn (3) that the information matrix is equal to

$$I(\theta) = \begin{pmatrix} \dfrac{n}{\lambda^2} & \dfrac{n-m}{a\lambda} \\ \dfrac{n-m}{a\lambda} & \dfrac{n-m}{a^2} \end{pmatrix}. \tag{15}$$

The same result can be obtained by using eqn (4). The log-likelihood of the kth observation is equal to

$$L_k(x_k;\theta) = \log\lambda - \lambda x_k \quad \text{for} \quad k \le m \tag{16}$$

$$= \log a + \log\lambda - a\lambda x_k \quad \text{for} \quad k > m,$$

and thus

$$\frac{\partial L_k}{\partial \lambda} = \frac{1}{\lambda} - x_k \quad \text{for} \quad k \leq m$$

$$= \frac{1}{\lambda} - ax_k \quad \text{for} \quad k > m,$$

(17)

and

$$\frac{\partial L_k}{\partial a} = 0 \qquad \qquad \text{for} \quad k \leq m$$

$$= \frac{1}{a} - \lambda x_k \quad \text{for} \quad k > m.$$

(18)

Substitution of these expressions in eqn (4) and calculation of the expected values gives the same results as given in eqn (15).

Suppose that we want to test the hypothesis H_0: $\lambda = 1$ and $a = 1$, i.e. that $X_1,..., X_n$ are identically distributed with parameter 1. Under this hypothesis, the information matrix (eqn (15)) is equal to

$$I(\theta_0) = \begin{pmatrix} n & n-m \\ n-m & n-m \end{pmatrix}$$

(19)

and either of the statistics of eqns (5) or (6) can be used.

In this example, the information matrix can be calculated explicitly. Usually, however, this will not be so easy. In such cases, substitution of the MLEs in eqns (3) and (4) gives asymptotically equivalent results. Using eqn (3) would lead to

$$I(\hat{\theta}) = \begin{pmatrix} \dfrac{n}{\hat{\lambda}^2} & \displaystyle\sum_{i=m+1}^{n} x_i \\[3ex] \displaystyle\sum_{i=m+1}^{n} x_i & \dfrac{n-m}{\hat{a}^2} \end{pmatrix}$$

(20)

By taking the inverse of the information matrix, the asymptotic (co)-variances of the MLEs can be found. For instance, from eqn (15) it can be shown that the asymptotic variance of $(\hat{\lambda} - \lambda)\sqrt{n}$ is equal to:

$$\frac{n\lambda^2 a}{na - (n - m)\lambda} = \frac{\lambda^2}{1 - (1 - \dfrac{m}{n})\dfrac{\lambda}{a}}.$$

(21)

Hence, and approximate $(1 - \alpha)$ confidence interval for λ is

$$\hat{\lambda} - \frac{\hat{\lambda}}{\sqrt{n\{1-(1-\frac{m}{n})\frac{\hat{\lambda}}{\hat{a}}\}}} z_{\frac{1}{2}\alpha} < \lambda < \hat{\lambda} + \frac{\hat{\lambda}}{\sqrt{n\{1-(1-\frac{m}{n})\frac{\hat{\lambda}}{\hat{a}}\}}} z_{\frac{1}{2}\alpha} .$$

(22)

Asymptotic results are derived conditionally on the observed initial act. Furthermore, it is assumed that the processes are observed for a fixed total time T. Asymptotic results hold for $T \rightarrow \infty$. Since the methods are all likelihood based, they can also be applied when observation is stopped at a 'stopping time', in particular after a fixed number of bouts (see section 1.5). For convenience it is assumed that, when the process is stopped after a fixed time T, the last incompletely observed bout length is neglected. However, this last bout length may also be included. Estimation and test procedures can be adjusted accordingly.

In sections 7.2 and 7.3 we describe procedures for CTMCs. The properties of estimators of parameters are treated in subsections 7.2.1–7.2.4. In section 7.3 we describe how the effects of various conditions on behavioural processes that are accurately described by a CTMC can be examined. Generalizations for semi-Markov chains are given in sections 7.4 and 7.5. Note that, when a behavioural process is described by a semi-Markov chain, some of the acts may still have exponentially distributed bout lengths. This subset of acts can be treated as normal Markov states, and can be studied with the methods of sections 7.2 and 7.3. For the remaining set of acts, the generalized methods of sections 7.4 and 7.5 must be used.

7.2 ESTIMATION IN CTMCs

Consider a record of a behavioural process that can be described by a CTMC. The behaviour might have been observed either for a fixed time period, T, or until a fixed number of bouts, N, had been recorded. Let N_{AB} denote the number of observed transitions from A to B and let S_A be the sum of the bout lengths of A. When the first and/or last bout are incompletely observed, they may be left out of the calculation of S_A. Alternatively, they may be included, which implies that they are considered as randomly censored observations. Conditional on the observed initial act, the log-likelihood of the process is

$$L(\mathfrak{R};\{\alpha\}) = \sum_{A} \sum_{B \neq A} \{N_{AB}\log\alpha_{AB} - \alpha_{AB}S_A\},$$

(7.1)

where $\mathfrak{R}$ denotes the complete record and $\{\alpha\}$ the vector of transition rates. The summation is taken over all different observed transitions.

The estimators of parameters given in the subsequent subsection are derived by maximizing (7.1). The estimators are all asymptotically normally distributed and strongly consistent (i.e. their value converges almost surely to the true value of the parameter). Furthermore, it can be proved that the log-transformed estimators of transition and termination rates are also asymptotically normally distributed. Asymptotic properties hold for either T or N tending to infinity (Billingsley, 1974).

We consider several types of parameters: transition rates, termination rates, transition probabilities, and transition rates towards groups of acts. We also describe the properties of log-transformed estimates of transition and termination rates, since these are more convenient for studying the effects of covariates on behavioural processes (subsection 7.3.1). The log-transformation implies that effects are assumed to be multiplicative rather than additive, which is more reasonable, since transition rates cannot be negative.

7.2.1 Transition rates

A CTMC is completely characterized by the distribution of its initial state and the set of its transition rates. Analysis is done conditional on the initial state, so that the remaining characterizing parameters are the transition rates. From (7.1) the maximum likelihood estimator (MLE) of the transition rate α_{AB} from act A to act B can be derived as

$$\hat{\alpha}_{AB} = \frac{N_{AB}}{S_A} , \tag{7.2}$$

(section 1.2 and eqn (1.5) and Billingsley, 1974, Chapters 7 and 8). From Billingsley's results it follows that for large observation time, T, or for a large total number of observed bouts, N,

$$\frac{\hat{\alpha}_{AB} - \alpha_{AB}}{\alpha_{AB}} \sqrt{N_{AB}} , \tag{7.3}$$

has approximately a standard normal distribution. From this result we can derive a $(1 - \alpha)$ confidence interval for α_{AB}:

$$\frac{\hat{\alpha}_{AB}}{1 + \dfrac{z_{\frac{1}{2}\alpha}\hat{\alpha}_{AB}}{\sqrt{N_{AB}}}} < \alpha_{AB} < \frac{\hat{\alpha}_{AB}}{1 - \dfrac{z_{\frac{1}{2}\alpha}\hat{\alpha}_{AB}}{\sqrt{N_{AB}}}} , \tag{7.4}$$

where $z_{1/2\alpha}$ is listed in Table A1.

The covariance between $\hat{\alpha}_{AB}$ and $\hat{\alpha}_{CD}$ is zero for $A \neq C$. The covariance between $\hat{\alpha}_{AB}$ and $\hat{\alpha}_{AD}$ ($B \neq D$) can be approximated by

$$-\frac{\alpha_{AB}\alpha_{AD}}{N_A^2}, \tag{7.5}$$

where N_A denotes the number of occurrences of A. Consequently, the correlation coefficient goes to zero when N_A tends to infinity. Thus, the MLEs of the transition rates are asymptotically independent. Hence, in large samples the rates can be analysed independently. As mentioned, it is often convenient to consider the log-transformed transition rates. It can be shown that for large T (or N)

$$\sqrt{N_{AB}} \; (\log\hat{\alpha}_{AB} - \log\alpha_{AB}) \tag{7.6}$$

has approximately a standard normal distribution. With respect to skewness, the distributions of the log-transformed estimators converge faster to a standard normal distribution than those of the estimators $\hat{\alpha}_{AB}$. Therefore a confidence interval based on this large-sample result is more accurate:

$$\hat{\alpha}_{AB}\exp\left[-\frac{z_{1/2\,\alpha}}{\sqrt{N_{AB}}}\right] < \alpha_{AB} < \hat{\alpha}_{AB}\exp\left[\frac{z_{1/2\,\alpha}}{\sqrt{N_{AB}}}\right]. \tag{7.7}$$

As in the untransformed case, the correlations disappear for large sample sizes:

$$Cov(\log\hat{\alpha}_{AB}, \log\hat{\alpha}_{AD}) = -\frac{1}{N_A^2}. \tag{7.8}$$

Hence, in large samples the log-transformed rates can also be analysed separately.

7.2.2 Termination rates

As mentioned in Chapter 1, eqn (1.5), the transition rate from A to B can be expressed as the product of the transition probability from A to B (the chance that a switch from A to B is made, given that A is terminated) and the termination rate of A (the chance per time unit that A is terminated):

$$\alpha_{AB} = p_{AB}\lambda_A. \tag{7.9}$$

An alternative complete characterization is, therefore, the set of termination rates together with the transition probabilities. The log-likelihood of the observed process can therefore also be considered as a function of these two types of parameters (by substituting (7.9) for α_{AB} in eqn (7.1)). By subsequent maximization it follows that the MLE of λ_A is

$$\hat{\lambda}_A = \frac{N_A}{S_A} \tag{7.10}$$

(see also Box 1.2). It can be shown (Billingsley, 1974) that the MLEs of the termination rates are independent of those of the transition probabilities. Furthermore, the MLEs of the different termination rates are independent, as are the MLEs of the transition rates. It can also be shown that

$$\frac{\hat{\lambda}_A - \lambda_A}{\lambda_A} \sqrt{N_A} \tag{7.11}$$

is asymptotically $N(0,1)$ distributed (Billingsley, 1974). It follows that a $(1 - \alpha)$ confidence interval of λ_A is

$$\frac{\hat{\lambda}_A}{1 + \dfrac{z_{\frac{1}{2}\alpha}\hat{\lambda}_A}{\sqrt{N_A}}} < \lambda_A < \frac{\hat{\lambda}_A}{1 - \dfrac{z_{\frac{1}{2}\alpha}\hat{\lambda}_A}{\sqrt{N_A}}} \;, \tag{7.12}$$

where $z_{\frac{1}{2}\alpha}$ can be found in Table A1.

The log-transformed estimators also have an asymptotic normal distribution. The distribution of

$$\sqrt{N_A}\,(\log\hat{\lambda}_A - \log\lambda_A) \tag{7.13}$$

tends to a standard normal distribution as T (or N) goes to infinity. A $(1 - \alpha)$ confidence interval for λ_A based on this result can be calculated by substituting $\hat{\lambda}_A$, λ_A, and N_A for $\hat{\alpha}_{AB}$, α_{AB}, and N_{AB}, respectively, in eqn (7.7).

7.2.3 Transition probabilities

Substituting (7.9) in (7.1) and maximizing over p_{AB} leads to the estimator:

$$\hat{p}_{AB} = \frac{N_{AB}}{N_A} \; . \tag{7.14}$$

As mentioned, MLEs of transition probabilities are independent of those of termination rates. Furthermore,

$$\sqrt{N_A}\,(\hat{p}_{AB} - p_{AB}) \tag{7.15}$$

is asymptotically $N(0,1)$ distributed. For $A \neq C$, $\hat{p}_{AB}$ and $\hat{p}_{CD}$ are independent. Since the sum of the transition probabilities from A to all other states is equal to one, they cannot be analysed separately. The covariance between $\hat{p}_{AB}$ and $\hat{p}_{AD}$ is approximately equal to

$$\frac{-p_{AB}p_{AD}}{N_A} . \qquad (7.16)$$

7.2.4 Transition rates towards groups of acts

Consider a group of acts $G = \{B_1,..., B_m\}$. The sum of the estimated transition rates from act A towards G

$$\alpha_{AG} = \sum_{i=1}^{m} \alpha_{AB_i} \qquad (7.17)$$

can be interpreted as the tendency to stop doing A in favour of G. If A is included in the group, the summation is taken over all other acts in G. In that case, α_{AG} is the tendency to stop A in favour of other acts in G. Since these parameters are chances per time unit, they can also be interpreted in terms of behavioural tendencies. They do not, however, characterize the process fully. Yet, especially when acts can be grouped in an ethologically relevant way, they provide a useful way to summarize information concerning a behavioural process. For instance, in section 8.2 we give an example of the analysis of interactive behaviour of rats, where transition rates towards the groups of contact-directed behaviour and non-contact-directed behaviour are studied.

The statistic

$$\frac{\hat{\alpha}_{AG} - \alpha_{AG}}{\alpha_{AG}} \sqrt{N_{AG}} \qquad (7.18)$$

is approximately standard normally distributed, where N_{AG} denotes the total number of transitions from A to one of the acts belonging to the group G. The corresponding confidence interval can be found by replacing B by G in eqn (7.4). The log-transformed analogue

$$\sqrt{N_{AG}} \, (\log\hat{\alpha}_{AG} - \log\alpha_{AG}) \qquad (7.19)$$

also has an asymptotic standard normal distribution. The confidence interval is calculated by substituting G for B in eqn (7.7). The (log-transformed) estimators of α_{AG} and α_{CP} are asymptotically independent for A $\neq$ C and/or $G \neq P$, where P is another group of acts, provided that G and P do not overlap.

7.3   EXAMINING THE EFFECTS OF COVARIATES ON CTMCS

As mentioned, one of the final goals of analysis will usually be to examine the effects of various conditions, so-called covariates, on behaviour. An obvious example is the influence of experimental circumstances. Another, usually ignored, possibility is the effect of time inhomogeneity. In Chapter 3 we describe methods for detecting changes in behaviour and getting an impression of the form of such changes. When changes are more or less abrupt, different periods can be defined. Subsequently, differences between the periods can be studied by considering the period as a qualitative covariate. When changes are gradual, and explicit models for the time dependency of behaviour can be formulated, observation time can be considered as a quantitative covariate. The effects of the behaviour of other individuals can also be analysed with the methods given in this section, by treating the acts of other individuals as a qualitative covariate. Examples are given below. Finally, we describe a way of examining whether certain acts that can be considered as Markov states can be grouped into a new Markovian state (so-called 'lumpability'; subsection 7.3.3).

The effects of different conditions on behavioural processes can be examined exploratively by combining the estimators of transition and termination rates by means of (weighted) summation. Such methods are described in subsection 7.3.1. More formal methods for estimating and testing effects are treated in subsequent subsections. For an analysis of the effects on transition probabilities we refer to Aitchison (1986) and Fienberg (1980). In subsection 7.3.2 it is shown how to generalize the log-likelihood function of the observed process, given in eqn (7.1), when transition rates are considered as functions of covariates. The effects of the covariates can then be estimated by likelihood maximization. Likelihood ratio tests of various ethologically relevant hypotheses (see above) are given in subsection 7.3.3. Such tests are asymptotically the most powerful. In complex situations, however, the methods of subsection 7.3.4 might be preferred, where transition or termination rates are considered as 'observations' in an ANOVA or regression analysis. These methods may be less powerful than the likelihood ratio tests of subsection 7.3.2, but the results are usually clearer.

7.3.1 Exploration

One way to get a first impression of the effects of covariates is to make tables or graphs of (log-transformed) estimates of the parameters given in subsection 7.2.2. However, the total number of parameters is often very

large. To get an overview of the effects, it is therefore usually more convenient to combine the parameter estimates of transition and termination
rates by means of (weighted) summation. The properties of such statistics
can be derived from the results in subsection 7.2.2. In this subsection we
give several such measures and describe how certain effects, likely to be
encountered in practice, can be discovered with them. Note, however, that
since these measures are in fact a summary of the total set of parameters,
they do not completely characterize CTMCs. Furthermore, they are usually
not chances per unit of time and therefore cannot be interpreted as behavioural tendencies (except under special 'lumpability' conditions, see
subsection 7.3.3). These measures, therefore, should only be used for
screening data. When the effects of covariates on these statistics are found,
there are several possible explanations. This should be examined further
with more refined methods such as those described in later subsections.

Properties of (weighted) sums of transition and termination rates

From the asymptotic normality (see Box 7.1) of the estimators of transition
and termination rates, it follows that (weighted) sums of these estimators
are also asymptotically normally distributed. Furthermore, it can be shown
that the log-transformed statistics, too, have an asymptotic normal distribution. The strong consistency property of the estimators implies that the
expectations of the statistics are all equal to their true values. Their approximate variances (given below) can be used for calculating confidence
intervals.

Let G be the group of acts $\{B_1,..., B_m\}$ and P the group $\{C_1,..., C_n\}$; then
the weighted sum of estimated transition rates from acts belonging to G to
acts belonging to P is

$$\hat{\xi}_{GP} = \sum_{i=1}^{m} \sum_{j=1}^{n} w_{ij} \hat{\alpha}_{B_i C_j},$$
(7.20)

where w_{ij} ($i = 1,..., m$; $j = 1,..., n$) denote the weight factors. If G and P
overlap, the summation is taken over all $B_i \neq C_j$. The approximate variance
of $\hat{\xi}_{GP}$ is equal to

$$Var(\hat{\xi}_{GP}) = \sum_{i=1}^{m} \sum_{j=1}^{n} w_{ij}^2 \frac{\alpha_{B_i C_j}^2}{N_{B_i C_j}}.$$
(7.21)

The variance of log $\hat{\xi}_{GP}$ is approximately

$$Var(\log \hat{\xi}_{GP}) = \frac{Var(\hat{\xi}_{GP})}{\xi_{GP}^2}.$$
(7.22)

The weighted sum of estimated termination rates of acts in G is

$$\hat{\gamma}_G = \sum_{i=1}^{m} w_i \hat{\lambda}_{B_i},$$ (7.23)

where $w_1,..., w_m$ are the weight factors. The variance of this statistic is approximately

$$Var(\hat{\gamma}_G) = \sum_{i=1}^{m} w_i^2 \frac{\lambda_{B_i}^2}{N_{B_i}},$$ (7.24)

and the variance of log $\hat{\gamma}_G$:

$$Var(\log \hat{\gamma}_G) = \frac{Var(\hat{\gamma}_G)}{\gamma_G^2}.$$ (7.25)

The variances given above are calculated for the case when the weight factors are deterministic. When stochastic weights are used, as in eqn (7.27) below, this is more difficult. However, since in a CTMC statistics of the form given in eqn (7.27) converge almost surely to constants when $T \rightarrow \infty$ (see e.g. Adke and Manjunath, 1984, section 5.5), the approximation by eqns (7.21) and (7.24) will usually be good enough to be used for the purpose of screening.

Detection of the global effects of covariates on behaviour

Covariates can have all sorts of effects on behaviour. Here, we will discuss certain global effects that are most likely to be encountered in practice, and ways of detecting them with the statistics given in the previous subsection.

One type of effect of, for instance, internal changes in animals, may be that behaviour accelerates, i.e. there is a quick alternation of acts, or slows down. To see whether this occurs, the summed termination rates, $\hat{\gamma}_G$ (with weight factors equal to one), can be studied, where G is the group of all behavioural categories. An example is given in Box 7.2.

Box 7.2   Detection of acceleration or slowing down of behaviour: dyadic interactions of male rats

The example used here is based on experiments carried out with the Etho-pharmacology Group, Leiden University. The results are discussed further in section 8.2. In each trial, two male rats of unequal weights were put together in an unfamiliar cage. In such situations, there is usually an abrupt change in behaviour after about 5 minutes. The first 5 minutes of the observation period is called 'period 1' and the subsequent period 'period 2'. Tables 1 and 2 show the estimated termination rates of several acts of the largest males in periods 1 and 2 respectively, for three rats. For the present, we ignore the behaviour of the smaller rats (see section 8.2). A description of the ethogram units is given in Box 8.1.

Table 1 Estimated termination rates of acts of the largest rats in period 1

Act	Rat no.		
	1	2	3
Rear	3.529	1.328	2.444
Explore	0.463	0.151	0.304
Self-groom	0.254	0.111	–
Sniff Opponent	0.403	1.787	0.508
Groom Opponent	0.645	0.381	0.550
Crawl over	1.775	0.800	2.723
Totals:	7.069	4.558	4.32

Table 2 Estimated termination rates of acts of the largest rats in period 2

Act	Rat no.		
	1	2	3
Rear	0.491	0.288	0.686
Explore	0.053	0.220	0.155
Self-groom	0.122	0.140	0.084
Sniff Opponent	0.332	0.443	0.986
Groom Opponent	0.342	0.464	0.493
Crawl Over	1.028	0.309	1.154
Totals:	2.368	1.864	3.558

The sums of the termination rates in the two periods are given in the last rows of the tables. As can be seen, these sums are much lower in period 2, which indicates that the alternation between acts is slower in period 1. This was confirmed by further investigations (Haccou *et al.*, 1988*a*).

To study changes in the tendency to perform a certain act, A, calculate the summed transition rate, $\hat{\xi}_{GA}$, where G is the group of all acts except A. The weight factors can be taken equal to one. Alternatively, the average tendency to start A can be calculated by taking the weight factors equal to $1/k$, where k is the total number of acts other than A. Changes in the tendency to perform a group of acts, G, can be investigated analogously by means of $\hat{\xi}_{PG}$, where P is the group of all acts not belonging to G.

The tendency to stop doing acts belonging to a group G can be studied by means of $\hat{\xi}_{GP}$ (with weight factors equal to one), where P is the group of all acts outside G. In a similar way, the tendency to stop G in favour of acts belonging to another group, P, can be analysed.

Differences in weighted sums can also give indications about global changes in behaviour. For instance, the average tendency to stop a group of acts G in favour of another group P can be studied by means of $\log \hat{\xi}_{GP} - \log \hat{\xi}_{PG}$, with weight factors equal to $1/m$ and $1/n$ respectively (where m is the number of acts in G and n the number of acts in P). An example of an application is given in Box 7.3.

Box 7.3   Comparison of tendencies to start and end a group
of acts: dyadic interactions of male rats

We consider once again the example of Box 7.2. Table 1 shows the transition tendencies towards the groups of respectively contact-directed and non-contact-directed behaviour in the two periods. These are mean transition tendencies of individuals (taken from Haccou *et al.*, 1988a), but for the purpose of illustration we will treat them as if they are based on one record.

The group of non-contact-directed acts consists of *Rear*, *Explore* and *Self-Groom*, and the contact-directed acts are *Sniff Opponent*, *Groom Opponent* and *Crawl over*. The first group will be called *NC* and the second group *C*. To calculate the mean tendency to start acts in *C*, the transition tendencies from all acts outside *C* towards acts in *C* are summed. As can be seen from Table 1, this leads to a total of 0.451 in period 1. Since there are two acts from which transitions towards *C* were made, the mean transition tendency is equal to 0.225. Analogously, mean tendencies from acts in *C* towards acts in *NC* are calculated, which gives 0.751 for period 1. The same is done in period 2. From Table 1 it can be seen that in both periods the mean tendency to stop contact-directed behaviour is much larger than the tendency to initiate it. To compare these differences for the two periods we calculate the differences between the logarithms of the two mean tendencies for both periods. For period 1 this gives: $\log 0.751 - \log 0.225 = 1.203$, and for period 2: $\log 0.546 - \log 0.053 = 2.332$. Apparently, the ratio of the mean tendencies from and towards contact acts is about three times larger in period 2 than in period 1.

Table 1 Transition tendencies (in s^{-1}) towards contact and non-contact behaviour in periods 1 and 2

	Period 1	Period 2
From non-contact acts to contact act		
Preceding act:		
Rear	0.301	0.092
Explore	0.150	0.040
Self-groom	–	0.027
Total:	0.451	0.159
Mean:	0.225	0.053
From contact acts to non-contact acts:		
Preceding act:		
Sniff Opponent	0.561	0.449
Groom Opponent	0.365	0.222
Crawl over	1.328	0.968
Total:	2.254	1.639
Mean:	0.751	0.546

It is easily seen how the statistics described can be used to study the effects of different experimental conditions, or to study global differences between the behaviour in different subphases during inhomogeneous observation periods. With only slight adjustments, the statistics can also be used to study social influences. In this case the behaviour of the other individual(s) is considered as the external influencing factor. For instance, suppose that we want to study the global effects of the behaviour of individual 2 on that of individual 1 in a dyadic interaction. Then the effects on the tendencies of individual 1 to terminate its current act can be studied by comparing:

$$\sum_i \hat{\lambda}_{(Y,Z_j)} = \sum_i \sum_l \hat{\alpha}_{(Y,Z_j)(Y,Z_l)}, \tag{7.26}$$

for the different acts Z_j of individual 2. This statistic is equal to $\hat{\xi}_{GP}$, with G as well as P equal to the group of all combinations of Y_i with Z_j ($i = 1,..., m$). An example is given in Box 7.4. The other statistics can be adjusted in a similar way. In eqn (7.26) it is implicitly assumed that the two individuals do not change their behaviour simultaneously. For adjustments in the case of simultaneous transitions, see section 1.2.

In the examples given above, the weight factors were chosen to be equal to one or, to calculate averages, equal to 1/(the number of acts in a group).

Box 7.4 Detection of global effects of acts of another individual: dyadic interactions of male rats

Until now, we have ignored the behaviour of the smaller rats in the example of Boxes 7.2 and 7.3. To study the effects of their behaviour on that of the larger rats, we divide it into three categories: *Immobile*, *Non-contact acts*, and *Contact acts*. See Box 8.1 for a more detailed description. The effect of the others' behaviour on the tendencies of the largest rat to terminate its behaviour can be studied by means of statistics as given in eqn (7.26). We will do this here for one individual, during the second period.

Table 1 gives the estimated termination tendencies of the largest rat for each of its acts, given that the other rat is performing non-contact-directed behaviour, together with the observed numbers of bouts. As can be seen, the numbers are widely different for different acts. Therefore, a weighted average is used instead of the sum of the termination rates. The weights given in eqn (7.27) were used. Their values are given in column 4 of Table 1. The weighted average is calculated by summing the products of the termination tendencies and their weights (column 5). In this case, the result

Table 1 Estimated termination tendencies (in s^{-1}) of the largest rat, when the other rat is doing non-contact-directed acts, during period 2

Act	$\hat{\lambda}$	N	$\sqrt{N}$	w	$w\hat{\lambda}$
Rear	0.295	25	5	0.166	0.049
Explore	0.139	64	8	0.265	0.037
Self-groom	0.079	10	3.162	0.105	0.008
Sniff Opponent	0.492	36	6	0.199	0.098
Groom Opponent	0.252	25	5	0.166	0.042
Crawl over	1.538	9	3	0.099	0.152
Totals:			30.162		0.386

is a weighted average termination tendency of 0.386 s^{-1}. In the same way, the weighted averages of termination tendencies during *Immobile* and *Contact-directed acts* were calculated, which led to values of 0.217 and 0.375 respectively. Apparently, the larger rat's tendencies to terminate current behaviour are on average much lower when the other rat is *Immobile* than during the other two types of behaviour of the other individual. The average termination tendencies during *Non-contact acts* and *Contact acts* of the other rat do not differ much.

The approximate variance of the average termination tendency during non-contact acts of the other rat can be calculated by means of eqn (7.24). This is done by taking the squared values of column 5 in Table 1, dividing by the corresponding numbers of acts (column 2) and summing the results, which gives a variance of 0.00303. A 95% confidence interval can be calculated as $(\hat{\lambda}/(1 + 1.96\,\hat{\sigma}), \hat{\lambda}/(1 - 1.96\,\hat{\sigma}))$ which gives $(0.348, 0.433)$. The confidence intervals of the other two averages are $(0.191, 0.250)$ for acts during *Immobile* and $(0.325, 0.443)$ for acts during *Contact acts*. This confirms that the average termination tendency during *Immobile* is lowered, and does not differ much between the two other groups of acts of the other individual.

Other weights may also be used. For instance, $\hat{\gamma}_G$ can be calculated with weights w_i such that $\Sigma w_i = 1$. A reason for attaching different weights to the termination rates can be that one wants a higher sensitivity to detect changes in termination rates in some of the acts rather than in others. Another possibility is that the different estimates are based on (widely) different numbers of bouts. In this case, for example, the following weights can be used when calculating $\hat{\gamma}_G$:

$$w_i = \frac{\sqrt{N_{B_i}}}{\sum_i \sqrt{N_{B_i}}} . \tag{7.27}$$

7.3.2 Expressing transition rates as functions of covariates

Up to now, we have considered the transition rates as the basic parameters of a CTMC. It is also possible, however, to express the transition rates as functions of a set of covariates and an (unknown) parameter vector θ. The estimate of θ gives an indication of the effects of the covariates. For instance, suppose that there are p covariates $z_1,..., z_p$, denoted by the vector z. A reasonable model for the effect of the covariates on α_{AB} is

$$\alpha_{AB}(\theta;z) = \mu_{AB}\exp[\beta_1 z_1 + ... + \beta_p z_p], \tag{7.28}$$

i.e. it is assumed that the covariates have multiplicative effects on the transition rates. In general, it is better to use multiplicative rather than additive effect models, since the transition rates cannot be negative.

Suppose that there are n records of behavioural processes, denoted by $\Re_1,..., \Re_n$. Let z_i be the vector of values of the set of covariates for the ith record. From (7.1) it follows that the log-likelihood of record i is equal to

$$L(\Re_i; \{\alpha(\theta; z_i)\}) = \sum_A \sum_{B \neq A} n_{AB,i} \log \alpha_{AB}(\theta; z_i) - \sum_A \left\{ \sum_{B \neq A} \alpha_{AB}(\theta; z_i) \right\} S_{A,i}.$$

(7.29)

$n_{AB,i}$ denotes the number of transitions from A to B and $S_{A,i}$ the total duration of A in the ith record. Note that $\theta = (\mu_{AB}, \beta_1,..., \beta_p)$. Since the data on different records/periods are assumed to be independent, the log-likelihood of the complete sample is

$$L(\Re_1,..., \Re_n; \{\alpha(\theta; z_1,..., z_n)\}) = \sum_{i=1}^n L(\Re_i; \{\alpha(\theta; z_i)\}).$$

(7.30)

Expressions like (7.28) are substituted for $\alpha_{AB}(\theta; z)$ in this equation. The MLE of θ is found by maximizing (7.30) over θ. This will usually require numerical techniques like the Newton-Raphson iteration (see Stoer and Bulirsch, 1980).

It might often be difficult to make *a priori* assumptions about the relation between transition rates and covariates. The model given in eqn (7.28) is fairly general. It may be more realistic to assume, however, that the covariates have different effects on the transition rates, so that $(\beta_1,..., \beta_n)$ also depend on A and B.

The method described usually leads to complicated numerical computations. In that case it may be preferable to apply ANOVA, regression analysis or a related analysis on the (log-transformed) estimators of transition or termination rates (see section 7.5).

Note that eqn (7.1) can be considered as a special case of eqn (7.29), with $\alpha_{AB}(\theta; z) = \alpha_{AB}$.

7.3.3 Likelihood ratio tests for CTMCs

Various hypotheses concerning CTMCs can be tested by means of likelihood ratio tests. The basic results are due to Billingsley (1974). In eqn (7.1) we give the log-likelihood of an observation of length T on a CTMC. In eqn (7.29) the log-likelihood is adjusted for the case when the transition rates are expressed as functions of other parameters. In the following, we use the notation of eqn (7.29), since it is more general and more convenient. Note that, since (7.1) is a special case of (7.29), the results apply to both cases. We first consider the case of one behavioural process. Let $\{\alpha(\theta;z)\}$ be the set of transition rates. The likelihood ratio test statistic for the hypothesis H_0: $\theta \in \Theta_0$ against H_1: $\theta \in \Theta_1$ (where Θ_0 is a subset of the closure of Θ_1)

$$\Lambda = 2[\max_{\theta \in \Theta_1} L(\Re;\{\alpha(\theta;z)\}) - \max_{\theta \in \Theta_0} L(\Re;\{\alpha(\theta;z)\})], \qquad (7.31)$$

and where $L(\Re;\{\alpha(\theta;z)\})$ is the log-likelihood of the process, as given in eqn (7.29). In the case of n independent observations of processes, the log-likelihood given in eqn (7.30) is substituted for $L(\Re;\{\alpha(\theta;z)\})$ in eqn (7.31). It is assumed that the total set of positive transitions does not depend on θ. Hence, whether or not a transition is possible is not related to θ.

Let the dimension of the parameter space Θ_1 be equal to r and that of Θ_0 equal to c (with $c < r$); then Λ has asymptotically a chi-squared distribution with $r - c$ degrees of freedom under the null hypothesis if certain regularity conditions are met (conditions 3.1 and 8.1–8.3 of Billingsley, 1974, Chapters 3 and 8). In our case, conditions 8.1 and 8.3 are met automatically provided that there are no instantaneous acts (point events). When there are instantaneous acts, some minor adjustments are needed (see section 7.1 and Metz, 1981). The remaining conditions are the usual regularity conditions, i.e. the $\alpha_{AB}(\theta;z)$ should in some sense be sufficiently smooth functions of θ and the parameter spaces Θ_0 and Θ_1 should not be restricted in odd ways (see e.g. Wilks, 1962, Chapter 13, for a precise formulation).

These results can be used to derive tests for a large number of relevant hypotheses, some of which will be treated in the next subsections. As mentioned in section 7.1, the tests can also be used when a fixed number of bouts is sampled or when the process has been observed until another type of stopping time. Furthermore, the information of the last, possibly censored, bout length may also be included in eqn (7.29).

Lumpability

If two or more states of a CTMC are lumped the resulting process is usually not Markovian. Under special conditions, however, it is: that is, the case when the states meet the so-called strong lumpability condition. As stated before, in a CTMC the chance per unit of time to switch to an arbitrary group of states $G = \{B_1,..., B_m\}$ only depends on the present state. Thus we can define the transition rate *from* A towards G, α_{AG}, by

$$\alpha_{AG} = \frac{Pr\{\text{state belongs to } G \text{ in } (x + dx,x)\,|\,\text{state is A on } x\}}{dx} \qquad (7.32)$$

(subsection 7.2.2). However, in general the rate of the reversed transition from G to A cannot be defined, since the chance per unit of time to enter

A when the process is in G usually also depends on the previous history of the process. Under the strong lumpability condition a group of acts $\{B_1,..., B_m\}$ of which each act can be considered as a Markov state, can be represented by a Markov state itself. It can be proved that the acts $B_1,..., B_m$ are strongly lumpable if and only if the transition rates from all the B_i to acts outside the group G are equal:

$$\alpha_{B_1,A} = ... = \alpha_{B_m,A} \quad \text{for all A} \notin G. \tag{7.33}$$

A similar results holds for the discrete time case. In ethological terms this means that as long as an animal performs a behavioural category belonging to G, the distinction between $B_1,..., B_m$ is not necessary with respect to its tendency to start any of the other (groups of) acts. Consider for instance the example given in Box 4.1 on mother-infant interactions in rhesus monkeys. When an infant sits on its mother its tendencies to take its mother's nipple or to leave its mother are not affected by grooming by the mother. Since the only other behavioural acts are *Off mother* and *On nipple*, *On mother with grooming* and *On mother without grooming* can be grouped into one category *On mother*. This example, taken from Metz *et al.* (1983), clearly shows that, besides leading to a simpler model, strong lumpability also has a clear ethological interpretation: apparently, and rather unexpectedly, grooming by the mother does not lead to a difference in the rate at which contact is broken. Strong lumpability of acts can be tested by means of the likelihood ratio principle.

Suppose that we want to test whether the group of acts $\{B_1,..., B_m\}$ are strongly lumpable. Let P be the group of remaining acts $\{C_1,..., C_r\}$ to which direct transitions from each of the B_i ($i = 1,..., m$) are possible. Note that strong lumpability of $B_1,..., B_m$ implies that there are no acts outside $G = \{B_1,..., B_m\}$ that can only be reached directly from a subset of the B_i. The likelihood ratio test statistic for testing strong lumpability of $B_1,..., B_m$ then becomes

$$\Lambda = 2[\sum_{i=1}^{m} \sum_{j=1}^{r} \{N_{B_iC_j}(\log\frac{N_{B_iC_j}}{S_{B_i}} - \log\frac{N_{GC_j}}{S_G})\}] , \tag{7.34}$$

where the N denote the number of transitions from the act(s) belonging to G to one of the acts C and S_G denote the total time the lumped act G is displayed. Under the null hypothesis of lumpability, the test statistic Λ has asymptotically a chi-squared distribution with $(m - 1)r$ degrees of freedom, i.e. reject the null hypothesis of lumpability if $\Lambda > \chi^2_{(m-1)r}(\alpha)$. See Table A2 for the critical values and Box 7.5 for an application of this test.

Box 7.5   Testing for lumpability: solitary rat behaviour

We will use the data on solitary rat behaviour given in Box 2.10 to illustrate
the test on lumpability of eqn (7.34). For convenience, part of the data is
summarized again here, in Tables 1 and 2. We will test whether *Turn*,
Shake, *Yawn* and *Squat* are lumpable. It is assumed that, although not all

Table 1 Total durations of acts in G

Act	total duration (s)
Turn	22.08
Shake	9.57
Yawn	2.12
Squat	42.52
Total:	76.29

Table 2 Transition frequencies from acts in G to other acts

To: From:	*Care*	*Walk*	*Sniff*	*Rear*
Turn	5	1	9	0
Shake	0	6	1	0
Yawn	0	0	1	0
Squat	4	2	8	0
Totals:	9	9	19	0

possible transitions from these acts to other acts were observed in this
particular case, all transitions are possible in principle. Table 1 shows the
total durations of the acts in this group (S_{Bi} in eqn (7.34)). The sum of these
durations, S_G, is equal to 76.29 s. Table 2 shows the transition frequencies
of all acts in the group to acts outside, N_{BiCj}, and the sums of the frequencies, N_{GCj}.

From these tables and eqn (7.34) it follows that the likelihood ratio test statistic is

$$\Lambda = 2[5\{\log(\frac{5}{22.08}) - \log(\frac{9}{76.29})\} + 1\{\log(\frac{1}{22.08}) - \log(\frac{9}{76.29})\}$$

$$+ 9\{\log(\frac{9}{22.08}) - \log(\frac{19}{76.29})\} + \dots +$$

$$+ 8\{\log(\frac{8}{42.52}) - \log(\frac{19}{76.29})\}] = 26.70. \tag{1}$$

m is equal to four in this case, and r is three (since *Rear* never occurs as a following act), so the number of degrees of freedom is nine. As can be seen from Table A2, the 5% critical value is 16.91, so it is concluded that the acts are not lumpable.

Effects of behaviour of other individuals

For simplicity let us consider an interaction between two individuals. Suppose that no simultaneous transitions can be made by the two individuals and that we want to test whether the behaviour of individual 2 affects the tendencies of individual 1 to make transitions between a subset of acts performed by individual 1. Let $\wp$ be the set of all possible transitions between these acts of individual 1. Furthermore, let G be the set of all different acts of individual 2. It is assumed that, if a transition between a pair of acts, say A and B, belonging to $\wp$ is possible, it can occur in combination with each of the acts of individual 2. The likelihood ratio test statistic of the hypothesis described above is based on

$$\Lambda = 2\sum_{\wp} [\{\sum_{X \in G} N_{(A,X)(B,X)} \log \frac{N_{(A,X)(B,X)}}{S_{AX}}\} - N_{AB} \log \frac{N_{AB}}{S_A}], \tag{7.35}$$

where $N_{(A,X)(B,X)}$ is the number of transitions between act A and act B made by individual 1 while individual 2 does X, and N_{AB} is the total number of transitions between A and B by individual 1, irrespective of what individual 2 is doing. S_{AX} is the total time spent on the combination of acts A and X and S_A the total time individual 1 performed act A. Under the null hypothesis that the behaviour of individual has no effect on the transition tendencies of individual 1 between the acts belonging to $\wp$ the test statistic has an asymptotic chi-squared distribution with $p(k - 1)$ degrees of freedom, where p is the number of possible transitions between acts from $\wp$ and k is the total number of different acts performed by individual 2.

When there are accidental simultaneous transitions by both individuals, it is best to incorporate these observations in the test statistic of eqn (7.35) by treating them as if individual 1 made the transition first.

If we want to test whether it makes a difference if individual 2 does any of the acts of a subset of acts, irrespective of the effects of the rest of the acts of individual 2, the statistic Λ defined in eqn (7.35) can be modified accordingly. In that case we sum only over the acts belonging to that subset and redefine N_{AB} and S_A accordingly (i.e. G is the considered subset of acts rather than all different acts of individual 2). In Box 7.6 an example of the application of this test is given.

Box 7.6 Testing the effects of behaviour of other individuals: the effects of mealybug defence on the tendency of wasps to parasitize

We illustrate the test on interactive behaviour (eqn (7.42)) with data on the interaction between mealybugs and parasitic wasps. This example is based on the as yet unpublished results of Dr J.W.A.M. Pijls (Ecology Department, Leiden University).

When a parasitic wasp encounters a mealybug, she inspects the host by drumming with her antennae. Mealybugs react to this in two different ways: they either show no extraordinary activity or they defend themselves by *Flipping* the abdomen or shooting drops of honeydew. After *Drumming* on a mealybug, a wasp can *Preen* her ovipositor, *Sit still*, *Search* in the nearby area or decide to *Oviposit* in the inspected mealybug. We wondered whether the reaction of the mealybug to detection by the wasp had any effect on the wasp's behaviour, including the decision to parasitize.

Table 1 Comparison of total and mean bout lengths (in s) of wasp acts for different mealybug reactions

Mealybug defence:	Absent		Vigorous		Complete record	
Bout length:	Total	Mean	Total	Mean	Total	Mean
Drum	88.5	4.97	73.1	7.31	161.6	5.76
Preen	87.0	6.69	110.2	11.02	197.2	8.57
Ovip.	37.1	9.27	30.3	7.57	67.4	8.42
Search	3.1	1.55	2.7	1.35	5.8	1.45
Still	19.4	6.47	59.7	11.94	79.1	9.89

A first impression of the effect can be obtained by comparing the mean bout lengths of different acts of the wasp for the two different types of reaction of the mealybug (see Table 1). It seems that the *Preen* and *Drum* bouts are prolonged if the mealybug defends itself and that the *Oviposition* attempts are terminated more quickly by the wasp. Moreover, the periods of *Sitting still* are longer.

There is no apparent difference between the two transition matrices of acts of the wasp (Tables 2 and 3).

Table 2 Transition matrix of wasp acts in the absence of mealybug defence

To: From:	*Drum*	*Preen*	*Ovip.*	*Search*	*Still*
Drum	–	7	4	4	3
Preen	9	–	0	4	0
Ovip.	0	4	–	0	0
Search	0	2	0	–	0
Still	2	0	0	1	–

Table 3 Transition matrix of wasp acts during vigorous mealybug defence

To: From:	*Drum*	*Preen*	*Ovip.*	*Search*	*Still*
Drum	–	3	4	1	2
Preen	4	–	0	3	3
Ovip.	0	4	–	0	0
Search	1	1	0	–	0
Still	3	2	0	0	–

A formal test of the effect of the reaction of the mealybug is based on the statistics of eqn (7.42). The transition matrix for the complete record (irrespective of the reaction of the mealybug) is given in Table 4. For ease of computation, the test statistic Λ can be rewritten as

$$\Lambda = 2\sum_{\wp}\left[\sum_X \{N_{(A,X)(B,X)}\log N_{(A,X)(B,X)} - N_{(A,B)(B,X)}\log S_{AX}\} \right.$$
$$\left. - N_{AB}\log N_{AB} + N_{AB}\log S_A\right], \tag{1}$$

Table 4 Transition matrix for the complete record

To: From:	*Drum*	*Preen*	*Ovip.*	*Search*	*Still*
Drum	–	10	8	5	5
Preen	13	–	0	7	3
Ovip.	0	8	–	0	0
Search	1	3	0	–	0
Still	5	2	0	1	–

Substitution of the data listed in Tables 1–4 yields

$$
\begin{aligned}
\Lambda = 2[\ & (7\log7 + 4\log4 + 4\log4 + 3\log3 + 9\log9 + 4\log4 + 4\log4 + 2\log2 \\
& + 2\log2 + 1\log1) \\
& + (3\log3 + 4\log4 + 1\log1 + 2\log2 + 4\log4 + 3\log3 + 3\log3 + 4\log4 \\
& + 1\log1 + 1\log1 + 3\log3 + 2\log 2) \\
& - \{(7 + 4 + 4 + 3)\log88.5 + (9 + 4)\log87.0 + 4\log37.1 + 2\log3.1 \\
& + (2 + 1)\log19.4\} \\
& - \{(3 + 4 + 1 + 2)\log 73.1 + (4 + 3 + 3)\log110.2 + 4\log30.3 \\
& + (1 + 1)\log2.7 + (3 + 2)\log79.1\} \\
& - (10\log10 + 8\log8 + 5\log5 + 5\log5 + 13\log13 + 7\log7 + 3\log3 \\
& + 8\log8 + 1\log1 + 3\log3 + 5\log5 + 2\log2 + 1\log1) \\
& + \{(10 + 8 + 5 + 5)\log161.6 + (13 + 7 + 3)\log197.2 + 8\log67.4 \\
& + (1 + 3)\log5.8 + (5 + 2 + 1)\log 79.1)] \\
= \ & 2(61.646 + 32.591 - 164.364 - 126.019 - 135.382 + 339.602) \\
= \ & 2 \times 8.04 = 16.08.
\end{aligned}
$$

Under the null hypothesis Λ is asymptotically chi-squared distributed with $p(k - 1)$ degrees of freedom. In this example, $p = 13$ and $k = 2$. From Table A2 it follows that the 5% critical value at 13 degrees of freedom is equal to 22.4. Hence, H_0 is not rejected. On the basis of this test it would be concluded that the reaction of the mealybug does not affect the wasp's behaviour. It should be realized, however, that the chi-squared approximation is presumably poor in this case, since the sample is small.

The test can also easily be generalized to test the effects of interactions when there are more than two individuals involved. For instance, suppose that we want to test the same hypothesis as before, but besides the two individuals under consideration there are other animals as well. The object is to study the effect of individual 2 on individual 1 irrespective of the acts of other individuals, denoted by P, Q, R,.... Then eqn (7.35) still applies if we replace X by all the combinations of acts of individual 2 with the (combination of) acts P, Q, R,... of the others. The summation in eqn (7.35) is over all P, Q, R,... during which transitions between (A,X) and (B,X) can occur. Suppose that there are q such combinations of acts P, Q, R,...; then the number of degrees of freedom of the asymptotic chi-squared distribution of the test statistic is equal to $p(k - 1)q$.

Testing for other influencing factors

Suppose we have k observational records of (large) lengths T_i ($i = 1,..., k$), either on different animals or on the same animal under different conditions. A test whether a set of transition rates α_{AB} between acts of a subset $\mathcal{P}$ is equal for the k processes can be based on

$$\Lambda = 2\sum_{i=1}^{k} \sum_{A,B \in \mathcal{P}} \left\{ N_{AB}(i) \left(\log \frac{N_{AB}(i)}{S_A(i)} - \log \frac{\sum_{j=1}^{k} N_{AB}(j)}{\sum_{j=1}^{k} S_A(j)} \right) \right\}, \qquad (7.36)$$

where $N_{AB}(i)$ is the number of transitions between acts A and B in the ith process and $S_A(i)$ the total time act A is displayed in this process. Under the null hypothesis this test statistic has an asymptotic chi-squared distribution with $p(k - 1)$ degrees of freedom, where p is the number of possible transitions between acts belonging to $\mathcal{P}$. It is assumed that all these transitions can occur in each of the k processes. Thus, if, on the grounds of previous screening (see subsection 7.3.1), it is suspected that under certain experimental conditions some transitions are impossible, the corresponding act(s) should not be included in $\mathcal{P}$. See Box 7.7 for an example.

Box 7.7   Analysing the effect of different breeding conditions

Suppose we are interested in the effect of different breeding conditions of juvenile mammals held in captivity. A number of k_1 juveniles are grown up solitarily, k_2 together with their mother and k_3 with their mother and at least one brother or sister. The behaviour of each juvenile is recorded during 1 hour every month until the juveniles are 1 year old. Several questions can be investigated by applying the test of eqn (7.36): are there differences between the individuals in each group; are there differences between the groups; is the behaviour altered in the course of the year; are there differences in the development of the behaviour between the groups, especially in the contact-directed behaviour, etc.? All these questions can be answered by applying this test and by combining the (independent) test results by summing the chi-squared test statistics calculated for the appropriately chosen subset of records (subsection 6.2.1).

Combination of replicas

In the tests of lumpability, eqn (7.34), and of the effects of behaviour of other individuals, eqn (7.35), only one observation of the process as a whole is considered. Furthermore, in the test for the effects of an influencing factor, eqn (7.36), we considered only one observation under each (experimental) condition. However, in many practical situations there will be replicas. In that case it might be best to combine records. The advantage of combining two or more records is that it leads to larger data sets so that, if the records can be considered as realizations of the same process, the parameter estimates as well as the approximation of the distributions of test statistics are more accurate and the tests have more power. If the records are essentially different, however, combining them may lead to increased variability and a possibly substantial loss of power when testing hypotheses. Combining records can especially be considered seriously if the replicas concern one and the same individual.

When records are combined, the numbers of transitions between acts, N_{AB} in eqns (7.1) and (7.29), are calculated by summing the observed numbers of transitions between A and B over all records. S_A is calculated by summing the time doing act A over all records. These values can then be substituted in the test statistics.

If the replicas concern different individuals or, more generally, if there is not sufficient evidence that there is no difference between the several records, it is preferable to treat the records separately. Parameter estimates of different records can be combined at a later stage, by taking (weighted) averages. The methods described in section 6.2 can be used for the combination of test results.

7.3.4 ANOVA and regression on transition and termination rates

Since the (log-transformed) estimators of transition and termination rates are asymptotically normally distributed (section 7.2), the estimates can be used as 'observations' in ANOVA or regression methods. The covariates are qualitative or quantitative factors. The effects on different transition or termination rates can be studied separately, because of the asymptotic independence of the estimators (section 7.2). It is preferable to use the log-transformed estimates instead of the estimates themselves as the input to the analysis. This amounts to assuming that influencing factors have multiplicative rather than additive effects. This is more appealing since we are used to thinking of behavioural tendencies in terms of multiplicative effects. For instance, compare 'the animal's tendency to stop A in favour of B becomes twice as large' to '... increases by four'. Furthermore, multiplicative effects are scale invariant and do not have to be constrained to ensure that the rates stay positive. More importantly, the variances of the log-transformed estimators of transition and termination rates are not related to their expected values (Box 3.1). From a statistical point of view, an additional advantage is that the log-transformation, by reducing the skewness, improves the approximation by a normal distribution.

One, slight, drawback is that the standard ANOVA and regression techniques need to be generalized, since the variances of (log-transformed) estimators depend on the numbers of observed transitions (or acts, in the case of termination rates), and therefore are unequal. Moreover, there are usually unequal numbers of observations for the different conditions, and, besides, often not all transitions between (combinations of) acts occur in each behavioural record. This leads to so-called 'empty cells' when under certain (experimental) conditions a transition does not occur in any of the records. Estimation and test procedures have to be adjusted in these respects (see e.g. Scheffé, 1959). Note that in almost all textbooks on ANOVA as well as in software packages, it is assumed that the observations have equal variances.

The adjusted procedures for one-way and two-way layout ANOVA, as well as for regression analysis, are given in the subsequent subsections. We only consider fixed effects. The effects of influencing factors can also be assumed to be random instead of fixed (Scheffé, 1959, Chapters 7 and 8). For instance, the objective of an analysis might be to make inferences about a large population, and the group of individuals used in the experiment is considered as a random sample from that population. In that case, the effect of individual differences should be assumed to be a random variable rather than a fixed quantity (e.g. LaMotte, 1983). Random- or mixed-effect ANOVA (where some effects are fixed and others random) is easily derived from fixed-effects ANOVA (see e.g. Sokal and Rohlf, 1981).

In the following, we will describe methods for studying log-transformed transition rates between acts. The same methods can be used to study log-transformed termination or transition rates towards groups of acts (subsections 7.2.2 and 7.2.4). In this case, N_{AB} should be replaced by N_A or N_{AG} respectively. An example of an ANOVA on log-transformed transition rates towards groups of acts can be found in section 8.2.

The one-way layout

Suppose that a behavioural process is studied under I different conditions and that a number of replicas are made. Let Y_{ij} denote the log-transformed estimator of a certain transition rate, α_{AB}, from act A to act B in the jth replica of the ith experimental condition ($i = 1,..., I$; $j = 1,..., n_i$). Furthermore, let N_{ij} be the number of transitions from A to B in the corresponding behavioural record. For large observation times, Y_{ij} is approximately normally distributed with variance $1/N_{ij}$ (subsection 7.2.1, eqn (7.6)). It is assumed that the different observations, and therefore the Y_{ij} are independent. The proposed ANOVA model is

$$Y_{ij} = \mu + \alpha_i + \varepsilon_{ij}, \tag{7.37}$$

where ε_{ij} is a normally distributed error term with zero expectation and variance θ^2/N_{ij}. Note that α_i denotes the main effect of the factor at level i. This notation will also be followed in succeeding subsections. Since the α_i are not uniquely determined we impose the side condition that

$$\sum_{i=1}^{I} a_i \alpha_i = 0, \tag{7.38}$$

for a chosen system of weights $a_1,..., a_I$ with

$$\sum_{i=1}^{I} a_i = 1. \tag{7.39}$$

A convenient choice for the weight factors is to take $a_I = 1$ and the others equal to zero. This would imply that $\alpha_I = 0$ and the other effects are considered relative to the effects of level I of the factor studied. An alternative choice is to take all weights to be equal or proportional to n_i.

Let μ_i be the sum of μ and α_i ($i = 1,..., I$); then the following model is equivalent to the one given in eqn (7.37)

$$Y_{ij} = \mu_i + \varepsilon_{ij}. \tag{7.40}$$

We will describe how to estimate the μ_i and how to test whether the factor has significant effects, using representation (7.40). The estimates of the parameters in (7.37) can be calculated from those in (7.40) as follows:

$$\hat{\mu} = \sum_{i=1}^{I} a_i \hat{\mu}_i, \tag{7.41}$$

and

$$\hat{\alpha}_i = \hat{\mu}_i - \hat{\mu}. \tag{7.42}$$

The factor θ^2 is incorporated as a safeguard against a slight misspecification. If the asymptotic approximation of the distribution of Y_{ij} holds good and if, furthermore, the relation between the expectation of Y_{ij} and the influencing factors assumed in the ANOVA model is correct, θ^2 is approximately equal to one. Thus, θ^2 can be considered as a correction factor for small departures from the approximation of the variance of Y_{ij} by $1/N_{ij}$ and/or for misspecifations of the model. For instance, in the model of eqn (7.40) it is assumed that, given a certain condition, the transition rates of all replicas have the same expectation. This assumption may not always be correct. If, for example, different individuals were used for the replicas, the expected values of transition rates may differ, since not all individuals react similarly to the same condition. In that case θ^2 accounts for the additional variation (where it is assumed that individuals 'on average' react similarly to the influencing factor). θ^2 is considered as an unknown parameter.

Data: The log-transformed estimates of transitions rates:

$$y_{1,1}, y_{1,2}, ..., y_{1,n_1}$$
$$y_{2,1}, y_{2,2}, ..., y_{2,n_2}$$

.   .

.   .

.   .

$$y_{I,1}, y_{I,2}, ..., y_{I,n_I}$$

with the corresponding numbers of observed transitions:

$$N_{1,1}, N_{1,2}, ..., N_{1,n_1}$$
$$N_{2,1}, N_{2,2}, ..., N_{2,n_2}$$

.   .

.   .

.   .

$$N_{I,1}, N_{I,2}, ..., N_{I,n_I}.$$

Assumptions:

A1: $Y_{ij} = \mu_i + \varepsilon_{ij}$, $i = 1,..., I$ and $j = 1,..., n_i$; ε_{ij} is normally distributed with zero expectation and variance θ^2/N_{ij}.

A2: Y_{ij} and $Y_{i'j'}$ are independent; hence the covariances between ε_{ij} and $\varepsilon_{i'j'}$, $Cov(\varepsilon_{ij}, \varepsilon_{i'j'}) = 0$ if $i \neq i'$ and/or $j \neq j'$.

A3:   Under H_0, $\mu_1 = ... = \mu_I$ (i.e. the factor does not affect the transition rates), and under H_1 there is at least one inequality.

Procedure: The so-called least-squares estimates of the μ_i are found by minimizing the following expression:

$$\sum_{i=1}^{I}\sum_{j=1}^{n_i}(y_{ij} - \mu_i)^2 N_{ij} \tag{7.43}$$

over all possible choices for the μ_i. This leads to

$$\hat{\mu}_i = \frac{\sum_{j=1}^{n_i} y_{ij} N_{ij}}{\sum_{j=1}^{n_i} N_{ij}}, \tag{7.44}$$

where θ^2 is estimated by

$$\hat{\theta}^2 = \frac{SS_1}{v_2}, \text{ with } v_2 = (\sum_{i=1}^{I} n_i) - I, \tag{7.45}$$

and SS_1 is the minimum value of expression (7.43) calculated by substituting $\hat{\mu}_i$ for μ_i in (7.43).

The test statistic is defined by

$$F = \frac{SS_0 - SS_1}{SS_1} \frac{v_2}{v_1}, \tag{7.46}$$

where SS_0 denotes the minimum sum of squares if all the μ_i are equal:

$$SS_0 = \sum_{i=1}^{I}\sum_{j=1}^{n_i}\left\{y_{ij} - \frac{\sum_{i=1}^{I}\sum_{j=1}^{n_i} y_{ij} N_{ij}}{\sum_{i=1}^{I}\sum_{j=1}^{n_i} N_{ij}}\right\}^2 N_{ij}, \tag{7.47}$$

$v_1 = I - 1$, and v_2 is as defined in eqn (7.45).

Under the null hypothesis, F has an F-distribution with v_1 and v_2 degrees of freedom. (See Scheffé, 1959, Chapter 3.) Hence, H_0 is rejected when $F > F(v_1, v_2; \alpha)$. See Table A4 for the critical values.

Multiple comparison method: If the test based on F gives a significant result, the differences between the expected values of transition rates under the different conditions can be studied by means of so-called contrasts:

$$\hat{\psi} = \sum_{i=1}^{I} w_i \hat{\mu}_i, \text{ where } \sum_{i=1}^{I} w_i = 0. \tag{7.48}$$

For instance, to study the difference between the effects of the first and second condition we can choose $w_1 = 1$, $w_2 = -1$ and $w_3 = \ldots = w_I = 0$.

Since Y_{ij} and $Y_{i'j'}$ are independent for $i \neq i'$ and/or $j \neq j'$, it follows from (7.44) that the $\hat{\mu}_i$ are independent and normally distributed with expectation μ_i and variance

$$Var(\hat{\mu}_i) = \frac{\theta^2}{\sum_{j=1}^{n_i} N_{ij}} .$$

(7.49)

Accordingly, the variance of $\hat{\psi}$ is:

$$Var(\hat{\psi}) = \sum_{i=1}^{I} w_i^2 Var(\hat{\mu}_i) = \sum_{i=1}^{I} w_i^2 \frac{\theta^2}{\sum_{j=1}^{n_i} N_{ij}} ,$$

(7.50)

which can be estimated by substituting $\hat{\theta}^2$ (eqn (7.45)) for θ^2.

The hypothesis $\psi = 0$ is rejected when

$$\frac{\hat{\psi}^2}{Var(\hat{\psi})(I-1)} > F(\nu_1, \nu_2; \alpha),$$

(7.51)

where $F(\nu_1, \nu_2; \alpha)$ is the same critical value as used in testing H_0.

It can be proved that if the test based on F (eqn (7.46)) gives a significant result there is at least one combination $w_1, ..., w_I$ for which (7.51) holds, and vice versa (Scheffé, 1959, section 3.4).

Residual analysis: The fit of the model in (7.40) can be tested by examining the residuals:

$$r_{ij} = \frac{(y_{ij} - \hat{\mu}_i)\sqrt{N_{ij}}}{\sqrt{\hat{\theta}^2}} .$$

(7.52)

If the model fits, the residuals have a Student distribution with ν_2 (eqn (7.45)) degrees of freedom (Table A3). Note that the residuals, in this case and in the subsequent cases, are not independent.

The two-way layout without interactions

Sometimes there may be more than one factor affecting the behaviour of an animal. For instance, when the effects of experimental conditions on interactive behaviour is studied, the acts of another animal as well as the experimental circumstances may influence an animal's behaviour. This can be studied by a two-way layout ANOVA, where one of the factors is the behaviour of the other animal and the other factor is the experimental condition (an example is given in section 8.2).

We denote the log-transformed estimator of the transition rate from A to B by Y_{ijk}, where

i is the level of factor 1, $i = 1,..., I$;

j is the level of factor 2, $j = 1,..., J$;

k is the index of the replica per combination of factor levels, $k = 1,..., K_{ij}$.

Note that the number of replicas may differ since the transition studied does not necessarily occur in every observation period. According to eqn (7.6) Y_{ijk} is approximately normal with variance $1/N_{ijk}$ (= the number of observed transitions from A to B in the ijkth record). It is assumed that the different Y_{ijk} are independent. The simplest two-way layout ANOVA model is

$$Y_{ijk} = \mu + \alpha_i + \beta_j + \varepsilon_{ijk}, \tag{7.53}$$

where the α_i represent the influence of factor 1 and the β_j the effect of factor 2. The error term ε_{ijk} is supposed to be normal with zero expectation and variance θ^2/N_{ijk}. Because of non-uniqueness, the side-conditions

$$\sum_{i=1}^{I} a_i \alpha_i = \sum_{j=1}^{J} b_j \beta_j = 0 \tag{7.54}$$

are imposed on the main effects, where $a_1,..., a_I$ and $b_1,..., b_J$ are chosen weight factors, with

$$\sum_{i=1}^{I} a_i = \sum_{j=1}^{J} b_j = 1. \tag{7.55}$$

A convenient choice, which we will use here, is $a_I = b_J = 1$, which implies that $\alpha_I = \beta_J = 0$, and the other effects are estimated relative to the combination of the last levels of both factors. Whether this choice is preferable to others depends on the possibilities for a sound ethological interpretation. Note that, once the main effects have been estimated under a certain system of weights, their values under other systems are easily calculated.

Just as in the case of the one-way layout, the factor θ^2 is introduced to enlarge the flexibility of the model and to account for small misspecifications. Its expected value is equal to one.

Data: For every combination of levels i and j of the two factors there are K_{ij} log-transformed transition rates:

$$y_{ij1}, y_{ij2},..., y_{ij,K_{ij}},$$

with corresponding number of observed transitions:

$$N_{ij1}, N_{ij2},..., N_{ij,K_{ij}}.$$

Assumptions:

A1: $Y_{ijk} = \mu + \alpha_i + \beta_j + \varepsilon_{ijk}$, $i = 1,..., I$, $j = 1,..., J$, $k = 1,..., K_{ij}$, ε_{ijk} is normally distributed with zero expectation and variance θ^2/N_{ijk}, $\alpha_I = \beta_J = 0$.

A2: Y_{ijk} and $Y_{i'j'k'}$ are independent;, hence the covariances $Cov(\varepsilon_{ijk}, \varepsilon_{i'j'k'}) = 0$ if $i \neq i'$ and/or $j \neq j'$ and/or $k \neq k'$.

A3:   Under $H_{0,A}$, $\alpha_1 = ... = \alpha_I = 0$ (factor 1 has no effect).

A4:   Under $H_{0,B}$, $\beta_1 = ... = \beta_J = 0$ (factor 2 has no effect).

A5:   Under H_1 there is at least one inequality in assumption A3 or A4.

Procedure: It follows from Scheffé (1959, section 4.4) that the vector $A = \alpha_1,..., \alpha_{I-1}$ can be calculated by

$$\hat{A} = V(\alpha)^{-1}L(\alpha), \tag{7.56}$$

where $V(\alpha)$ is a symmetric matrix of dimension $(I - 1)(I - 1)$ with elements

$$V_{ix}(\alpha) = \delta_{ix}s_i - \sum_{j=1}^{J} \frac{g_{ij}g_{xj}}{t_j}, \tag{7.57}$$

$i = 1,..., I - 1$, where $\delta_{ix} = 1$ if $i = x$ and zero otherwise,

$$g_{ij} = \sum_{k=1}^{K_{ij}} N_{ijk}, \tag{7.58}$$

$$s_i = \sum_{j=1}^{J} g_{ij} \text{ and } t_j = \sum_{i=1}^{I} g_{ij}. \tag{7.59}$$

In eqn (7.56) $L(\alpha)$ is a vector of length $(I - 1)$ with elements

$$L_i(\alpha) = S_i - \sum_{j=1}^{J} \frac{g_{ij}T_j}{t_j}, \quad i = 1,..., I - 1, \tag{7.60}$$

where

$$S_i = \sum_{j=1}^{J} \sum_{k=1}^{K_{ij}} y_{ijk}N_{ijk} \text{ and } T_j = \sum_{i=1}^{I} \sum_{k=1}^{K_{ij}} y_{ijk}N_{ijk}. \tag{7.61}$$

The β_i are estimated analogously by

$$\hat{B} = V(\beta)^{-1}L(\beta), \tag{7.62}$$

where $V(\beta)$ is a symmetric matrix of dimension $(J - 1)(J - 1)$ with elements

$$V_{jx}(\beta) = \delta_{jx}t_j - \sum_{i=1}^{I} \frac{g_{ij}g_{ix}}{s_i}, \tag{7.63}$$

$j = 1,..., J - 1$, where $\delta_{jx} = 1$ if $j = x$ and zero otherwise, and $L(\beta)$ is a vector of length $(J - 1)$ with elements

$$L_j(\beta) = T_j - \sum_{i=1}^{I} \frac{g_{ij}S_i}{S_i}, \quad j = 1,..., J - 1. \tag{7.64}$$

Subsequently, μ can be estimated by

$$\hat{\mu} = \frac{\sum_{i=1}^{I} S_i - \sum_{i=1}^{I-1} s_i \hat{\alpha}_i - \sum_{j=1}^{J-1} t_j \hat{\beta}_j}{\sum_{i=1}^{I} s_i}. \tag{7.65}$$

Define the sum of squares under the assumptions of H_1 as

$$SS_1 = \sum_{i=1}^{I} \sum_{j=1}^{J} \sum_{k=1}^{K_{ij}} (y_{ijk} - \hat{\mu} - \hat{\alpha}_i - \hat{\beta}_j)^2 N_{ijk}. \tag{7.66}$$

Under $H_{0,A}$ and $H_{0,B}$, the sums of squares are respectively:

$$SS_{0,A} = \sum_{i=1}^{I} \sum_{j=1}^{J} \sum_{k=1}^{K_{ij}} (y_{ijk} - \frac{T_j}{t_j})^2 N_{ijk}, \tag{7.67}$$

$$SS_{0,B} = \sum_{i=1}^{I} \sum_{j=1}^{J} \sum_{k=1}^{K_{ij}} (y_{ijk} - \frac{S_i}{s_i})^2 N_{ijk}. \tag{7.68}$$

The test statistic F for testing $H_{0,A}$ (factor 1 does not affect the transition rate under consideration) is

$$F = \frac{SS_{0,A} - SS_1}{SS_1} \frac{v_2}{v_1}. \tag{7.69}$$

The number of degrees of freedom is equal to

$$v_1 = I - 1 \text{ and } v_2 = (\sum_{i=1}^{I} \sum_{j=1}^{J} K_{ij}) - (I + J - 1). \tag{7.70}$$

To test $H_{0,B}$ (factor 2 has no effect) replace $SS_{0,A}$ by $SS_{0,B}$ in (7.69) and change v_1 to $J - 1$ in (7.70).

The parameter θ^2 is estimated by

$$\hat{\theta}^2 = \frac{SS_1}{v_2}. \tag{7.71}$$

Under the null hypotheses, F has an F-distribution with v_1 and v_2 degrees of freedom. Either null hypothesis is rejected when $F > F(v_1, v_2; \alpha)$. See Table A4 for the critical values.

Multiple comparison method: If either of the tests gives a significant result, multiple comparison can be applied by means of contrasts (as in the previous subsection). For instance, if factor 1 gives a significant result, the differences between levels can be examined by means of

$$\hat{\psi} = \sum_{i=1}^{I-1} w_i \hat{\alpha}_i, \text{ with } \sum_{i=1}^{I} w_i = 0. \tag{7.72}$$

Let w be the vector $(w_1,..., w_{I-1})'$; then it can be shown that the variance of $\hat{\psi}$ is

$$Var(\hat{\psi}) = \theta^2 w' V(\alpha)^{-1} w, \tag{7.73}$$

which is estimated by substituting (7.71) for θ^2. The hypothesis $\psi = 0$ is rejected when

$$\frac{\hat{\psi}^2}{V\hat{a}r(\hat{\psi})(I-1)} \geq F(v_1, v_2; \alpha), \tag{7.74}$$

where $F(v_1, v_2; \alpha)$ is the critical value used for testing $H_{0,A}$. A multiple comparison of the effects of levels of factor 2 can be performed analogously. In this case $V(\alpha)$ is replaced by $V(\beta)$ in eqn (7.73) and I by J in (7.74). Furthermore, $F(v_1, v_2; \alpha)$ in (7.74) is then the critical value for testing $H_{0,B}$.

Residual analysis: When the model of eqn (7.53) fits, the residuals

$$r_{ijk} = \frac{(y_{ijk} - \hat{\mu} - \hat{\alpha}_i - \hat{\beta}_j)\sqrt{N_{ijk}}}{\sqrt{\hat{\theta}^2}} \tag{7.75}$$

have a Student distribution with v_2 (eqn (7.70)) degrees of freedom (Table A3).

Two-way layout with interactions

When there are two influencing factors, a more general model than the previous one can be used:

$$Y_{ijk} = \mu_{ij} + \varepsilon_{ijk}, \tag{7.76}$$

where ε_{ijk} has the same distribution as before. In this model it is assumed that there are interactive effects between the factors. When there are no empty cells, the model of (7.76) can also be represented as

$$Y_{ijk} = \mu + \alpha_i + \beta_j + \gamma_{ij} + \varepsilon_{ijk}, \tag{7.77}$$

and main effects can be estimated and tested under this model as well as the one given in (7.53). Which of the models (the one in (7.53) or that of

(7.77)) is used depends on the assumptions that can be made about interactive effects. When there are no prior clues it is best to test for main effects under the more general model of eqn (7.77). The presence of interactions can be determined by testing the simpler model of (7.53) against the model given in (7.76). If it is known that the γ_{ij} are zero (either from the analysis of previous experiments or from the interpretation of the γ_{ij}) main effects can be estimated and tested under the model of eqn (7.53). Alternatively, the decision about which procedures are used for testing main effects may be based on the outcome of the test of (7.53) against (7.76). If the null hypothesis is rejected, the more general model of (7.76) can be used. If not, the simpler model may be used. The outcome of such an analysis, however, strongly depends on the power of the test of (7.53) against (7.76) (Scheffé, 1959, section 4.1). Therefore, it is advisable to test for main effects under (7.76) unless there are strong indications that the interactions are zero. However, sometimes one is forced to assume that there are no interactive effects, since under (7.76) main effects cannot be estimated and tested when there are empty cells.

We will first consider the estimation of μ_{ij} and the test of model (7.53) against (7.76). Note that this estimation and test procedure can also be performed when there are empty cells. We denote the set of non-empty cells by D.

Data: As in the two-way layout without interactions.

Assumptions:
A1: $Y_{ijk} = \mu_{ij} + \varepsilon_{ijk}$, $i = 1,..., I$, $j = 1,..., J$, $k = 1,..., K_{ij}$; ε_{ijk} is normally distributed with zero expectation and variance θ^2/N_{ijk}.
A2: Y_{ijk} and $Y_{i'j'k'}$ are independent if $i \neq i'$ and/or $j \neq j'$ and/or $k \neq k'$.
A3: Under H_0, $\mu_{ij} = \mu + \alpha_i + \beta_j$ for $i = 1,..., I$ and $j = 1,..., J$ and under H_1 there is at least one inequality.

Procedure: For non-empty cells the parameters μ_{ij} are estimated by

$$\hat{\mu}_{ij} = \frac{\sum\limits_{k=1}^{K_{ij}} Y_{ijk} N_{ijk}}{g_{ij}}, \tag{7.78}$$

where g_{ij} is as defined in (7.58). The sum of squares under the assumptions of H_1 is equal to

$$SS_2 = \sum\limits_{(i,j) \in D} \sum\limits_{k=1}^{K_{ij}} (Y_{ijk} - \hat{\mu}_{ij})^2 N_{ijk}. \tag{7.79}$$

The test statistic is

$$F = \frac{SS_1 - SS_2}{SS_2} \frac{v_2}{v_1} , \tag{7.80}$$

where SS_1 is defined in eqn (7.66) and the numbers of degrees of freedom are equal to

$$v_1 = p - I - J + 1 \quad \text{and} \quad v_2 = \sum_{(i,j) \in D} K_{ij} - p , \tag{7.81}$$

where p is the number of non-empty cells. Under H_0 the test statistic of eqn (7.81) has an F-distribution with v_1 and v_2 degrees of freedom (see Table A4 for the critical values). The parameter θ^2 is estimated by

$$\hat{\theta}^2 = \frac{SS_2}{v_2} . \tag{7.82}$$

Residual analysis: When the model of eqn (7.76) fits, the residuals

$$r_{ijk} = \frac{(y_{ijk} - \hat{\mu}_{ij})\sqrt{N_{ijk}}}{\sqrt{\hat{\theta}^2}} \tag{7.83}$$

have Student's distribution with v_2 (eqn (7.81)) degrees of freedom (see Table A3).

When there are no empty cells, we can define

$$\mu_{ij} = \mu + \alpha_i + \beta_j + \gamma_{ij} . \tag{7.84}$$

As before, the conditions given in eqns (7.54) and (7.55) are imposed on the parameters α_i and β_j ($i = 1,..., I$, $j = 1,..., J$). In addition we have the side conditions

$$\sum_{i=1}^{I} a_i \gamma_{ij} = \sum_{b=1}^{J} b_j \gamma_{ij} = 0 \tag{7.85}$$

for all j and i, respectively. Under these conditions the parameters are defined as

$$\mu = \sum_{i=1}^{I} \sum_{j=1}^{J} a_i b_j \mu_{ij} \tag{7.86}$$

$$\alpha_i = \sum_{j=1}^{J} b_j \gamma_{ij} - \mu \tag{7.87}$$

$$\beta_i = \sum_{i=1}^{I} a_i \gamma_{ij} - \mu \tag{7.88}$$

$$\gamma_{ij} = \mu_{ij} - \mu - \alpha_i - \beta_j. \tag{7.89}$$

The parameters are estimated by substituting $\hat{\mu}_{ij}$ (cf. eqn (7.78)) in eqns (7.86)–(7.89). We will now describe a procedure for testing main effects of the factors. Note, however, that if the test based on (7.80) gives a significant result, performing a test for main effects of either of the factors is a bit strange, since a significant interactive effect implies that both factors affect the behaviour.

Data: See above.

Assumptions:

A1: $Y_{ijk} = \mu + \alpha_i + \beta_j + \gamma_{ij} + \varepsilon_{ijk}$, $i = 1,..., I$, $j = 1,..., J$, $k = 1,..., K_{ij}$; ε_{ijk} is normally distributed with expectation zero and variance θ^2/N_{ijk}. (This assumption is equivalent to A1 above).

A2:   See above.

A3:   Under $H_{0,A}$, $\alpha_1 = ... = \alpha_I = 0$ (factor 1 has no effect).

A4:   Under $H_{0,B}$, $\beta_1 = ... = \beta_J = 0$ (factor 2 has no effect).

A5:   Under H_1 there is at least one inequality in assumptions A3 or A4

Procedure: The hypotheses $H_{0,A}$ and $H_{0,B}$ can be tested by procedures similar to the one given by Scheffé (1959, section 4.4). Define

$$W_{A,i} = \left(\sum_{j=1}^{J} \frac{b_j^2}{g_{ij}} \right)^{-1} \quad \text{and} \quad W_{B,j} = \left(\sum_{i=1}^{I} \frac{a_i^2}{g_{ij}} \right)^{-1} \tag{7.90}$$

$$\hat{A}_i = \sum_{j=1}^{J} b_j \hat{\mu}_{ij} \quad \text{and} \quad \hat{B}_j = \sum_{i=1}^{I} a_i \hat{\mu}_{ij} \tag{7.91}$$

$$\bar{A} = \frac{\sum_{i=1}^{I} W_{A,i} \hat{A}_i}{\sum_{i=1}^{I} W_{A,i}} \quad \text{and} \quad \bar{B} = \frac{\sum_{j=1}^{J} W_{B,j} \hat{B}_j}{\sum_{j=1}^{J} W_{B,j}}. \tag{7.92}$$

Furthermore, define the sums of squares

$$SS_A = \sum_{i=1}^{I} W_{A,i} (\hat{A}_i - \bar{A})^2 \quad \text{and} \quad SS_B = \sum_{j=1}^{J} W_{B,j} (\hat{B}_j - \bar{B})^2. \tag{7.93}$$

A test of $H_{0,A}$ against H_1 can be based on

$$F = \frac{SS_A}{SS_2} \frac{v_2}{v_1}, \qquad (7.94)$$

with

$$v_1 = (I - 1) \quad \text{and} \quad v_2 = \sum_{i=1}^{I} \sum_{j=1}^{J} K_{ij} - IJ. \qquad (7.95)$$

Under $H_{0,A}$, F has an F-distribution with v_1 and v_2 degrees of freedom. $H_{0,B}$ can be tested in the same way, by replacing SS_A by SS_B in eqn (7.94). Furthermore, v_1 in (7.95) is then equal to $J - 1$.

Multiple comparison: If $H_{0,A}$ is rejected, the contrast

$$\hat{\psi} = \sum_{i=1}^{I} w_i \hat{\alpha}_i, \quad \text{where} \quad \sum_{i=1}^{I} w_i = 0, \qquad (7.96)$$

can be calculated. The variance of $\hat{\psi}$ is

$$Var(\hat{\psi}) = \theta^2 \sum_{i=1}^{I} \sum_{j=1}^{J} \frac{w_i^2 b_j^2}{g_{ij}}, \qquad (7.97)$$

with g_{ij} as defined in (7.58), and can be estimated by substituting the estimate of θ^2 (eqn (7.82)). The hypothesis $\psi = 0$ is rejected when

$$\frac{\hat{\psi}^2}{V\hat{a}r(\hat{\psi})^2(I - 1)} \geq F(v_1, v_2, \alpha), \qquad (7.98)$$

where $F(v_1, v_2; \alpha)$ is the critical value used for testing $H_{0,A}$. A multiple comparison of effects of levels of factor 2 can be performed analogously. In that case, b_j is replaced by a_i in eqn (7.97) and $I - 1$ by $J - 1$ in eqn (7.98). Furthermore, in (7.98) the critical value for testing $H_{0,B}$ should be used.

Linear regression analysis

The ANOVA methods given in the previous subsections can be used to examine the effects of factors whose values can be grouped into a limited number of classes. Sometimes, however, influencing factors may have a continuous range of values. Examples are the concentration of a chemical substance, or the temperature. In such cases it is preferable to apply regression methods.

Let Y_i be the log-transformed estimator of the transition rate from A to B and let x_i be the value of the covariate studied in record i. The simplest type of regression model is

$$Y_i = a + bx_i + \varepsilon_i, \qquad (7.99)$$

where the ε_i are mutually independent and approximately normally distributed with variance θ^2/N_i. The question of whether or not the variable x affects the transition rate corresponds to testing the hypothesis $b = 0$ against $b \neq 0$.

Data: n log-transformed estimates of transition rates $y_1,..., y_n$ with corresponding numbers of observed transitions $N_1,..., N_n$ and values of the covariable $x_1,..., x_n$.

Assumptions:

A1: $Y_i = a + bx_i + \varepsilon_i$, $i = 1,..., n$; the ε_i are normally distributed with zero expectation and variance θ^2/N_i.

A2: Y_i and Y_j are independent; hence the covariance between ε_i and ε_j is zero if $i \neq j$.

A3: Under H_0 there is no effect of x, and thus $b = 0$.

Procedure: The weighted least-squares estimates of the unknown parameters a and b are found by minimizing

$$\sum_{i=1}^{n} (y_i - a - bx_i)^2 N_i, \tag{7.100}$$

for all possible values of a and b. This results in

$$\hat{b} = \frac{(\sum N_i)(\sum N_i x_i y_i) - (\sum N_i x_i)(\sum N_i y_i)}{(\sum N_i)(\sum N_i x_i^2) - (\sum N_i x_i)^2}, \tag{7.101}$$

and

$$\hat{a} = \frac{\sum N_i y_i - \hat{b} \sum N_i x_i}{\sum N_i}, \tag{7.102}$$

where all summations run from $i = 1$ to n. Under the null hypothesis the parameter a is estimated by

$$a^* = \frac{\sum N_i y_i}{\sum N_i}. \tag{7.103}$$

The test statistic F is defined by

$$F = \frac{SS_0 - SS_1}{SS_1} \frac{v_2}{v_1}, \tag{7.104}$$

where the sums of squares SS_0 and SS_1 are equal to, respectively,

$$\sum (y_i - a^*)^2 N_i \quad \text{and} \quad \sum (y_i - \hat{a} - \hat{b}x_i)^2 N_i \tag{7.105}$$

and the numbers of degrees of freedom $v_1 = 1$ and $v_2 = n - 2$.

Under the null hypothesis, F has an F-distribution with v_1 and v_2 degrees of freedom. H_0 is rejected when $F > F(v_1, v_2; \alpha)$. See Table A4 for the critical values.

A one-sided version of this test may also be applied when the objective is to test whether the transition rate increases with x (i.e. $b > 0$) or decreases with x ($b < 0$). Such a test is based on Student's distribution with $n - 2$ degrees of freedom: in the right-sided case, reject H_0 if $t = \sqrt{F}$ exceeds $t_{n-2}(\alpha)$. See Table A3 for the critical values.

A $(1 - \alpha)$ confidence interval for the slope parameter b is given by

$$\hat{b} - t_{n-2}(\tfrac{1}{2}\alpha)s_{\hat{b}} < b < \hat{b} + t_{n-2}(\tfrac{1}{2}\alpha)s_{\hat{b}}. \tag{7.106}$$

The standard deviation of $\hat{b}$ is estimated by

$$s_{\hat{b}} = \sqrt{\frac{\hat{\theta}^2}{\sum N_i x_i^2 - \frac{(\sum N_i x_i)^2}{\sum N_i}}}, \quad \text{with } \hat{\theta}^2 = \frac{SS_1}{n - 2}. \tag{7.107}$$

Residual analysis: When the model of (7.99) fits, the residuals

$$r_i = \frac{(y_i - \hat{a} - \hat{b}x_i)\sqrt{N_i}}{\sqrt{\hat{\theta}^2}} \tag{7.108}$$

have Student's distribution with $v_2 = n - 2$ degrees of freedom (Table A3).

Further remarks

In this subsection, we treated only the one-way and two-way layout ANOVA models and the simplest linear regression model. Generalizations of other such models, e.g. higher-way layout ANOVA or ANCOVA (analysis of covariance), are also possible. Furthermore, when the covariances between the different log-transformed transition rates (subsection 7.2.1, eqn (7.8)) are considered too large (this can happen if certain acts occur too infrequently), generalizations of multivariate procedures (e.g. Mardia *et al.*, 1979) might be used. It may be difficult, however, to derive these.

As mentioned, the incorporation of the factor θ^2 in the ANOVA model can, to a certain degree, account for variation between the replicas. When observations are matched, another way to deal with differences between replicas is to incorporate an extra factor in the ANOVA model. This may be a fixed effect or, as has been mentioned already, a random effect. Another, more heuristic, approach is to use a jack-knife method, i.e. the analysis is performed several times while one or more arbitrary observations are left out. This approach is especially useful if there are too few observations to incorporate an extra factor in the ANOVA and one wants to check whether or not the results are due to one or more 'extreme' observations.

Note that the number of 'observations' per cell (for instance n_i in the one-way layout) is random since it cannot be assured that the transitions between acts that are studied occur in all behavioural records. The ANOVA is carried out conditionally on the realized values of these numbers. This can be done safely as long as they are independent of the influencing factor(s) studied. This should be checked beforehand.

The proposed procedure is to carry out separate analyses on each transition rate studied. It is not always advisable to analyse all possible transition rates between acts in this way, either because there are too few observations or because we want a coarser overview of the effects. In those cases we can analyse transition tendencies to groups of acts that intuitively belong together and/or are interesting in the light of the research topic. In addition, graphical representations of main effects often provide a handy way to gain insight in a behavioural process and the effects of various factors on the process. (Examples can be found in sections 8.2 and 8.3.) When, in addition, a simultaneous test of the effects of, for example, experimental treatment is needed, one of the methods given in section 6.2 can be used to derive a combination test.

If the log-transition rate of one or more transitions is a non-linear function of a covariable, there are several ways to analyse the data. The simplest way is to assume a polynomial of degree 2 or higher rather than a linear relation. The procedures for estimation and testing are essentially the same as in the linear case. Useful results and examples of such procedures when the variances of the observations are unequal can be found in Wetherill (1986, section 9.7) and in Sen and Srivastava (1990). An important advantage of this approach is that it can be generalized easily when there is more than one covariable.

If a polynomial regression is not adequate, likelihood methods as described in subsection 7.3.2 can be applied, provided that the model is specified up to a number of unknown parameters. If it is not desired to make strong assumptions concerning the distributions of the log-transformed estimates, so-called 'generalized linear models' can be applied. See McCullagh and Nelder (1983) for a monograph on this topic.

7.4   PARAMETER ESTIMATION IN SEMI-MARKOV CHAINS

Whereas all transition rates are constant in CTMCs, transition rates in semi-Markov chains may depend on the residence times in the states. In behavioural processes this means that the tendency to stop behavioural category A and start B may change during a bout, and is therefore a function of the bout length x. A CTMC can thus be considered as a special case of a semi-Markov chain in which $\alpha_{AB}(x) = \alpha_{AB}$ for all A and B.

The possible dependency on time has several consequences. In CTMCs each transition rate could be represented by the product of the termination rate and the transition probability (eqn (7.9)). In semi-Markov chains, this is not always the case, since the forms of, for example, $\alpha_{AB}(x)$ and $\alpha_{AC}(x)$ (B $\neq$ C) as functions of x may differ. This is most likely to happen in interactive behaviour. For instance, when the model is applied to a dyadic interaction A represents a combination of acts, say *Groom* by the first animal and *Explore* by the other animal. In this case it may happen that *Groom* has a minimum duration, m, whereas *Explore* does not. Suppose that after the minimum duration, *Groom* is terminated with a constant probability per time unit and *Explore* has a constant termination rate right from the start. Then

$$\alpha_{(Groom,Explore)(X,Explore)}(x) = 0 \quad \text{for} \quad x \leq m$$
$$= \text{constant} \quad \text{for} \quad x > m, \tag{7.109}$$

whereas

$$\alpha_{(Groom,Explore)(Groom,Y)}(x) = \text{constant}, \tag{7.110}$$

for all possible following acts X of animal 1 and Y of animal 2. For such cases, the representation of the transition rates as products of termination rates and transition probabilities

$$\alpha_{AB}(x) = \lambda_A(x)p_{AB}, \tag{7.111}$$

is not possible. This implies that it is not always possible also to define transition rates towards groups of acts, as in (7.17). In those cases where the representation (7.111) can be used, however, termination rates, transition probabilities and transition rates towards groups of acts can be defined and interpreted analogously to the case of CTMCs (see section 7.2).

As mentioned before (section 4.4) transition rates from a certain act to different other acts can be considered as 'competing risks' that do not necessarily operate independently. On the basis of the observed bout lengths and transitions between acts it is not possible, however, to distinguish models with dependent and independent transition rates (David and

Moeschberger, 1978). Since, moreover, models with independent competing risks are the most parsimonious, we will assume that the rates are independent (as was also implicitly done for CTMCs). An alternative attitude is to consider the transition rates as descriptive statistics of behaviour rather than interpret them as behavioural tendencies.

The forms of the $\alpha_{AB}(x)$ as functions of x are not always known beforehand. For those transition rates that can be expressed explicitly as functions of x, maximum likelihood estimation is possible (subsection 7.4.1). If this is not so, estimated survivor functions (subsection 4.7.1) can give an impression of the forms of $\alpha_{AB}(x)$. In subsection 7.4.2 we discuss this topic further.

7.4.1 Maximum likelihood estimation

We consider one record of behaviour. (For convenience, we will use a slightly different notation than in previous chapters.) The observation consists of the sequence of acts $A_1,..., A_n$ and the corresponding bout lengths $x_1,..., x_n$. The last bout length may be incompletely observed. When the process is considered as a semi-Markov chain, the log-likelihood of the complete record (which, as before, will be denoted by $\Re$) is

$$L(\Re;\{\alpha(x)\}) = \sum_{i=1}^{n-1} \log\alpha_{A_iA_{i+1}}(x_i) - \sum_{i=1}^{n} \int_0^{x_i}\lambda_{A_i}(s)\,ds, \qquad (7.112)$$

where

$$\lambda_{A_i}(x) = \sum_{B \neq A_i} \alpha_{A_iB}(x). \qquad (7.113)$$

Note that, unless the transition rates can be represented as in (7.111), $\lambda_{Ai}(x)$ cannot be interpreted as the tendency to terminate A_i. When all transition rates are constant, (7.112) reduces to (7.1). In (7.112) the last, incompletely observed, bout is considered as a censored observation. Alternatively, this bout may be excluded and the second summation taken over $i = 1$ to $n - 1$. The MLEs have the asymptotic properties listed in Box 7.1 (Fleming and Harrington, 1991).

Transition rates

Note that the transition rates are functions of x involving a parameter (vector) θ. The log-likelihood of eqn (7.112) is maximized over θ. For instance, the termination rate of the Weibull distribution might be used as a model for a monotonically increasing or decreasing transition rate:

$$\alpha_{AB}(x) = \mu\rho(\mu x)^{\rho-1} \qquad (7.114)$$

(see eqn (4.4)), which contains two parameters, μ and ρ. Such functions are substituted in (7.112), which is subsequently maximized to derive the MLEs of the parameters involved.

For groups of transition rates that involve different sets of parameters, the likelihood function factorizes. This implies that the sets of parameters involved can be estimated and analysed separately. For instance, suppose that there is a group of acts *G* that can be considered as proper Markov states; then the transition rates from acts belonging to *G* to other acts can be estimated by (7.2). Furthermore, the methods of analysis given insections 7.2 and 7.3 can be applied to this subset of acts. As another example, suppose that only the set of transition rates from acts belonging to *G* can be fully specified as functions of *x*. For the rest of the acts, the form of the transition rates is unknown, but it is known that they do not depend on parameters involved in the transition rates from acts in *G*. In that case, the estimates of parameters of transition rates from acts in *G* to all other acts can be calculated by maximizing

$$L^*(\mathfrak{R};\{\alpha(x)\}) = \sum_{i=1}^{n-1} \sum_{\wedge A_i \in G} \log \alpha_{A_i A_{i+1}}(x_i) - \sum_{i=1}^{n} \sum_{\wedge A_i \in G} \int_0^{x_i} \lambda_{A_i}(s) \mathrm{d}s. \quad (7.115)$$

The methods given in subsection 7.4.2 can then be used to get an impression of the forms of transition rates from acts outside *G*.

Usually, transition rates from different behavioural categories to other acts will not share common parameters. However, this is not necessarily always the case. To return to the previous example of dyadic interactions, consider the group of all behavioural categories involving *Groom* by individual 1. All transitions from *Groom* + X to Y + X (where X denotes an arbitrary act of individual 2 and Y any act of individual 1 other than grooming) involve the same minimum duration, since this is inherent to *Groom*. The corresponding transition rates thus have a common parameter. Note that for transition rates from *Groom* + X to *Groom* + Z, this is not so, since these transitions are not subject to the constraint of the minimum duration of *Groom*.

Transition rates from a certain behavioural category towards other acts can also be estimated separately by likelihood maximization, as long as they do not have common parameters. Suppose that $\alpha_{AB}(x)$ does not share any parameters with other transition rates. Let x_j $(j = 1,..., m)$ be the observed durations of act A and let δ_j be equal to one when the *j*th bout of A was followed by B and zero otherwise. Then $\alpha_{AB}(x)$ is estimated by maximizing

$$L(x_1,...,x_m;\delta_1,...,\delta_m;\alpha_{AB}(x)) = \sum_{j=1}^{m} \{\delta_j \log \alpha_{AB}(x_j) - \int_0^{x_j} \alpha_{AB}(s) \mathrm{d}s\}.$$

$$(7.116)$$

To return once again to our last example, this implies that transition rates from *Groom* + X corresponding to initiatives of individual 1 can be estimated separately from those corresponding to initiatives of individual 2.

Termination rates and transition probabilities

If the representation given in eqn (7.111) is possible, termination rates and transition probabilities can be defined and estimated separately. The estimators of transition probabilities (and their asymptotic properties) are equal to those for CTMCs (see subsection 7.2.3). From (7.111) and (7.112) it follows that termination rates can be estimated by maximizing

$$L(\Re;\{\lambda(t)\}) = \sum_{i=1}^{n-1} \log\lambda_{A_i}(x_i) - \sum_{i=1}^{n} \int_0^{x_i} \lambda_{A_i}(s)ds. \qquad (7.117)$$

Similar remarks as made for transition rates apply here.

7.4.2 Non-parametric estimation of cumulative transition rates

As mentioned, it may not always be possible to specify fully the form of all transition rates beforehand. In this case, an impression of the way $\alpha_{AB}(x)$ changes as a function of x can be obtained by means of log-survivor plots (subsection 4.7.1). As before, let $x_1,..., x_m$ be the observed bout lengths of A. To examine the transition rate to B, all bouts followed by acts other than B are considered as randomly censored observations (section 4.4), and the Kaplan-Meier estimator of the survivor function of the set of bout lengths is calculated (subsection 4.7.1). We will denote this function by $\bar{F}_{AB,n}(x)$. Since

$$-\log\bar{F}_{AB}(x) = \int_0^x \alpha_{AB}(s)ds, \qquad (7.118)$$

minus the logarithm of the estimated survivor function is an estimator of the cumulative transition rate from A to B. Plotting this function against x gives an impression of the form of $\alpha_{AB}(x)$ (see e.g. Fig. 4.9).

The estimated log-survivor function can also be used to test the fit of a specified model for $\alpha_{AB}(x)$, by making use of the relationship in eqn (7.118). An example is given in section 8.3.

For different acts A, the estimators $\bar{F}_{AB,n}(x)$ are independent. Asymptotic properties of the estimators, including the asymptotic covariance functions between $\bar{F}_{AB,n}(x)$ and $\bar{F}_{AC,n}(x)$ (B $\neq$ C), are given in Fleming and Harrington (1991). Note that the functions $\bar{F}_{AB}(x)$ are equal to the so-called 'cause-specific hazard functions' in survival analysis (e.g. Kalbfleisch and Prentice, 1980, Chapter 7). The relation with survival analysis is further explained in the next section.

7.5   EXAMINING THE EFFECTS OF COVARIATES ON SEMI-MARKOV CHAINS

As mentioned in subsection 7.4.1, the effects of covariates on subsets of behavioural categories that can be considered as proper Markov states can be examined with the methods given in section 7.3. We will consider here the remaining set of behavioural categories, with transition rates that are not constant during bouts. Note that behavioural categories with (nearly) fixed minimum durations and otherwise constant termination rates (subsection 1.3.1) can be analysed in the same way as proper Markov states, after the minimum duration has been subtracted from the bout lengths (e.g. section 8.2). As with CTMCs, covariates may be, for example, different periods (in the case of time inhomogeneity), experimental conditions, or acts of other individuals.

7.5.1 Exploration

For transition rates that can be fully specified as functions of x (e.g. eqn (7.114)), the parameter estimates can be studied for different values of the influencing covariates. Combining transition rates through (weighted) summation, as with CTMCs (subsection 7.3.1), is in general not possible. When the representation given in (7.111) is possible, however, the parameters of termination rates and/or transition rates towards groups of acts may be studied.

For transition rates where the dependency on time cannot be specified beforehand, plots of the estimated cumulative transition rates (subsection 7.4.2) against x can be made for different values of covariates. When the effects of a covariate are multiplicative, plots of $\log(-\log \bar{F}_{AB,n}(x))$ for different values of the covariate are approximately parallel (e.g. Kalbfleisch and Prentice, 1980, section 4.5).

7.5.2 Formal estimation methods and tests based on survival analysis

When transition rates can be fully specified (e.g. eqn (7.114)), the MLEs of the parameters involved can be estimated for different records. Subsequently, the effects of covariates on these parameters can be analysed with ANOVA or regression methods, making use of the asymptotic normality of the MLEs (Box 7.1). An example of such an analysis can be found in section 8.3. However, when $\alpha_{AB}(x)$ contains more than one parameter, the asymptotic covariances between the MLEs may be too large to use univariate methods. In that case, multivariate analogues will have to be used, which may be rather tedious.

An alternative way to analyse the effects of covariates on transition rates is to specify regression models for the effects and estimate the regression parameters directly through maximum likelihood estimation. To do this, survival analysis methods can be applied. These are a large set of methods originally developed for analysing so-called 'survival' or 'failure' times: the times from start of observation until the occurrence of a certain event, such as death or the breakdown of a system by a certain cause. Observations consist of the times until occurrence of the event, or until occurrence of a censor (section 4.4). Different survival and censor times are assumed to be independent. There also may be different types of events studied, e.g. breakdown by different causes. In this case, the different causes are 'competing risks' and an occurrence of one type of event acts as a censor for occurrences of other types. There are many textbooks on survival analysis (e.g. Kalbfleisch and Prentice, 1980, Fleming and Harrington, 1991) and the methods are implemented in several computer packages (e.g. SAS, BMDP, SPSS). We will give here both an outline of how behavioural processes that can be described by semi-Markov chains can be analysed by means of these methods and some examples of specific methods that might be of interest in an ethological context. For an extensive treatment we refer to the literature cited above.

When a behavioural process is adequately described by a semi-Markov chain, the starts of acts are 'renewal points', i.e. from those points onwards, past events no longer affect the probabilities of future events. This means that each bout can be considered as an independent observation. As a consequence, the start of each bout can be considered as the start of an observation in survival analysis. The bout lengths correspond to the survival times and the succeeding acts are competing risks. The transition rates then correspond to 'cause-specific hazard rates'.

Transition rates from one behavioural category to others

Suppose that the transition rates from behavioural category A to other categories do not share any parameters with transition rates from other acts and can thus be analysed separately (subsection 7.4.1). Note that, besides parameters describing the dependency on x, parameters describing the dependency on covariates also have to be different. Let there be p covariates, represented by a vector $z = (z_1,..., z_p)'$. The data considered are the observed bout lengths of A, $x_1,..., x_m$, the observed following acts, $B_1,..., B_m$, and the corresponding values of the covariates $z_1,..., z_m$. If certain bouts were the last observed ones, the following act may be unknown. In this case the corresponding bout lengths are considered as censored observations.

Let D be the group of different acts in $B_1,..., B_m$. For each act $B \in D$ the transition rate from A to B is a function of the bout length x, the values of the covariates, z, and a vector of unknown parameters θ. The log-likelihood of the observations is

$$L(x_1,...,x_m; B_1,...,B_m; z_1,...,z_m; \theta)$$

$$= \sum_{B \in D} \sum_{i=1}^{m} [\{\delta_{B_i,B} \log \alpha_{AB}(x_i; z_i; \theta)\} - \int_0^{x_i} \alpha_{AB}(s; z_i; \theta) ds], \quad (7.119)$$

where $\delta_{B_i,B}$ is zero if $B_i \neq B$ and one otherwise. If $\alpha_{AB}(x; z; \theta)$ is completely specified, θ can be estimated by maximizing (7.119). This will usually have to be done by a numerical method such as Newton–Raphson iteration (e.g. Box 8.5). A frequently considered model for a monotonically changing rate is the Weibull (see also (4.4) and (7.114)) regression model:

$$\alpha_{AB}(x; z; \theta) = \rho(\mu x)^{\rho-1} \exp[z_1\beta_1 + ... + z_p\beta_p]. \quad (7.120)$$

In (7.120) it is assumed that the covariates have a multiplicative effect on the transition rates, which is more plausible than an additive effect (subsection 7.3.4). The parameters ρ, μ, and/or $\beta_1,..., \beta_p$ may be assumed to be different for different following acts B.

When there are stochastic time-lags, one of the models of subsection 1.3.1 may be used, for example the model given in eqn (1.16). In this model it is assumed that the transition rates have a point of inflection, which is a measure of the duration of the time-lag, and become constant for large x. It is usually reasonable to assume that the time-lag, and therefore the point of inflection, does not depend on the following act, whereas the asymptotic constant values of the transition rate do. Both values may be affected by the covariates (see e.g. section 8.3), which leads to the following model:

$$\alpha_{AB}(x; z; \theta) = \frac{\lambda_{AB}(z; \theta)(1 - \exp[-x/\beta(z; \theta)])}{1 + \exp[1 - x/\beta(z; \theta)]}. \quad (7.121)$$

If it is assumed that the effects of covariates on the point of inflection and the asymptotic transition rates are different, $\lambda_{AB}(z; \theta)$ and $\beta(z; \theta)$ may involve different parameters θ_i. If only some of the transitions are subject to a time-lag (cf. the example given in subsection 7.4.1), the model in (7.121) may be used for the corresponding subset of B, whereas for other acts B it can be assumed that the transition rates are constant in time.

It may not always be possible to define the relationship between transition rates and x explicitly. In that case, a very flexible model is the so-called 'proportional hazards model' (Cox, 1975). In this model it is assumed that there is an unspecified 'base-line' hazard rate, which depends on x. The covariates have multiplicative effects on the base-line hazard. Within the context of this book it can be assumed, for instance, that

$$\alpha_{AB}(x;z;\theta) = \lambda_{AB}(x)\exp[z_1\beta_1 + ... + z_p\beta_p]. \tag{7.122}$$

In (7.122), $\lambda_{AB}(x)$ corresponds to the base-line hazard. Kalbfleisch and Prentice (1980, Chapter 7) give methods for estimating $\beta_1,..., \beta_m$, irrespective of the value of $\lambda_{AB}(x)$, by maximization of the so-called 'partial likelihood'. In addition, the cumulative base-line hazard

$$\int_0^x \lambda_{AB}(s)ds \tag{7.123}$$

can be estimated. Plots of these functions against time can be used to get an impression of the form of the transition rates. Methods for checking the correctness of the assumption of multiplicative effects of the covariates can also be found in Kalbfleisch and Prentice (1980, Chapter 4). It may also be assumed that the transition rates for different following acts are proportional (as in eqn (7.111)), by including the following act as an extra, nominal covariate. In that case, the base-line hazard only depends on A. This is further explained in Box 7.8.

Box 7.8   Considering following acts as a nominal variable

Let x be the bout length of act A and let there be p covariates $z_1,..., z_p$. Furthermore, suppose that A can be followed by either of the acts B, C or D. We assume that the transition rates from A to following acts are proportional (as in eqn (7.111)). Let F denote the following act (i.e. F = B, C or D) and define two dummy variables, say y_1 and y_2, with the following values:

F:	B	C	D
y_1	1	0	0
y_2	0	1	0

To include the following act as a nominal variable the model of eqn (7.122) is adjusted as follows:

$$\alpha_{AF}(x;z;\theta) = \lambda_A(x)\exp[z_1\beta_1 + \ldots + z_p\beta_p + y_1\gamma_1 + y_2\gamma_2]. \tag{1}$$

Substitution of the values of y_1 and y_2 gives

$$\alpha_{AB}(x;z;\theta) = \lambda_A(x)\exp[z_1\beta_1 + \ldots + z_p\beta_p + \gamma_1]$$

$$\alpha_{AC}(x;z;\theta) = \lambda_A(x)\exp[z_1\beta_1 + \ldots + z_p\beta_p + \gamma_2] \tag{2}$$

$$\alpha_{AD}(x;z;\theta) = \lambda_A(x)\exp[z_1\beta_1 + \ldots + z_p\beta_p],$$

so γ_1 and γ_2 reflect the differences (or, more precisely, ratios) in values of transition rates towards respectively B and C, and that of the transition rate towards D. The parameters γ_1 and γ_2 can be estimated in the same way as the β_i ($i = 1,\ldots, p$) (Kalbfleisch and Prentice, 1980).

Simultaneous analysis of transition rates from several acts

When transition rates from certain acts share parameters, it will be necessary to analyse them simultaneously. Let G be the group of acts whose transition rates share parameters. The data are the observed bout lengths of all acts belonging to G, $x_1,\ldots, x_m$, the corresponding acts $A_1,\ldots, A_m$, the observed following acts, $B_1,\ldots, B_m$, and the corresponding values of the vectors of covariates $z_1,\ldots, z_m$. If certain bouts were the last observed ones and the following act is unknown, the corresponding bout lengths are considered as censored observations. As before, let D be the group of different following acts. The log-likelihood of this set of observations is

$$L(x_1,\ldots,x_m;A_1,\ldots,A_m;B_1,\ldots,B_m;z_1,\ldots,z_m;\theta)$$

$$= \sum_{A\in G}\sum_{B\in D}\sum_{i=1}^{m}[\{\delta_{A_i,A}\delta_{B_i,B}\log\alpha_{AB}(x_i;z_i;\theta)\} - \int_0^{x_i}\alpha_{AB}(s;z_i;\theta)ds],$$

$$\tag{7.124}$$

where, as before, δ_{XY} is zero if $X \neq Y$ and one otherwise. As in the preceding subsection, models for the transition rates can be defined and substituted in (7.124) to estimate the parameters. Alternatively, a semi-parametric model like the proportional hazards model may be used (e.g. the model given in (7.122)).

8 EXAMPLES OF ANALYSES BASED ON CONTINUOUS TIME MARKOV CHAIN MODELLING

8.1 INTRODUCTORY REMARKS

In this chapter we give some examples of practical applications of analyses based on CTMCs and their generalizations. The examples are partially based on our collaborative projects with experimental research groups (i.e. the Ethopharmacology Group, Leiden University, and the Primate Research Centre, TNO, Rijswijk, The Netherlands), the rest are derived from the literature.

We do not use all the formal methods given in previous chapters in all instances. One reason for this is that in a number of cases, at the time the research was carried out the most appropriate methods were not available. For instance, the non-parametric procedure for determining the number of change points in a sequence of bouts (subsection 3.3.1) has only recently been developed. Another reason is that we want to indicate here the possibilities of the type of analysis outlined in this book in an ethologically relevant context. In this chapter, the emphasis therefore lies on a general approach rather than on detailed step-by-step analyses.

In subsections 8.2.1–8.2.3 we describe the process of modelling dyadic interactions between male rats, starting with the original ethogram and finishing with a set of behavioural categories that can be described accurately with a CTMC model (where some categories have minimum durations). In subsection 8.2.4 we describe how the model was used to study the effects of brain stimulation and of an opponent's behaviour simultaneously (see also Haccou *et al.*, 1988a).

In section 8.3 we consider the mother-infant interaction of rhesus monkeys. Under normal conditions, this interaction can be modelled accurately by a three-state Markov chain (see Dienske and Metz, 1977, Dienske *et al.*, 1980) at a certain level of resolution. Conflicts between mother and infant can be modelled by means of a 'point event' (subsection 1.3.1). In subsection 8.3.2 we describe how this model can be used as a basis to study the behaviour at other levels of resolution. Haccou *et al.* (1988*c*) studied the effects of amphetamine on the infant's behaviour. To this end, a semi-Markov model incorporating time-lags was formulated, which made it possible to analyse the effects on minimum durations as well as transition tendencies. The results of this analysis are given in subsection 8.3.3.

In section 8.4 we consider the application of a function of a Markov chain for modelling the mating behaviour of barbs (Putters *et al.*, 1984).

8.2 DYADIC INTERACTION OF MALE RATS

The results described in this section come from a collaborative project with Dr M.R. Kruk and W. Meelis (Ethopharmacology Group, Leiden), Dr E.T. van Bavel, and K.M. Wouterse. The purpose of the project was to investigate the relation between neural mechanisms and the interactive behaviour of male rats. From previous studies it was known that above a certain threshold electrical stimulation at a specific site in the hypothalamus induces violent attack (Kruk *et al.*, 1979, 1983, 1984). In the set-up used in such experiments, both rats are outside their home cage. Fighting does not normally occur in such cases. These observations led to the question of whether the neural mechanisms involved in attack behaviour could also be involved in the regulation of interactive behaviour in the absence of fighting. To investigate this, interactive behaviour at subthreshold stimulation intensities was compared with normal interactive behaviour. A further question concerns how increased stimulation establishes the change from 'normal' to violent behaviour. To this end, behaviour at various subthreshold stimulation intensities was compared. The behavioural records were analysed by means of a CTMC model, which made it possible to analyse the effects of stimulation intensity as well as of opponent behaviour on the behaviour of the stimulated rats.

8.2.1 Experimental set-up and the original ethogram

The test animals were five male rats (strain CPB/WE-zob), weighing about 400 g each (these will be termed the 'Experimentals'). Their 'Opponents' were naive rats (strain Wistar) weighing about 200 g. Electrodes were implanted in the Experimentals. These did not hinder their movements (Kruk *et al.*, 1979; the experiments were performed under the licence of the Pharmacological Laboratory and were reviewed by the Committee for Animal Experimentation of the University of Leiden).

Trials took place outside the home cage of the rats. In each trial, an Experimental was put in the cage first. Five minutes after this, stimulation was started. After a further 5 min, a naive Opponent was placed in the cage. Stimulation continued for 25 min, at intensities of 0, 25, 50 and 75% of the threshold at which violent attacks are induced. Each Experimental received each stimulation condition once, in a balanced randomized order.

The behavioural interactions were recorded on video. Behavioural records were later made from the videotapes (played at one-third normal speed). The original ethogram is described in Box 8.1.

Box 8.1 Original ethogram for dyadic interactions of male rats

Experimental animals:

Walk	
Lie down	Lying, with front paws stretched, with at most slight head motions
Scan	Head is moved slowly to and fro, while neck is kept stretched out
Sniff	Sniffing ground, objects lying on the ground, or cage
Rear	Standing on hind paws sniffing air
Self-groom	
Crawl over	Crawling over Opponent, leaving drops of urine on its fur
Groom Opponent	Grooming of the neck and back region of the Opponent with the incisors, while the front paws are placed on the Opponent
Sniff anogenital	Sniffing Opponent in anogenital region
Sniff side	Sniffing side of Opponent
Sniff oral	Sniffing oral region of Opponent

Opponents:

Walk	
Scan	
Sniff	See description above
Self-groom	
Immobile	Sitting still with at most slight head motion
Sniff electrode	Sniffing electrode implanted in Experimental
Sniff Experimental	

8.2.2 Data exploration

At first, only the behaviour of the Experimentals was analysed, in order to obtain global indications about the presence of time inhomogeneity and the definition of behavioural categories. At a later stage, the behaviour of the Opponents was included in a more detailed analysis. An advantage of this approach is that initially fewer behavioural categories have to be dealt with, since the Opponents' acts are disregarded. It should be borne in mind, however, that deviations from exponentiality in the bout lengths of acts of the Experimentals may be caused by reactions to changes in behaviour of the Opponents (Box 4.14).

The data exploration involved several stages at which diverse aspects of the behaviour were examined. Here we discuss only a selection of the results which gave the most important indications for the formulation of the final model given in subsection 8.2.3:

1. A global inspection of the bar plots showed that the behavioural category *Lie down* occurred only during the last 5 minutes of the trials. This indicates clear time inhomogeneity.

2. Bout lengths of all acts in the ethogram (except *Lie down*, since this occurred only seldomly) were tested for exponentiality with Darling's test against mixtures of exponentials (cf. subsection 4.6.4, eqns (4.42) and (4.43)). The test statistics were calculated for each record separately. Subsequently, per current intensity, the test statistics of different records were combined by means of the 'sum test' given in subsection 6.2.1 (eqn (6.1)). Since at this stage we were interested in finding deviations in the direction of a function of a Markov chain which is not a semi-Markov chain, the right-sided test was used. The results are given in Table 8.1. As can be seen, there was a relatively large proportion of significant deviations from exponentiality in the direction of a mixture in the bout lengths of *Scan*, *Groom Opponent* and *Self-groom*.

3. Whether these deviations could be due to time inhomogeneity (we only tested for the presence of one change point) was investigated by means of cumulative bout length plots (subsection 3.2.2) and the likelihood ratio change point test (subsection 3.3.1). There were not many significant change points in *Groom Opponent* and *Self-groom*. Whereas 15 of the 20 records showed significant change points in the duration of *Scan*, this could not explain the deviations from exponentiality. Table 8.2 shows the results of tests for exponentiality on the bout lengths occurring before and after the estimated change point. As can be seen, there are still relatively many deviations.

4. *Scan* bouts were split up according to the following act. Table 8.3 shows the mean durations of *Scan* bouts and those of *Scan* bouts followed by *Self-groom*. Apparently, *Scan* bouts followed by *Self-groom* are longer than other *Scan bouts*.

5. It also appeared that the chance that *Scan* was followed by *Self-groom* depends on the preceding act. When *Scan* was preceded by *Self-groom* the

Table 8.1 Results of Darling's test for exponentiality (eqns (4.42) and (4.43)) of the bout lengths of acts of the Experimentals given in Box 8.1. The test results of different individuals were combined with the sum test (eqn (6.1))

Act	Current intensity (% of threshold)			
	0	25	50	75
Walk	−4.47	−4.75	−5.63	−4.90
Scan	2.54*	10.35*	3.38*	0.95
Sniff	4.28*	0.34	−2.75	−3.32
Rear	−2.43	−2.39	−4.05	−3.65
Self-groom	3.11*	4.00*	1.17	0.06
Crawl over	−2.93	−1.52	−1.95	−1.06
Groom Opponent	3.59*	5.52*	−0.25	−0.73
Sniff anogenital	−1.01	−0.31	−0.86	−0.86
Sniff side	–	–	−0.90	−2.10
Sniff oral	−0.61	−1.52	−1.89	−0.80

* $p < 0.05$ (right-sided test). Under H_0 the test statistic has a $N(0,1)$ distribution. The critical value is 1.645 (see Table A1).

Table 8.2 Results of Darling's test for exponentiality (eqns (4.42) and (4.43)) of the bout lengths of *Scan* for records in which significant changes occurred

Rat no.	Period[+]	Current intensity (% of threshold)			
		0	25	50	75
1	A	–	−1.03	–	−1.89
	B	–	1.19	–	1.37
2	A	−1.50	−1.04	−1.17	–
	B	4.34*	4.20*	0.89	–
3	A	−2.11	8.12*	0.64	−0.80
	B	0.41	2.74	−1.10	−1.73
4	A	−1.55	−2.19	−2.97	−2.08
	B	−0.78	1.48	−0.22	2.49*
5	A	–	−1.14	−1.39	–
	B	–	2.65*	5.61*	–

+ A: before estimated change point, B: after estimated change point.
* $p < 0.05$, right-sided test. Under H_0 the test statistic has a $N(0,1)$ distribution. The critical value is 1.645 (see Table A1).

Table 8.3 Mean durations of *Scan* bouts split up according to whether or not they were followed by *Self-groom*

Rat no.	Current intensity (% of threshold)*							
	0		25		50		75	
	A	B	A	B	A	B	A	B
1	1.8	2.3	1.3	2.5	1.1	2.8	1.3	2.6
2	0.5	2.6	0.9	2.9	1.7	2.7	1.6	2.4
3	1.1	1.8	0.6	2.9	1.8	2.7	0.5	2.4
4	1.0	2.0	1.5	2.7	1.2	2.1	1.3	2.0
5	1.1	2.0	1.5	3.3	1.0	2.1	1.2	1.9

Mean durations are given in seconds.
* Per intensity: column A = bouts not followed by *Self-groom*, column B = bouts followed by *Self-groom*.

Table 8.4 Relative transition frequencies of *Scan* to *Self-groom*, split up according to whether or not the previous act was *Self-groom*

Rat no.	Current intensity (% of threshold)*							
	0		25		50		75	
	A	B	A	B	A	B	A	B
1	0.44	0.03	0.54	0.08	0.66	0.04	0.76	0.01
2	0.64	0.01	0.75	0.07	0.85	0.02	0.88	0.01
3	0.50	0.04	0.67	0.02	0.82	0.01	0.60	0.02
4	0.36	0.04	0.67	0.02	0.82	0.01	0.60	0.02
5	0.47	0.07	0.65	0.05	0.62	0.13	0.74	0.03

* Per current intensity: column A = relative transition frequency to *Self-groom* from *Scan* bouts preceded by *Self-groom*, column B = relative transition frequency to *Self-groom* from *Scan* bouts not preceded by *Self-groom*.

chance that *Self-groom* followed was much higher than when *Scan* was not preceded by *Self-groom* (see Table 8.4). Apparently, we had a situation as described in section 5.1 (see also Fig. 5.2): *Scan* bouts interrupting

Self-groom had to be considered as belonging to a different behavioural state than *Scan* bouts that do not interrupt *Self-groom*.

6. Similar results to those given in steps 4 and 5 were found for *Scan* bouts that interrupt *Groom Opponent*. Thus, one option was to distinguish different categories of *Scan* bouts: (a) those preceded by *Self-groom*, (b) those preceded by *Groom Opponent*, and (c) those preceded by other acts. A more parsimonious solution, however, is to include *Scan* bouts in other acts. Thus, a sequence of bouts like *Self-groom Scan Self-groom* would be considered as one *Self-groom* bout.

7. We thus ended up with new behavioural categories, which include different forms of *Scan*. Furthermore, in the first instance the different forms of sniffing at the Opponent (i.e. anal, oral or side) were grouped into one category *Sniff Opponent*. Also, *Walk* and *Sniff cage* were grouped into a category called *Explore*. The resulting set of behavioural categories is described in Box 8.2.

Box 8.2   Resulting set of behavioural categories of Experimental animals after data exploration

Descriptions of the original ethogram units, which we refer to here, were given in Box 8.1.

Explore	Consists of *Walk*, *Scan* interruptions and *Sniff*
Lie down	Lying, with front paws stretched, with at most slight head motions
Rear	Standing on hind paws sniffing air
Self-groom	Grooming own fur, with *Scan* interruptions
Crawl over	Crawling over Opponent, leaving drops of urine on its fur
Groom Opponent	Grooming of the neck and back region of the Opponent with the incisors, while the front paws are placed on the Opponent. Includes *Scan* interruptions
Sniff Opponent	*Sniff anogenital*, *Sniff side*, *Sniff oral*, including *Scan* interruptions

---

8. Investigation of cumulative bout length plots and change point estimation in the bout lengths of the resulting categories led to the impression that at the beginning of the trials there was a shift from a relatively intensive interaction with Opponents to a more quiet stage.

9. The records were provisionally divided into two parts (on the basis of the results of step 8), and it was tested whether the bout lengths of the

behavioural categories of Box 8.2 were exponentially distributed. The results for both periods are given in Table 8.5. The test results of different individuals were combined with the sum test (eqn (6.1)). As can be seen, there are still many deviations in the direction of mixtures.

Table 8.5 Results of Darling's test on exponentiality (eqns (4.42) and (4.43)) of the bout lengths of acts of the Experimentals given in Box 8.2. Records split up into two periods A: period 1 (from the start of observation until the first significant change or until 5 minutes) and B: period 2 (starting 5 minutes after the start of observation). The test results of different individuals were combined with the sum test (eqn (6.1))

Act	Period	Current intensity (% of threshold)			
		0	25	50	75
Explore	A	0.64	1.75*	−0.33	3.44*
	B	8.37*	6.48*	4.60*	3.61*
Rear	A	−1.81	0.51	−1.52	−0.39
	B	−1.02	−2.06	−1.52	−2.40
Self-groom	A	2.49*	1.24*	0.007	1.50
	B	11.27*	3.40*	2.85*	1.25
Crawl over	A	−1.03	−1.23	−1.61	−
	B	−2.79	−	−1.29	−0.41
Groom Opponent	A	3.09*	0.04	−1.77	−1.08
	B	4.29*	2.19*	0.79	3.84
Sniff Opponent	A	0.05	0.76	−1.45	−2.03
	B	−1.33	−0.75	−0.61	0.10

* $p < 0.05$ (right-sided test).

8.2.3 Model formulation and goodness-of-fit testing

The next step was to include the behaviour of the Opponents. The behavioural repertoire of the Opponents was much smaller than that of the Experimentals (see Box 8.1). Furthermore, the Experimental rats were never *Immobile*. The reasons for this are presumably because the Opponents were much smaller than the Experimentals. Moreover, the Experimentals were confronted with other rats several times, while for each trial a naive Opponent was used. Since we were primarily interested in the Experimentals' behaviour, and since we wanted to keep the number of different behavioural categories as small as possible, we distinguished initially only three behavioural categories in the Opponents' acts, namely *Immobile*, *Non-contact acts* and *Contact acts*. A detailed description of these categories is provided in Box 8.3.

Box 8.3 Set of behavioural categories of Opponents used in further analysis

Sniff Experimental	Sniffing any part of the Experimental's body
Non-contact acts	All acts, except *Immobile*, not aimed at body contact with the Experimental
Immobile	

The reason for this (initial) choice of categories is because we expected that the distinction would be relevant for the Experimentals' behaviour. If the resulting categories of combinations of acts of the Experimental and Opponent showed deviations from exponentiality in the direction of mixtures, the behaviour of the Opponent could be considered in more detail at a later stage.

Since the data exploration showed that *Lie* bouts occur only during the last 5 minutes of the trials, and since we had the impression that the occurrence of this act may be due to fatigue caused by the prolonged electrical stimulation, we decided to analyse only the first 20 minutes of each record. Thus, *Lie* no longer occurs in the behavioural repertoire.

Definition of periods

As already mentioned, the last 5 minutes of the records were excluded from further analysis. Since the data exploration indicated that relatively early in the records there was a shift with regard to interactive behaviour, we looked at cumulative plots of bouts of the group of all *Non-contact acts* of the Experimentals. These plots all showed clear and abrupt change points (see e.g. Fig. 8.1). A non-parametric change point procedure (Pettitt, 1979) was used to estimate the place of change points and to test for them. This procedure is equivalent to the non-parametric test of one against zero change points, given in subsection 3.3.1. The estimated places of changes and an indication of which ones are significant (according to the corrected critical values given by Meelis *et al.*, 1991) are given in Table 8.6.

Most records showed a significant change within the first 5 minutes. There is only one significant change point after 5 minutes (rat no. 4, current intensity 0). For this record, however, the cumulative bout length plot

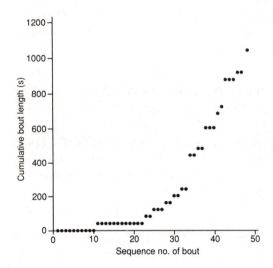

Figure 8.1 Cumulative bout length plot of *Non-contact* acts of an Experimental (one observation period).

Table 8.6 Estimated places of change points in the bout lengths of the combined behavioural category *Non-contact acts* of the Experimentals (in minutes after start of observation). The significance of changes was tested with Pettitt's test (Pettitt, 1979), using the corrected critical values given by Meelis *et al.* (1991)

Rat no.	Current intensity (% of threshold)			
	0	25	50	75
1	2.27*	2.58	0.60*	1.93
2	9.45	1.72	1.53*	1.62*
3	3.00*	1.93*	1.38*	3.03*
4	9.12*	3.45*	0.08	2.12*
5	4.35	5.53*	4.92	3.45

* $p < 0.05$.

indicated that there was another change point earlier in the record. This makes the results of the estimation and test procedure less reliable, since they are based on the assumption that there is at most one change point. Furthermore, the plot did not indicate a very clear change at the estimated point. We therefore used 5 minutes as the criterion to split up the records. Accordingly, there were two periods: (1) the period up to a significant change point or until 5 minutes after the start of the observation, and (2) the period from 5 to 20 min after the start of the observation. These periods were analysed separately. In this way it was ensured that period 2 was fairly homogeneous, and thus could be used to test the initial fit of the model. Once an accurate model was formulated for period 2 (which contains a large amount of data) it could be tested whether the same model could be used in period 1 (for which only a relatively small amount of data is available).

We also tested for time inhomogeneity in the behaviour of the Opponents, but found no apparent changes.

The fit of the model in periods

The test given in subsection 5.2.1 (eqn (5.4)) was used to test for deviations of first order sequential dependency. Table 8.7 gives the results for period 2. There were not enough data to test this for period 1. As can be seen from Table 8.7, only one of the 20 tests gives a significant result. As argued in section 6.1, a small number of deviations can be expected, even when the null hypothesis is true. It can thus be concluded that the results do not indicate that there are deviations from first order sequential dependency, and the set of behavioural categories can be modelled as states in a semi-Markov chain.

Moran's test (eqns (4.44) and (4.46)) was used to find out whether the bout lengths of the combined categories were exponentially distributed. Per

Table 8.7 Results of the test on first-order sequential dependency (see subsection 5.2.1, eqn (5.4)), for period 2

Rat no.	Current intensity (% of threshold)			
	0	25	50	75
1	−0.50	0.01	0.27	0.18
2	−0.17	−1.06	0.33	0.40
3	3.10*	1.26	0.19	0.58
4	0.45	−0.74	0.25	1.95
5	0.99	0.92	1.54	1.13

* $p < 0.05$ (right-sided test). Under H_0 the test statistic has a standard normal distribution, the critical value is 1.645 (see Table A1).

behavioural category and current intensity, the test statistics for different individuals were combined with the sum test (subsection 6.2.1, eqn (6.1)). Results for period 2 are given in Table 8.8. In this instance, a two-sided version of the test is used to see whether there are deviations in the direction of increasing termination rates. As can be seen, there is only one significant deviation in the direction of mixtures left (*Explore + Immobile*, at 50% stimulation intensity). This confirms the hypothesis that the behavioural categories can be considered as semi-Markov states. The results furthermore indicate that for some categories, i.e. *Rear*, and *Sniff Opponent*, the termination rate increases during the bout. Inspection of the log-survivor plots indicated that this was due to time-lags (see e.g. Fig. 8.2). When bout lengths in period 1 were analysed, it appeared that *Explore* also had a minimum duration, of about 0.3 s. Bout length distributions of *Crawl over* also showed deviations in the direction of an increasing termination rate (see Table 8.8). There were insufficient observations of *Crawl over* to investigate this thoroughly, so in this analysis this was excluded from further consideration.

The minimum durations for *Rear* (0.4 s) and *Sniff Opponent* (0.25 s) were subtracted from the bout lengths. After the corrections for minimum durations, tests for exponentiality of bouts in periods 1 as well as 2 did not give a higher number of significant deviations than would be expected under the null hypothesis.

It was thus concluded that the behaviour can be described adequately with a semi-Markov chain model. The behavioural categories consisting of

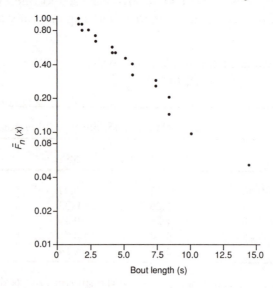

Figure 8.2 Log-survivor plot of *Rear + Non-contact*.

Table 8.8 Results of Moran's test on exponentiality (eqn (4.46)) for bouts of the combinations of acts of Experimentals and Opponents (see Boxes 8.2 and 8.3), in period 2. The test results of different individuals were combined with the sum test (eqn (6.1))

Acts		Current intensity (% of threshold)			
Experimental	Opponent	0	25	50	75
Explore	*Imm*	−0.73	1.94	2.13*	0.88
	NC	−1.36	−0.69	1.27	−0.34
	Sniff E	−0.48	−1.38	−2.28*	−1.20
Rear	*Imm*	−2.40*	−2.76*	−2.71*	−3.97*
	NC	−4.09*	−2.31*	−3.90*	−6.87*
	Sniff E	−2.64*	−1.37	−1.24	−1.61
Self-groom	*Imm*	−1.48	1.38	−	−1.55
	NC	1.82	0.67	−0.43	−0.97
	Sniff E	−1.24	−1.31	−1.19	0.35
Crawl over[+]	*NC*	−1.34	−2.35*	−	−
	Sniff E	−0.62	−	−2.22*	−0.65
Groom O[+]	*NC*	−1.11	−1.44	0.57	−2.52*
	Sniff E	−0.62	−0.05	−0.38	1.41
Sniff O[+]	*NC*	−4.04*	−3.18*	−1.66	−2.49*
	Sniff E	0.59	−0.53	−0.33	−

Imm = Immobile, NC = No-contact, Sniff E = Sniff Experimental, Groom O = Groom Opponent, Sniff O = Sniff Opponent.
+ These acts of the Experimentals nearly did not occur in combination with *Immobile* by the Opponents.
* $p < 0.05$, two-sided test. Under H_0 the test statistic has a $N(0,1)$ distribution. The critical values are −1.96 and 1.96.

combinations with *Self-groom, Crawl over* and *Groom Opponent* were considered as proper Markov states, i.e. with exponentially distributed bout lengths. The combinations with *Explore, Rear* and *Sniff Opponent* were modelled as semi-Markov states with time-lags. Note, however, that the minimum durations of these categories are due to the Experimentals, not the Opponents. Thus, the behavioural categories were considered as semi-Markov states, where the Experimentals' transition tendencies change during a bout (see subsection 1.3.1, eqn (1.9)), whereas the Opponents' tendencies remain constant. Methods for estimating the minimum bout durations are explained in Box 8.4.

Box 8.4   Modelling and estimation of time-lags in acts of
              Experimentals

Let $\lambda_{(A,B;Exp)}(x)$ be the chance per time unit that an Experimental stops doing
A while the Opponent is doing B, and, analogously, let $\lambda_{(A,B;Opp)}(x)$ be the
chance per time unit that an Opponent stops doing B while the Experimental
does A (where x denotes the bout duration of (A,B)). These parameters are
called the 'termination tendencies' of the behavioural category (A,B) of,
respectively, the Experimental and the Opponent. The Experimental's
termination tendency of (A,B) is equal to

$$\lambda_{(A,B;Exp)}(x) = \sum_D \alpha_{(A,B)(D,B)}(x), \tag{1}$$

where D denotes the set of all other acts of the Experimental. The termination tendency of (A,B) of the Opponent is defined similarly.

It is assumed that act A of the Experimental has a fixed minimum dura-
tion, τ, whereas the Opponent can terminate act B at any time. $\lambda_{(A,B;Exp)}$ is
thus equal to zero for x less than τ after which it has a constant value, say
μ_1. $\lambda_{(A,B;Opp)}(x)$ is constant, say μ_2, for all values of x (Fig. 1). Let $x_1,..., x_n$
be the bout lengths of the behavioural category (A,B) and let $\delta_i = 1$ when
the ith bout was terminated by the Experimental and 0 when the bout was
terminated by the Opponent. The estimator of τ, $\hat{\tau}$, is the minimum of

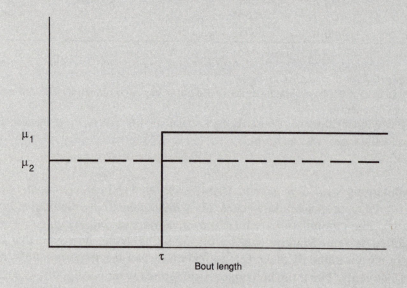

Figure 1 Solid line: Termination tendency of Experimental; Dashed line:
Termination tendency of Opponent.

those x_i for which $\delta_i = 1$. Note that $\hat{\tau}$ is not necessarily equal to the minimum x_i since bouts terminated by the Opponent can be smaller. To estimate μ_1, leave out all x_i smaller than or equal to $\hat{\tau}$ and subtract $\hat{\tau}$ from the remaining bout lengths (see subsection 1.3.1). The remaining data set is $y_1,...,y_m$. Then $\hat{\mu}_1$ is m/S_1, where S_1 is the sum of the y_i that were terminated by the Experimental. τ can also be estimated by taking the minimum of the bout lengths of A by the Experimental (irrespective of the Opponent's behaviour). μ_1 is estimated as described above, with this estimator of τ. The estimator of the Opponent's termination tendency, $\hat{\mu}_2$, is calculated as usual, i.e. n/S_2, where S_2 is the sum of the x_i that were terminated by the Opponent.

Once $\hat{\mu}_1$ is calculated, the transition rate from, say (A,B), to (C,B) is estimated by

$$\hat{\alpha}_{(A,B)(C,B)} = \hat{\mu}_1 \frac{N_{(A,B)(C,B)}}{N_{(A,B)}}, \tag{2}$$

where $N_{(A,B)(C,B)}$ is the number of observed transitions from (AB) to (CB) and $N_{(A,B)}$ is the total number of (A,B) bouts.

8.2.4 Further analysis

The model was used to examine the effects of current intensity on the dyadic interaction. We will discuss here only the main results of the analysis of the Experimentals' behaviour. For further details see Haccou *et al.* (1988*a*).

Three types of effects can be considered:
(1) the main effects of the current intensity
(2) the main effects of the behaviour of Opponents
(3) the interactive effects of current intensity and Opponent behaviour.
These effects can all be estimated and tested by means of ANOVA on the transition rates (subsection 7.5.1). For instance, the effect of the Opponent's behaviour on the Experimental's tendency to stop *Explore* and start *Groom Opponent* can be examined by comparing the transition rates:

$\alpha_{(Explore,Immobile)(Groom\ Opponent,Immobile)}$, $\alpha_{(Explore,Non-contact)(Groom\ Opponent,Non-contact)}$, and $\alpha_{(Explore,Contact)(Groom\ Opponent,Contact)}$.

Note that the chance of exactly simultaneous transitions is zero. In practice, simultaneous transitions may be recorded, but hardly ever occur (this was discussed in section 1.2). The effects of current intensity are examined by comparing transition rates across treatments. An ANOVA model for the Experimental's tendency to make transitions from *Explore* to *Groom*

Opponent including both main effects and interactive effects would look like this:

$$\log(\alpha(j)_{(Explore,i)(GroomOpponent,i)}) = \mu + \alpha_i + \beta_j + \gamma_{ij} + \varepsilon_{ij}, \qquad (8.1)$$

where i denotes the Opponent's act and j the treatment (subsection 7.5.1).

In principle, the effects on all possible transition rates can be examined in this way. When the numbers of bouts are large enough, the estimators of the transition rates are approximately independent, and thus separate ANOVA tests can be carried out. In practice, however, there are too few data to do this. To get an overview of the effects, it is also preferable to combine parameters somehow into ethologically well-interpretable statistics. To be able to carry out an ANOVA, the resulting statistics, consisting of a combination of transition rates, should be asymptotically normally distributed.

The transition rates towards groups of acts meet all the criteria listed above (section 7.3). Within the context of the research it is sensible to divide the acts into two groups: *Contact acts* (*Crawl over*, *Groom Opponent* and *Sniff Opponent*) and *Non-contact acts* (*Explore*, *Rear*, *Self-groom*). The transition rate from a behavioural category towards a group of acts is calculated by summing the transition rates from that category towards all the acts in the group. If the category belongs to the group in question, the transition rates are summed over all the other acts in the group. For example, the transition rate from *Explore + Immobile* to *Non-contact acts + Immobile* is: $\alpha_{(Explore,Immobile)(Rear,Immobile)}$ plus $\alpha_{(Explore,Immobile)(Self-groom,Immobile)}$. The resulting set of parameters have similar asymptotic properties as ordinary transition rates (section 7.3), and thus can be analysed in the same way.

The ANOVA model given in eqn (8.1) was applied to the transition rates towards the two groups of acts. We found significant main effects of the Opponent's behaviour on the Experimental's tendencies to stop *Rear* and *Explore* in favour of *Contact acts*: these tendencies are lowest when the Opponent is immobile and highest when the Opponent sniffs at the Experimental. Similar effects, though non-significant, were found on the Experimental's tendencies to stop *Self-groom*, *Sniff Opponent* or *Groom Opponent* in favour of (other) *Contact acts*. The two most intensive *Contact acts*, *Crawl over* and *Groom Opponent*, did not even occur in combination with *Immobile*. All these effects indicate that, the less the Opponent interferes with the Experimental, the lower the chance that the Experimental starts interfering with him. *Immobile* thus appears to be a very effective way of avoiding attention.

The current intensity significantly affected the Experimentals' tendencies to make transitions from *Rear* and *Self-groom* to other *Non-contact acts*.

Here, the effects are dissimilar: whereas the tendency to stop *Rear* in favour of other *Non-contact acts* increases, the tendency to stop *Self-groom* in favour of other *Non-contact acts* goes down with increasing stimulation. Figure 8.3 shows the estimated main effects of current intensity on all the Experimentals' transition tendencies towards *Non-contact acts* (except *Crawl over*, since there were too few data on this act). Tendencies to stop *Groom Opponent*, *Sniff Opponent* and *Rear* increase, whereas tendencies to stop *Explore* and *Self-groom* decrease. Apparently, the Experimentals become more self-directed at higher stimulation intensities. The only contradiction seems to be the effect on *Rear*. This is very interesting, since we found that, whereas *Rear* was considered as a *Non-contact act* by us, in some ways the Opponents reacted to *Rear* in a similar way as to *Contact acts* of the Experimentals (Haccou *et al.*, 1988a).

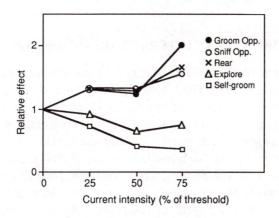

Figure 8.3 Estimated mean effects of current intensity on the Experimentals' transition tendencies towards *Non-contact* acts. Adapted from Haccou *et al.* (1988a) with permission from Academic Press Inc.

There were significant interactive effects of the current intensity and Opponent behaviour on the tendencies from *Crawl over* and *Sniff Opponent* to *Non-contact acts*, indicating that the Experimentals' reaction to the behaviour of the other rat is affected by the stimulation. There were too few observations to investigate this further for *Crawl over*, but inspection of the effects on transition rates from *Sniff Opponent* revealed that at high current intensities the reaction to *Immobile* becomes very strong: at the highest current intensity, on average Experimentals stopped sniffing the Opponent within 1 s when he was immobile.

The results suggest that activation of the neural system in the hypothalamic attack area makes the stimulated animal more self-directed, since tendencies to stop non-contact behaviour in general decrease, whereas tendencies to stop contact behaviour increase (Fig. 8.3). At the same time, however, stimulated rats react more to their Opponent's activities at high stimulation intensities. The results also illustrate how important it is to take proper account of individuals other than the test animal in experiments with interactive behaviour, since, apparently, the behaviour of the other animal may affect the test animal at least as strongly as the experimental treatment. When gross behavioural measures are used, experimental effects are liable to be overlooked. For instance, whereas there is a clear effect of the treatment on the Experimentals' tendencies to stop self-grooming, the total time spent self-grooming did not change with the current intensity. The reason for this is presumably the influence of the Opponents' behaviour on the Experimentals' tendencies to stop self-grooming. For further details see Haccou *et al.* (1988*a*).

8.3   MOTHER-INFANT INTERACTION OF RHESUS MONKEYS

The results presented here are based on a project together with Dr H. Dienske (Primate Centre, TNO, Rijswijk) and Dr E.G. Langeler. The aim was to investigate the effects of low doses of amphetamine on the infant's behaviour. The effects of psycho-active drugs are usually studied at relatively high dosages, which grossly exceed the levels used in therapeutic applications in man. The high levels used are partially due to difficulties with the detection of low-dose effects. Gross behavioural measures usually fail to detect such effects, especially when the drug changes the time structure rather than the form of behaviour. By formulating a detailed model for the time structure, it is, however, possible to study low-dose effects.

8.3.1 Experimental set-up

We studied 12 mother-infant pairs. The infants were about 3 months old. All trials were done with singly housed mother-infant pairs in a test cage. The animals were accustomed to the cage as well as the presence of observers. The infants were given subcutaneous injections with either amphetamine (0.2 mg/kg) or a placebo on two consecutive days. In half of the tests the amphetamine injection was given first and the placebo the following day. In the other half, the order was reversed. The observers did not know whether the placebo or amphetamine was given. Observation started immediately after injection and lasted for 2 h 47 min, during which behaviour was recorded continuously with a keyboard.

8.3.2 Levels of resolution

In the mother-infant interaction of rhesus monkeys several levels of resolution (time scales) can be distinguished (Dienske *et al.*, 1981). There is gross periodicity since the infants often alternate between awake and nursing periods during observation. Nursing periods consist of long bouts of nipple contact. In these periods, which last on average about 10 to 15 min, the infants sit very still on their mothers' lap, with their eyes closed, and suck milk. Nipple contact also occurs during awake periods, but then no milk is sucked. The distribution of *Nipple* bouts in awake periods differs markedly from the nursing periods. Awake periods and nursing periods therefore should be analysed separately. A schematic example is given in Fig. 8.4.

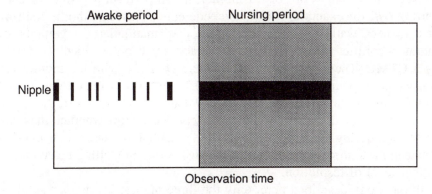

Figure 8.4 Periodicity in mother-infant interaction of rhesus monkeys.

At the next level of resolution there is an alternation between fully awake and drowsy subphases during awake periods (Fig. 8.5). In drowsy phases, which often occur just before and after nursing periods, infants have a high tendency to initiate body contact and a low tendency to end it. Such phases are often characterized by a high frequency of conflicts with the mother. Drowsy and fully awake phases should be distinguished before further analysis, since bout distributions of acts differ between these phases. The methods described in Chapter 3 can be used to detect the various phases.

Within homogeneous periods, the mother-infant interaction of rhesus monkeys is (under normal conditions) accurately described by a CTMC with three states (Dienske and Metz, 1977, Dienske *et al.*, 1980; see also Boxes 1.5 and 4.1): *Off* (no body contact between mother and infant), *On* (the infant sits on its mother's lap or back, usually grabbing her fur with its hands, without having the nipple in its mouth), and *Nipple* (the infant sits on its mother's lap with the nipple in its mouth).

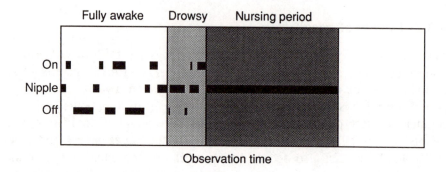

Figure 8.5 Time inhomogeneity in awake periods of rhesus infants.

When the infant is on- or off-mother, it can perform a variety of acts. During *Off*, for example, it can have side contact with its mother (sitting next to her), walk, explore, seize its mother, manipulate its genitals, or initiate a conflict with its mother. These acts are also described accurately by a CTMC (Dienske *et al.*, 1981), where conflicts with the mother are considered as point events (subsection 1.3.1). Acts occurring during *On* and *Nipple* can also be considered as Markov states. Note, however, that since the bout length distributions of such acts depend on whether they are performed during *On*, *Off*, or *Nipple*, it is essential to make the distinction between these three behavioural categories before modelling behaviour at a finer level of resolution.

In the study described here, only the three highest levels of resolution were considered. The distinction between awake and nursing periods can easily be made on observational grounds. Since an accurately fitting model was available for homogeneous periods, parametric methods could be used for detection and definition of drowsy and fully awake periods (subsection 3.1.1).

8.3.3 Generalization of the original model to a semi-Markov model

As mentioned in subsection 8.3.2, the behavioural categories *Off*, *On* and *Nipple* under normal conditions are accurately described by states in a Markov model. In this case, however, we found that the model did not fit well, since there were relatively long time-lags (subsection 1.3.1). To take this into account, the model given in eqn (1.16) was used (see subsection 1.3.1 and Haccou *et al.*, 1988*c*, for a discussion of the choice of a model for time-lags). It was thus assumed that the termination rates of the three behavioural categories all had forms such as in Fig. 1.4. Each termination rate is therefore characterized by two parameters: the point of inflection, β (a measure for the duration of the time-lag), and the asymptotic termination rate, λ (the long-term tendency to terminate a bout).

Per behavioural category, the MLEs of β and λ are calculated by means of Newton-Raphson iteration (Box 8.5). Once these parameters are estimated, the fit of the model is tested graphically as follows. Since the survivor function is

$$\bar{F}(x) = \exp[-\int_0^x \lambda(s)ds], \tag{8.2}$$

it follows that

$$-\log\bar{F}(x) = \int_0^x \lambda(s)ds. \tag{8.3}$$

From eqn (1.16) it follows that

$$\int_0^x \lambda(s)ds = \lambda x - \lambda\beta\,\frac{e+1}{e}\log\left(1+e^{1-\frac{x}{\beta}}\right) + \lambda\beta\,\frac{e+1}{e}\log(1+e). \tag{8.4}$$

So a plot of

$$\frac{-\log\bar{F}_n(x) + \hat{\lambda}\hat{\beta}\left(\frac{e+1}{e}\right)\log\left(1+e^{1-\frac{x}{\beta}}\right) - \hat{\lambda}\hat{\beta}\left(\frac{e+1}{e}\log(1+e)\right)}{\hat{\lambda}} \tag{8.5}$$

against x (where $\bar{F}_n(x)$ is the empirical survivor function, eqn (4.12)) should give approximately a straight line through the origin, with a slope of 45°.

Box 8.5 Estimation of parameters of the time-lag model of eqn (1.16)

The probability density of x is equal to

$$f(x) = \lambda(x)\exp[-\int_0^x \lambda(s)ds]. \tag{1}$$

From eqn (1.16) it can be derived that

$$\int_0^x \lambda(s)ds = \lambda x + \lambda\beta\,\frac{e+1}{e}\log\frac{1+e^{1-\frac{x}{\beta}}}{1+e}. \tag{2}$$

It follows that the log-likelihood of a series of observations $x_1,...,x_n$ is equal to

$$L(x_1,...,x_n;\lambda,\beta) = \sum_{i=1}^{n} \log f(x_i)$$

$$= n\log\lambda + \sum_{i=1}^{n}\log(1 - e^{-\frac{x_i}{\beta}}) - \sum_{i=1}^{n}\log(1 + e^{1-\frac{x_i}{\beta}})$$

$$-\sum_{i=1}^{n}\lambda x_i - \lambda\beta\left(\frac{e+1}{e}\right)\sum_{i=1}^{n}\log\frac{1 + e^{1-\frac{x_i}{\beta}}}{1 + e}. \tag{3}$$

The parameters λ and β are estimated by maximizing this function. Since the likelihood equations cannot be solved explicitly, this is done by Newton-Raphson iteration (Rao, 1973, Stoer and Bulirsch, 1980). To do this, the first and second partial derivatives with respect to λ and β are needed. These are equal to

$$\frac{\partial L}{\partial\lambda} = \frac{n}{\lambda} - \sum_{i=1}^{n}x_i + \beta(e+1)\sum_{i=1}^{n}\log\frac{1 + e}{1 - e^{1-\frac{x_i}{\beta}}} \tag{4}$$

$$\frac{\partial L}{\partial\beta} = \lambda(e+1)\sum_{i=1}^{n}\log\frac{e+1}{1 + e^{1-\frac{x_i}{\beta}}} - \frac{1}{\beta^2}\sum_{i=1}^{n}\frac{x_i}{e^{\frac{x_i}{\beta}} - 1}$$

$$-\left(\frac{1}{\beta^2} + \frac{\lambda}{\beta}(e+1)\right)\sum_{i=1}^{n}\frac{x_i}{e^{\frac{x_i}{\beta}-1} + 1}. \tag{5}$$

$$\frac{\partial^2 L}{\partial\lambda^2} = -\frac{n}{\lambda^2} \tag{6}$$

$$\frac{\partial^2 L}{\partial\beta^2} = \frac{1}{\beta^2}\left[\sum_{i=1}^{n}\frac{x_i}{e^{\frac{x_i}{\beta}} - 1}\{2 - \frac{x_i}{1 - e^{-\frac{x_i}{\beta}}}\}\right]$$

$$+ \frac{1}{\beta^2}\left[\sum_{i=1}^{n}\frac{x_i}{e^{\frac{x_i}{\beta}} + 1}\{2 - \left(\frac{1}{\beta} + \lambda(e+1)\right)\frac{x_i}{1 + e^{1-\frac{x_i}{\beta}}}\}\right] \tag{7}$$

$$\frac{\partial^2}{\partial\lambda\partial\beta} = (e+1)\sum_{i=1}^{n}\log\frac{e+1}{1+e^{1-\frac{x_i}{\beta}}} + \frac{1}{\beta}\sum_{i=1}^{n}(1+e^{1-\frac{x_i}{\beta}})(x_i e^{1-\frac{x_i}{\beta}}). \tag{8}$$

To calculate the maximum likelihood estimators two start values are chosen, denoted by λ_0 and β_0. Subsequently, the next iteration values of the estimators are calculated by

$$\begin{pmatrix}\lambda_1\\\beta_1\end{pmatrix} = \begin{pmatrix}\lambda_0\\\beta_0\end{pmatrix} - \begin{pmatrix}\dfrac{\partial^2 L}{\partial\lambda^2} & \dfrac{\partial^2 L}{\partial\lambda\partial\beta}\\[2mm]\dfrac{\partial^2 L}{\partial\lambda\partial\beta} & \dfrac{\partial^2 L}{\partial\beta^2}\end{pmatrix}^{-1}\begin{pmatrix}\dfrac{\partial L}{\partial\lambda}\\[2mm]\dfrac{\partial L}{\partial\beta}\end{pmatrix}. \tag{9}$$

To calculate λ_2 and β_2, λ_1 and β_1 are substituted for λ_0 and β_0 in (9), etc. This is repeated until the difference between the current and the previous values of λ_i and β_i is small enough (relative to the values of λ_i and β_i). Furthermore, the partial derivatives should be small. An example of a criterion for this is

$$\sqrt{\left(\frac{\partial L}{\partial\lambda}\right)^2 + \left(\frac{\partial L}{\partial\beta}\right)^2} < 10^{-5}. \tag{10}$$

Good starting values for λ_0 and β_0 are respectively $1/\bar{x}$ and the minimum bout duration.

The improvement in fit with respect to the original model can be seen by plotting $-\log \bar{F}_n(x)/\hat{\lambda}$, where $\hat{\lambda}$ is the MLE of the termination rate under the pure Markov model (= $1/\bar{x}$, see Box 1.2), against x. Following the same line of argument as above, it can be proved that, if the termination rate is constant, such a plot should give an approximately straight line through the origin with a slope of 45°. Figure 8.6 shows the plots of the survivor function of one data set transformed in the two alternative ways. As can be seen, the fit of the semi-Markov model is considerably better. The same conclusion could be drawn for the other data sets.

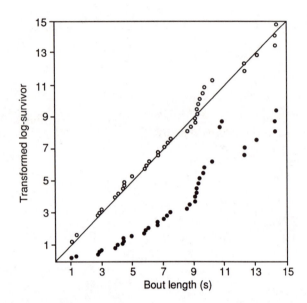

Figure 8.6 Log-survivor of one set of bouts, transformed in two different ways:
(1) according to eqn. (8.5):   ∘
(2) $-\log \bar{F}_n(x)/\bar{x}$:   •
When the model of eqn (1.16) fits, transformation (1) should resemble the line drawn.
When the pure Markov model fits, transformation (2) should resemble the line drawn.

8.3.4 Further analysis

We did not find any effects of amphetamine either on the duration of awake
and nursing periods, or on the drowsy and fully awake phases. Here, we
will therefore only discuss the results at the level of the behavioural categories *On*, *Off* and *Nipple*. For each of these categories, the parameters β and
λ were estimated by means of the method outlined in Box 8.5. Subsequently, the infant's tendencies to initiate transitions between the behavioural categories were estimated by multiplying the estimated asymptotic
termination rate, $\hat{\lambda}$, by the appropriate transition frequencies. For instance,
an infant's long-term tendency to switch from *On* to *Off* is estimated by

$$\hat{\alpha}_{On,Off;l} = \frac{N_{On,Off;l}}{N_{On}} \hat{\lambda}_{On}, \tag{8.6}$$

where $N_{(On,Off;l)}$ is the number of transitions from *On* to *Off* that were
initiated by the infant, $N_{(On)}$ is the total number of *On* bouts and $\hat{\lambda}_{(On)}$ is the
estimated asymptotic termination rate of *On*. The parameters were thus
estimated for each subphase. Next, the estimates of the different subphases

within each observation period were combined by taking weighted averages. Let $\hat{\xi}_i$ be the estimated value of a parameter, say $\alpha_{(On,Off;I)}$, in the ith subphase of an observation period, and suppose that there are M subphases. Then the weighted average of the estimates is calculated by

$$\hat{\xi} = \frac{\sum_{i=1}^{M} \frac{z_i \hat{\xi}_i}{\hat{\sigma}_i}}{\sum_{i=1}^{M} \frac{z_i}{\hat{\sigma}_i}} , \tag{8.7}$$

where $\hat{\sigma}_i$ is the estimated standard deviation of $\hat{\xi}_i$ and z_i is the length of the ith subphase. An example of such a calculation is given in Box 8.6. Since the weight factors are proportional to the length of the subphases, the combined estimates, $\hat{\xi}$, reflect the relative importance of the drowsy and fully awake subphases in an observation period.

Box 8.6 Calculation of weighted averages of parameter estimates in the subphases and their asymptotic variances

Table 1 shows the duration of subphases in one observation period and the estimated values of $\lambda_{(On)}$ in the subphases. From these values, the infant's transition tendency towards *Off* is calculated by multiplying the values of $\hat{\lambda}$ by the fraction of transitions from *On* to *Off* initiated by the infant (column 3). The results are given in column 4. Column 5 shows the estimated asymptotic variances of $\hat{\lambda}_{(On)}$. These are equal to the left upper element of the inverse information matrix (i.e. the matrix in eqn (8), Box 8.5), when the estimated value of $\lambda_{(On)}$ is substituted in the equations. When the Newton-Raphson iteration procedure is carried out, these values are

Table 1
Estimated values of λ_{On} and $\alpha_{(On,Off;I)}$ in the subphases of one observation period and their estimated variances

Length of subphase(s)	$\hat{\lambda}_{(On)}$ (s^{-1})	$N_{(On,Off;I)}/N_{(On)}$	$\hat{\alpha}_{(On,Off;I)}$ (s^{-1})	$Var(\hat{\lambda}_{(On)})$	$Var(\hat{\alpha}_{(On,Off;I)})$
1055	0.1551	0.25	0.0388	1.503×10^{-3}	9.394×10^{-5}
806	0.1263	0.50	0.0631	3.068×10^{-4}	7.670×10^{-5}
2260	0.1256	0.69	0.0856	2.921×10^{-4}	1.391×10^{-5}

calculated also. The variances of the transition rates are estimated from these values by multiplying them by $(N_{(On,Off;l)} / N_{(On)})^2$. The results are given in column 6.

Table 2 shows the data needed for calculating the weighted average of the $\hat{\alpha}_{(On,Off;l)}$ according to eqn (8.7). The z_i are equal to the durations of the subphases (column 1 of Table 1), the $\hat{\xi}_i$ correspond to the $\hat{\alpha}_{(On,Off;l)}$ and the σ_i are equal to the square root of the variances given in column 6 of Table 1. The weighted average, $\hat{\xi}$, is equal to $21207.700/392549.160 = 0.0540$ (s^{-1}). The estimated variance of $\hat{\xi}$ is equal to

$$Var\hat{\xi} = \frac{\sum_{i=1}^{M}\left(\frac{z_i}{\hat{\sigma}_i}\right)^2 \hat{\sigma}_i^2}{\left(\sum_{i=1}^{M}\frac{z_i}{\hat{\sigma}_i}\right)^2} = \frac{\sum_{i=1}^{M}z_i^2}{\left(\sum_{i=1}^{M}\frac{z_i}{\hat{\sigma}_i}\right)^2}. \tag{1}$$

From Table 2 it follows that in this case the estimated variance of the weighted average is equal to $6870261/(392549.160)^2 = 4.458\times10^{-5}$.

Table 2 Calculation of weighted averages of the transition rates in Table 1 and their variances

i	z_i	$\hat{\xi}_i$	$\hat{\sigma}_i$	$z_i/\hat{\sigma}_i$	$z_i\hat{\xi}_i/\hat{\sigma}_i$	z_i^2
1	1055	0.0388	0.009690	108875.13	4224.355	1113025
2	806	0.0631	0.008756	92051.165	5808.428	649636
3	2260	0.0856	0.011794	191622.86	16402.917	5107600
Sum:	4121			392549.16	21207.700	6870261

The log-transformed $\hat{\xi}$ are asymptotically normally distributed (Chapter 7). The effects of inter-individual differences and experimental conditions were examined with an ANOVA model:

$$\log(\hat{\xi}_{ij}) = \log(\xi) + \mu_i + \beta_j + \varepsilon_{ij}, \tag{8.8}$$

where ε_{ij} are $N(0,\theta(Var\hat{\xi})/\hat{\xi}^2)$ distributed (it is shown in Box 8.6 how $Var\hat{\xi}$ is calculated), μ_i denotes the effect of inter-individual differences and β_j the effect of experimental treatment. The model given in (8.8) was used for each of the parameters. Tables 8.9 and 8.10 show the results of the effects of amphetamine. As can be seen, time-lags increased significantly in all three behavioural categories, with a factor of about 4 to 6. The drug

Table 8.9 Effects of amphetamine on time-lags

Behavioural category	$\exp(\gamma_1 - \gamma_0)^+$	t^{++}	p
On	4.50	2.24 (10)	0.025–0.05
Nipple	5.77	2.30 (10)	0.025–0.05
Off	5.03	2.30 (10)	0.025–0.05

+ Factor with which the time-lag changes when amphetamine is administered.
++ Student's t statistic (degrees of freedom).

Table 8.10 Effects of amphetamine on the infants' transition tendencies

From	To	$\exp(\gamma_1 - \gamma_0)^+$	t^{++}	p
On	*Off*	0.19	-3.32 (9)	0.01 – 0.025
On	*Nipple*	2.00	2.43 (10)	0.025– 0.05
Off	*On*	1.96	1.69 (10)	0.10 – 0.25
Off	*Nipple*	4.04	2.03 (10)	0.05 – 0.10
Nipple	*On*	0.34	-2.40 (10)	0.025– 0.05
Nipple	*Off*	0.19	-1.47 (8)	0.10 – 0.25

+ Factor with which the time-lag changes when amphetamine is administered.
++ Student's t statistic (degrees of freedom).

thus appears to decrease the infants' ability to switch between acts. There were also significant effects on all transition tendencies except those from *Off* to *On*, *Off* to *Nipple* and *Nipple* to *Off*. Here, too, however, the *p*-values were very small. The effects on the transition tendencies all point towards a more mother-directed internal state of the infants: all tendencies towards *Nipple* increase, whereas all tendencies towards *Off* decrease. The tendency to switch from *Nipple* to *On* also decreases.

The main conclusions are that, at the dose examined, amphetamine affects two short-time processes in behaviour, namely the duration of time-lags of the behavioural states *On*, *Nipple* and *Off*, and the infants' tendencies to alternate between these states. Processes on longer time scales, however, were not altered: neither the alternation between nursing and awake periods nor the cycle of subphases during awake periods was affected.

8.4   MATING BEHAVIOUR OF BARBS

Putters *et al.* (1984) studied the temporal pattern of mating behaviour of the barb *Barbus nigrofasciatus* by means of a function of a Markov chain. Operative formal methods of analysis for such models still have to be developed. Here, we only give an outline of their approach and main results to give an impression of how such analyses can be carried out.

Barb courtship starts with the male taking a parallel posture to the female, slightly below her to the side, with his back towards her. From this posture the male may swim forward, away from the female, turn and circle around her, return to the initial courtship posture, or attempt to mate. During mating, the male slides alongside the female, tipping her head downwards and bending his tail over her back. With a brief jerk a few eggs and some sperm are then shed simultaneously. After mating, the male 'jumps' away from the female and resumes courtship after about 5 seconds.

A schematic example of the time structure of barb behaviour is given in Fig. 8.7. Putters *et al.* (1984) studied the distribution of intervals between

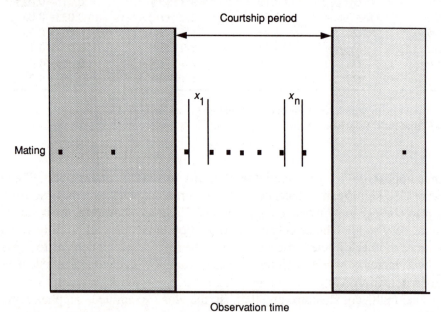

Figure 8.7 Schematic example of the time structure of courtship behaviour in barbs. Shaded regions indicate periods in which activities other than courtship predominate. Such periods were left out. Gaps between mating within the remaining ('courtship') periods were analysed, with the exception of the first 5 s, during which no mating can occur. The first and last considered gaps of the courtship period are indicated in the figure. They are denoted by x_1 and x_n respectively.

mating *per se* in periods in which the pattern is more or less stationary, leaving out periods during which other activities predominate. The selection of such periods was based on cumulative plots of gaps between courtship against observation time (subsection 3.2.2). Furthermore, the periods thus selected were checked for stationarity by comparing the first, middle and last segments of each period by means of a Kruskal-Wallis test. They also tested for short-term sequential dependencies among gaps between mating (subsection 5.4.1).

In the first 5 seconds of courtship periods no mating can occur, due to spatial separation. These initial subphases were removed before further analysis. The distribution of gaps between mating in the remaining part of the periods was examined using log-survivor plots.

8.4.1 Model formulation

The log-survivor plots of inter-mating intervals of *B. nigrofasciatus* appeared to be convex, indicating a decreasing termination rate (subsection 4.2.1 and section 4.3). This means that matings are clustered in time, since the longer the interval between mating, the less likely additional matings occur. Three models that could explain such clustering were fitted to the data: the so-called branching Poisson model (Cox and Lewis, 1978), a waning-attention model with hazard rate

$$\lambda(x) = \alpha e^{-\beta x} + \gamma \tag{8.9}$$

(Hauske, 1967, Metz, 1974), and a mixture of two exponentials (eqn (4.1)). The parameters of the models were estimated by maximization of the likelihood function (subsection 7.4.1). Only the fit of the mixture was satisfactory, indicating that the data could be described by a function of a Markov chain.

Inter-mating intervals of this species of barbs thus appear to consist of at least two behavioural states. It was impossible, however, to split the intervals up according to observable differences. The resulting minimal model is thus a three-state Markov chain in which two of the states are lumped together. Mating was considered as a 'point event' (subsection 1.3.1). Unfortunately, further analysis based on this model is difficult since transitions between the two lumped states cannot be identified. This implies that several alternative CTMC models can describe the observed behaviour. The different possibilities are given in Fig. 8.8. The values of the transition rates between the states and the transition probability from state I (mating) to the other states are subject to certain constraints. Putters *et al.* (1984) describe how the ranges of possible values can be calculated.

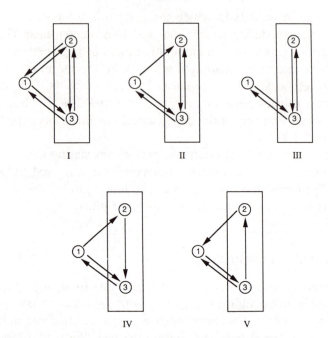

Figure 8.8 All possible alternative three-state models for mating in *B. nigrofasciatus*. State I corresponds to mating and is considered as a point event. States II and III cannot be distinguished on observational grounds.

8.4.2 Model identification

A thorough search for possible observable features to distinguish the two behavioural states did not meet with success. Another approach is to look at the mating behaviour of related species. It was found that in *B. tetratzona* mating intervals could be split up according to the female's willingness (indicated by her hovering near plants). This led to two sets of gaps which were both exponentially distributed. If the two sets are lumped, the same pattern as observed in *B. nigrofasciatus* arises. When the female is unwilling, no mating occurs. This strongly points in the direction of model III in Fig. 8.8. Based on this model, the male's mating rate conditional on female willingness, α_{23}, was estimated. It turned out that this parameter was almost constant for different males and courtship periods. This, too, is an argument in favour of model III.

Finally, the model was tested by experimental manipulation of the female's willingness. This was done in two ways. First, the female's willingness was enhanced by repeatedly exposing her to courting males without allowing mating to occur. Afterwards, the female was put in a tank with a male without interrupting mating. As predicted by the model, mating then initially occurred with exponentially distributed intervals, at the male rate, α_{23}. Second, females were exhausted by putting them in tanks with single males for a number of days. For such females, inter-mating intervals were still independent and mating occurred at a lower rate, as predicted.

APPENDIX I STATISTICAL TABLES

In this appendix the most important statistical tables referred to in the main text are included. Note that in some cases more extensive tables are available in the statistical literature. Moreover, many standard statistical software packages provide exact or approximate p-values for a number of well-known test statistics.

CONTENTS

A12. Non-parametric test for testing three against two abrupt changes in the mean boutlength. See subsection 3.3.1.

A13. Likelihood ratio test for one against zero change points for exponentially distributed boutlengths. See subsection 3.3.1.

A14. Likelihood ratio test for two against one change points for exponentially distributed boutlengths. See subsection 3.3.1.

A15. Likelihood ratio test for three or more against two change points for exponentially distributed boutlengths. See subsection 3.3.1.

A16. Likelihood ratio test for one against zero change points for normally distributed boutlengths. See subsection 3.3.1.

A17. Kolmogorov-Smirnov test for testing for exponentiality with unknown mean. See subsection 4.6.2.

A18. Improved large-sample approximation for the Kolmogorov-Smirnov test for testing for exponentiality with unknown mean. See subsection 4.6.2.

A19. The Cramér-von Mises test for testing for exponentiality with unknown mean. See subsection 4.6.2.

A20. The Shapiro-Wilk W test for goodness of fit for a sample of two-parameter exponentially distributed random variables. See subsection 4.6.2.

A21. Barlow's 'cumulative total time on test' for testing against increasing or decreasing termination rate. See subsection 4.6.3.

A22. Test of Matthews and Farewell for an abrupt change in the termination rate. See subsection 4.6.4.

A23. Chen's correlation goodness-of-fit test for censored data. See subsection 4.7.2.

A24. The Kruskal-Wallis non-parametric k-sample test. Table of the small-sample critical values for $k = 3$. See subsection 5.3.1.

A25. Spearman's rank correlation test. See subsection 5.4.2.

A26. Homogeneity test for the geometric distribution. See subsection 5.4.3.

A27. The coefficients $a_{i,n}$ for the Shapiro-Wilk test for departure from normality. See subsection 6.2.1.

A28. The Shapiro-Wilk test for departure from normality. See subsection 6.2.1.

Table A1 The standard normal distribution

x	0	1	2	3	4	5	6	7	8	9
0.0	5000	5040	5080	5120	5160	5199	5239	5279	5319	5359
0.1	5398	5438	5478	5517	5557	5596	5636	5675	5714	5753
0.2	5793	5832	5871	5910	5948	5987	6026	6064	6103	6141
0.3	6179	6217	6255	6293	6331	6368	6406	6443	6480	6517
0.4	6554	6591	6628	6664	6700	6736	6772	6808	6844	6879
0.5	6915	6950	6985	7019	7054	7088	7123	7157	7190	7224
0.6	7257	7291	7324	7357	7389	7422	7454	7486	7517	7549
0.7	7580	7611	7642	7673	7704	7734	7764	7794	7823	7852
0.8	7881	7910	7939	7967	7995	8023	8051	8078	8106	8133
0.9	8159	8186	8212	8238	8264	8289	8315	8340	8365	8389
1.0	8413	8438	8461	8485	8508	8531	8554	8577	8599	8621
1.1	8643	8665	8686	8708	8729	8749	8770	8790	8810	8830
1.2	8849	8869	8888	8907	8925	8944	8962	8980	8997	9015
1.3	9032	9049	9066	9082	9099	9115	9131	9147	9162	9177
1.4	9192	9207	9222	9236	9251	9265	9279	9292	9306	9319
1.5	9332	9345	9357	9370	9382	9394	9406	9418	9429	9441
1.6	9452	9463	9474	9484	9495	9505	9515	9525	9535	9545
1.7	9554	9564	9573	9582	9591	9599	9608	9616	9625	9633
1.8	9641	9649	9656	9664	9671	9678	9686	9693	9699	9706
1.9	9713	9719	9726	9732	9738	9744	9750	9756	9761	9767
2.0	9772	9778	9783	9788	9793	9798	9803	9808	9812	9817
2.1	9821	9826	9830	9834	9838	9842	9846	9850	9854	9857
2.2	9861	9864	9868	9871	9875	9878	9881	9884	9887	9890
2.3	9893	9896	9898	9901	9904	9906	9909	9911	9913	9916
2.4	9918	9920	9922	9925	9927	9929	9931	9932	9934	9936
2.5	9938	9940	9941	9943	9945	9946	9948	9949	9951	9952
2.6	9935	9955	9956	9957	9959	9960	9961	9962	9963	9964
2.7	9965	9966	9967	9968	9969	9970	9971	9972	9973	9974
2.8	9974	9975	9976	9977	9977	9978	9979	9979	9980	9981
2.9	9981	9982	9982	9983	9984	9984	9985	9985	9986	9986
3.0	9987	9987	9987	9988	9988	9989	9989	9989	9990	9990
3.1	9990	9991	9991	9991	9992	9992	9992	9992	9993	9993
3.2	9993	9993	9994	9994	9994	9994	9994	9995	9995	9995
3.3	9995	9995	9995	9996	9996	9996	9996	9996	9996	9997
3.4	9997	9997	9799	9997	9997	9997	9997	9997	9997	9998
3.5	9998	9998	9998	9998	9998	9998	9998	9998	9998	9998
3.6	9998	9998	9999	9999	9999	9999	9999	9999	9999	9999

Table of the standard normal cumulative distribution function $\Phi(x) = Pr\{X < x\}$.

Example: suppose that $x = 1.85$. Start with $x = 1.8$ (the row beginning with 9641). In the sixth column one can find the Φ-value, multiplied by 10^4, corresponding to $x = 1.85$. Hence, $\Phi(1.85) = 0.9678$ and the right-sided p-value is equal to 0.0322.

Appendix I Statistical tables

Table A2 The upper critical values of the chi-squared distribution

							α							
k	0.995	0.99	0.975	0.95	0.90	0.75	0.50	0.25	0.10	0.05	0.025	0.01	0.005	0.001
1	–	–	0.001	0.004	0.016	0.102	0.455	1.32	2.71	3.84	5.02	6.63	7.88	10.8
2	0.010	0.020	0.051	0.103	0.211	0.575	1.39	2.77	4.61	5.99	7.38	9.21	10.6	13.8
3	0.072	0.115	0.216	0.352	0.584	1.21	2.37	4.11	6.25	7.81	9.35	11.3	12.8	16.3
4	0.207	0.297	0.484	0.711	1.06	1.92	3.36	5.39	7.78	9.49	11.1	13.3	14.9	18.5
5	0.412	0.554	0.831	1.15	1.61	2.67	4.35	6.63	9.24	11.1	12.8	15.1	16.7	20.5
6	0.676	0.872	1.24	1.64	2.20	3.45	5.35	7.84	10.6	12.6	14.4	16.8	18.5	22.5
7	0.989	1.24	1.69	2.17	2.83	4.25	6.35	9.04	12.0	14.1	16.0	18.5	20.3	24.3
8	1.34	1.65	2.18	2.73	3.49	5.07	7.34	10.2	13.4	15.5	17.5	20.1	22.0	26.1
9	1.73	2.09	2.70	3.33	4.17	5.90	8.34	11.4	14.7	16.9	19.0	21.7	23.6	27.9
10	2.16	2.56	3.25	3.94	4.87	6.74	9.34	12.5	16.0	18.3	20.5	23.2	25.2	29.6
11	2.60	3.05	3.82	4.57	5.58	7.58	10.3	13.7	17.3	19.7	21.9	24.7	26.8	31.3
12	3.07	3.57	4.40	5.23	6.30	8.44	11.3	14.8	18.5	21.0	23.3	26.2	28.3	32.9
13	3.57	4.11	5.01	5.89	7.04	9.30	12.3	16.0	19.8	22.4	24.7	27.7	29.8	34.5
14	4.07	4.66	5.63	6.57	7.79	10.2	13.3	17.1	21.1	23.7	26.1	29.1	31.3	36.1
15	4.60	5.23	6.26	7.26	8.55	11.0	14.3	18.2	22.3	25.0	27.5	30.6	32.8	37.7
16	5.14	5.81	6.91	7.96	9.31	11.9	15.3	19.4	23.5	26.3	28.8	32.0	34.3	39.3
17	5.70	6.41	7.56	8.67	10.1	12.8	16.3	20.5	24.8	27.6	30.2	33.4	35.7	40.8
18	6.26	7.01	8.23	9.39	10.9	13.7	17.3	21.6	26.0	28.9	31.5	34.8	37.2	42.3
19	6.84	7.63	8.91	10.1	11.7	14.6	18.3	22.7	27.2	30.1	32.9	36.2	38.6	43.8
20	7.43	8.26	9.59	10.9	12.4	15.5	19.3	23.8	28.4	31.4	34.2	37.6	40.0	45.3
21	8.03	8.90	10.3	11.6	13.2	16.3	20.3	24.9	29.6	32.7	35.5	38.9	41.1	46.8
22	8.64	9.54	11.0	12.3	14.0	17.2	21.3	26.0	30.8	33.9	36.8	40.3	42.8	48.3
23	9.26	10.2	11.7	13.1	14.8	18.1	22.3	27.1	32.0	35.2	38.1	41.6	44.2	49.7
24	9.89	10.9	12.4	13.8	15.7	19.0	23.3	28.2	33.2	36.4	39.4	43.0	45.6	51.2
25	10.5	11.5	13.1	14.6	16.5	19.9	24.3	29.3	34.4	37.7	40.6	44.3	46.9	52.6
26	11.2	12.2	13.8	15.4	17.3	20.8	25.3	30.4	35.6	38.9	41.9	45.6	48.3	54.1
27	11.8	12.9	14.6	16.2	18.1	21.7	26.3	31.5	36.7	40.1	43.2	47.0	49.6	55.5
28	12.5	13.6	15.3	16.9	18.9	22.7	27.3	32.6	37.9	41.3	44.5	48.3	51.0	56.9
29	13.1	14.3	16.0	17.7	19.8	23.6	28.3	33.7	39.1	42.6	45.7	49.6	52.3	58.3
30	13.8	15.0	16.8	18.5	20.6	24.5	29.3	34.8	40.3	43.8	47.0	50.9	53.7	59.7
40	20.7	22.2	24.4	26.5	29.1	33.7	39.3	45.6	51.8	55.8	59.3	63.7	66.8	73.4
50	28.0	29.7	32.4	34.8	37.7	42.9	49.3	56.3	63.2	67.5	71.4	76.2	79.5	86.7
60	35.5	37.5	40.5	43.2	46.5	52.3	59.3	67.0	74.4	79.1	83.3	88.4	91.6	99.6
70	43.3	45.4	48.8	51.7	55.3	61.7	69.3	77.6	85.5	90.5	95.0	100.4	104.2	112.3
80	51.2	53.5	57.2	60.4	64.3	71.1	79.3	88.1	96.6	101.9	106.6	112.3	116.3	124.8
90	59.2	61.8	65.6	69.1	73.3	80.6	89.3	98.6	107.6	113.1	118.1	124.1	128.3	137.2
100	67.3	70.1	74.2	77.9	82.4	90.1	99.3	109.1	118.5	124.3	129.6	135.8	140.2	149.4

Table of the upper critical values $\chi^2_k(\alpha)$ of the chi-squared distribution:
$Pr\{\chi^2_k \geq \chi^2_k(\alpha)\} = \alpha$, where k denotes the number of degrees of freedom.

Table A3 The upper critical values of Student's distribution

				α			
k	0.200	0.100	0.050	0.025	0.010	0.005	0.001
1	1.376	3.078	6.314	12.706	31.820	63.656	318.294
2	1.061	1.886	2.920	4.303	6.965	9.925	22.327
3	0.978	1.638	2.353	3.182	4.541	5.841	10.214
4	0.941	1.533	2.132	2.776	3.747	4.604	7.173
5	0.920	1.476	2.015	2.571	3.365	4.023	5.893
6	0.906	1.440	1.943	2.447	3.143	3.707	5.208
7	0.896	1.415	1.895	2.365	2.998	3.499	4.785
8	0.889	1.397	1.860	2.306	2.896	3.355	4.501
9	0.883	1.383	1.833	2.262	2.821	3.250	4.297
10	0.879	1.372	1.812	2.228	2.764	3.169	4.144
11	0.876	1.363	1.796	2.201	2.718	3.106	4.025
12	0.873	1.356	1.782	2.179	2.681	3.055	3.930
13	0.870	1.350	1.771	2.160	2.650	3.012	3.852
14	0.868	1.345	1.761	2.145	2.624	2.977	3.787
15	0.866	1.341	1.753	2.131	2.602	2.947	3.733
16	0.865	1.337	1.746	2.120	2.583	2.921	3.686
17	0.863	1.333	1.740	2.110	2.567	2.898	3.646
18	0.862	1.330	1.734	2.101	2.552	2.878	3.610
19	0.861	1.328	1.729	2.093	2.539	2.861	3.579
20	0.860	1.325	1.725	2.086	2.528	2.845	3.552
21	0.859	1.323	1.721	2.080	2.518	2.831	3.527
22	0.858	1.321	1.717	2.074	2.508	2.819	3.505
23	0.858	1.319	1.714	2.069	2.500	2.807	3.485
24	0.857	1.318	1.711	2.064	2.492	2.797	3.467
25	0.856	1.316	1.708	2.060	2.485	2.787	3.450
26	0.856	1.315	1.706	2.056	2.479	2.779	3.435
27	0.855	1.314	1.703	2.052	2.473	2.771	3.421
28	0.855	1.313	1.701	2.048	2.467	2.763	3.408
29	0.854	1.311	1.699	2.045	2.462	2.756	3.396
30	0.854	1.310	1.697	2.042	2.457	2.750	3.385
35	0.852	1.306	1.690	2.030	2.438	2.724	3.340
40	0.851	1.303	1.684	2.021	2.423	2.704	3.307
45	0.850	1.301	1.679	2.014	2.412	2.690	3.281
50	0.849	1.299	1.676	2.009	2.403	2.678	3.261
60	0.848	1.296	1.671	2.000	2.390	2.660	3.232
70	0.847	1.294	1.667	1.994	2.381	2.648	3.211
80	0.846	1.292	1.664	1.990	2.374	2.639	3.195
90	0.846	1.291	1.662	1.987	2.368	2.632	3.183
100	0.845	1.290	1.660	1.984	2.364	2.626	3.174
200	0.843	1.286	1.652	1.972	2.345	2.601	3.131
500	0.842	1.283	1.648	1.965	2.334	2.586	3.107
1000	0.842	1.282	1.646	1.962	2.330	2.581	3.098

Table of the upper critical values $t_k(\alpha)$ of Student's distribution:
$Pr\{T \geq t_k(\alpha)\} = \alpha$, where k denotes the number of degrees of freedom.

Appendix I Statistical tables

Table A4 Upper critical values of the F-distribution

v_2 \ v_1	1	2	3	4	5	6	7	8	9	10	12	15	20	24	30	40	60	120	∞
1	161	199	216	225	230	234	237	239	241	242	244	246	248	249	250	251	252	253	254
2	18.5	19.0	19.2	19.2	19.3	19.3	19.4	19.4	19.4	19.4	19.4	19.4	19.4	19.5	19.5	19.5	19.5	19.5	19.5
3	10.1	9.55	9.28	9.12	9.01	8.94	8.89	8.85	8.81	8.79	8.74	8.70	8.66	8.64	8.62	8.59	8.57	8.55	8.53
4	7.71	6.94	6.59	6.39	6.26	6.16	6.09	6.04	6.00	5.96	5.91	5.86	5.80	5.77	5.75	5.72	5.69	5.66	5.63
5	6.61	5.79	5.41	5.19	5.05	4.95	4.88	4.82	4.77	4.74	4.68	4.62	4.56	4.53	4.50	4.46	4.43	4.40	4.36
6	5.99	5.14	4.76	4.53	4.39	4.28	4.21	4.15	4.10	4.06	4.00	3.94	3.87	3.84	3.81	3.77	3.74	3.70	3.67
7	5.59	4.74	4.35	4.12	3.97	3.87	3.79	3.73	3.68	3.64	3.57	3.51	3.44	3.41	3.38	3.34	3.30	3.27	3.23
8	5.32	4.46	4.07	3.84	3.69	3.58	3.50	3.44	3.39	3.35	3.28	3.22	3.15	3.12	3.08	3.04	3.01	2.97	2.93
9	5.12	4.26	3.86	3.63	3.48	3.37	3.29	3.23	3.18	3.14	3.07	3.01	2.94	2.90	2.86	2.83	2.79	2.75	2.71
10	4.96	4.10	3.71	3.48	3.33	3.22	3.14	3.07	3.02	2.98	2.91	2.85	2.77	2.74	2.70	2.66	2.62	2.58	2.54
11	4.84	3.98	3.59	3.36	3.20	3.09	3.01	2.95	2.90	2.85	2.79	2.72	2.65	2.61	2.57	2.53	2.49	2.45	2.40
12	4.75	3.89	3.49	3.26	3.11	3.00	2.91	2.85	2.80	2.75	2.69	2.62	2.54	2.51	2.47	2.43	2.38	2.34	2.30
13	4.67	3.81	3.41	3.18	3.03	2.92	2.83	2.77	2.71	2.67	2.60	2.53	2.46	2.42	2.38	2.34	2.30	2.25	2.21
14	4.60	3.74	3.34	3.11	2.96	2.85	2.76	2.70	2.65	2.60	2.53	2.46	2.39	2.35	2.31	2.27	2.22	2.18	2.13
15	4.54	3.68	3.29	3.06	2.90	2.79	2.71	2.64	2.59	2.54	2.48	2.40	2.33	2.29	2.25	2.20	2.16	2.11	2.07
16	4.49	3.63	3.24	3.01	2.85	2.74	2.66	2.59	2.54	2.49	2.42	2.35	2.28	2.24	2.19	2.15	2.11	2.06	2.01
17	4.45	3.59	3.20	2.96	2.81	2.70	2.61	2.55	2.49	2.45	2.38	2.31	2.23	2.19	2.15	2.10	2.06	2.01	1.96
18	4.41	3.55	3.16	2.93	2.77	2.66	2.58	2.51	2.46	2.41	2.34	2.27	2.19	2.15	2.11	2.06	2.02	1.97	1.92
19	4.38	3.52	3.13	2.90	2.74	2.63	2.54	2.48	2.42	2.38	2.34	2.23	2.16	2.11	2.07	2.03	1.98	1.93	1.88
20	4.35	3.49	3.10	2.87	2.71	2.60	2.51	2.45	2.39	2.35	2.28	2.20	2.12	2.08	2.04	1.99	1.95	1.90	1.84
21	4.32	3.47	3.07	2.84	2.68	2.57	2.49	2.42	2.37	2.32	2.25	2.18	2.10	2.05	2.01	1.96	1.92	1.87	1.81
22	4.30	3.44	3.05	2.82	2.66	2.55	2.46	2.40	2.34	2.30	2.23	2.15	2.07	2.03	1.98	1.94	1.89	1.84	1.78
23	4.28	3.42	3.03	2.80	2.64	2.53	2.44	2.37	2.32	2.27	2.20	2.13	2.05	2.01	1.96	1.91	1.86	1.81	1.76
24	4.26	3.40	3.01	2.78	2.62	2.51	2.42	2.36	2.30	2.25	2.18	2.11	2.03	1.98	1.94	1.89	1.84	1.79	1.73
25	4.24	3.39	2.99	2.76	2.60	2.49	2.40	2.34	2.28	2.24	2.16	2.09	2.01	1.96	1.92	1.87	1.82	1.77	1.71
26	4.23	3.37	2.98	2.74	2.59	2.47	2.39	2.32	2.27	2.22	2.15	2.07	1.99	1.95	1.90	1.85	1.80	1.75	1.69
27	4.21	3.35	2.96	2.73	2.57	2.46	2.37	2.31	2.25	2.20	2.13	2.06	1.97	1.93	1.88	1.84	1.79	1.73	1.67
28	4.20	3.34	2.95	2.71	2.56	2.45	2.36	2.29	2.24	2.19	2.12	2.04	1.96	1.91	1.87	1.82	1.77	1.71	1.65
29	4.18	3.33	2.93	2.70	2.55	2.43	2.35	2.28	2.22	2.18	2.10	2.03	1.94	1.90	1.85	1.81	1.75	1.70	1.64
30	4.17	3.32	2.92	2.69	2.53	2.42	2.33	2.27	2.21	2.16	2.09	2.01	1.93	1.89	1.84	1.79	1.74	1.68	1.62
40	4.08	3.23	2.84	2.61	2.45	2.34	2.25	2.18	2.12	2.08	2.00	1.92	1.84	1.79	1.74	1.69	1.64	1.58	1.51
60	4.00	3.15	2.76	2.53	2.37	2.25	2.17	2.10	2.04	1.99	1.92	1.84	1.75	1.70	1.65	1.59	1.53	1.47	1.39
120	3.92	3.07	2.68	2.45	2.29	2.17	2.09	2.02	1.96	1.91	1.83	1.75	1.66	1.61	1.55	1.50	1.43	1.35	1.25
∞	3.84	3.00	2.60	2.37	2.21	2.10	2.01	1.94	1.88	1.83	1.75	1.67	1.57	1.52	1.46	1.39	1.32	1.22	1.00

Table of the upper critical values $F(v_1,v_2;0.05)$ of the F-distribution with v_1 and v_2 degrees of freedom.

Table A5 Upper critical values of T_1

		α
N	0.05	0.01
2	0.9750	0.9950
3	0.8709	0.9423
4	0.7679	0.8643
5	0.6838	0.7885
6	0.6161	0.7218
7	0.5612	0.6644
8	0.5157	0.6152
9	0.4775	0.5727
10	0.4450	0.5358
12	0.3924	0.4751
15	0.3346	0.4069
20	0.2705	0.3297
24	0.2354	0.2871
30	0.1980	0.2412
40	0.1576	0.1915
60	0.1131	0.1371
120	0.0632	0.0759
∞	0	0

Test for one upper outlier in a sample of exponentially distributed random variables. See subsection 2.3.3. Adapted from Eisenhart *et al.* (1947).

Table A6 Upper critical values of T_2

	α	
N	0.05	0.01
3	0.989	0.998
4	0.944	0.977
5	0.900	0.949
6	0.863	0.922
7	0.833	0.898
8	0.809	0.878
9	0.788	0.860
10	0.770	0.845
12	0.740	0.820
14	0.717	0.799
16	0.698	0.782
18	0.682	0.768
20	0.668	0.755
30	0.621	0.711
40	0.590	0.683
50	0.569	0.662
60	0.553	0.647
70	0.540	0.633
80	0.529	0.623
90	0.520	0.615
100	0.512	0.601
150	0.483	0.579
200	0.466	0.560

Test for two upper outliers in a sample of exponentially distributed random variables. See subsection 2.3.3.

Table A7 Lower critical values of T_3

	α	
N	0.05	0.01
3	0.00844	0.00167
4	0.00424	0.000836
5	0.00255	0.000502
6	0.00170	0.000335
7	0.00122	0.000239
8	0.000913	0.000179
9	0.000710	0.000140
10	0.000568	0.000112
12	0.000388	0.0000761
14	0.000281	0.0000552
16	0.000213	0.0000419
18	0.000167	0.0000328
20	0.000135	0.0000264
30	0.0000589	0.0000116
40	0.0000329	0.00000644
50	0.0000209	0.00000410
100	0.00000518	0.00000102

Test for one lower outlier in a sample of exponentially distributed random variables. See subsection 2.3.3. Adapted from Barnett and Lewis (1980) with permission of John Wiley & Sons Limited.

Table A8 Lower critical values of T_4

N	α	
3	0.129	0.0577
4	0.0553	0.0240
5	0.0309	0.0132
6	0.0197	0.00844
7	0.0137	0.00584
8	0.0101	0.00428
9	0.00776	0.00328
10	0.00614	0.00259
12	0.00412	0.00173
14	0.00295	0.00124
16	0.00222	0.000933
18	0.00173	0.000726
20	0.00138	0.000581
30	0.000597	0.000249
40	0.000331	0.000138
50	0.000209	0.0000876
100	0.0000515	0.0000214

Test for two lower outliers in a sample of exponentially distributed random variables. See subsection 2.3.3.

Table A9 Upper critical values of T_5 and T_6

	α	
N	0.05	0.01
3	0.974	0.995
4	0.894	0.957
5	0.830	0.912
6	0.782	0.875
7	0.746	0.845
8	0.717	0.821
9	0.694	0.800
10	0.675	0.783
11	0.658	0.768
12	0.644	0.755
13	0.631	0.743
14	0.620	0.733
15	0.610	0.724
16	0.601	0.715
17	0.593	0.707
18	0.586	0.700
19	0.579	0.694
20	0.573	0.687
21	0.567	0.682

Test for one or two upper outlier(s) in a sample of two-parameter exponentially distributed random variables (displaced exponential). The critical values for T_5 are listed. See subsection 2.3.3. The test statistic T_6 has the same distribution with N replaced by $N - 1$. Adapted from Barnett and Lewis (1980) with permission of John Wiley & Sons Limited.

Table A10 Upper critical values of Λ_1

N	α			
	0.1	0.05	0.025	0.01
20	5.76	6.65	7.78	9.24
30	6.15	7.18	8.34	9.87
40	6.43	7.56	8.74	10.32
50	6.64	7.85	9.04	10.67
60	6.82	8.09	9.30	10.95
70	6.97	8.30	9.51	11.19
80	7.10	8.47	9.69	11.40
90	7.21	8.63	9.86	11.58
100	7.31	8.76	10.00	11.74
125	7.52	9.06	10.31	12.09
150	7.70	9.30	10.57	12.37
175	7.85	9.50	10.78	12.61
200	7.98	9.68	10.96	12.82

Non-parametric test for testing one against zero abrupt changes in the mean boutlength. See subsection 3.3.1. Adapted from Meelis *et al.* (1990) with permission of Elsevier Science Publishers BV.

Table A11 Upper critical values of Λ_2

N	α			
	0.1	0.05	0.025	0.01
20	4.61	5.14	5.77	6.37
30	5.37	6.04	6.77	7.54
40	5.91	6.68	7.47	8.37
50	6.33	7.18	8.02	9.01
60	6.67	7.58	8.47	9.54
70	6.96	7.92	8.85	9.99
80	7.21	8.22	9.18	10.37
90	7.44	8.48	9.46	10.71
100	7.63	8.71	9.72	11.02
125	8.05	9.21	10.27	11.66
150	8.40	9.61	10.72	12.19
175	8.69	9.96	11.10	12.63
200	8.94	10.25	11.42	13.02

Non-parametric test for testing two against one abrupt changes in the mean bout length. See subsection 3.3.1. Adapted from Meelis *et al.* (1990) with permission of Elsevier Science Publishers BV.

Table A12 Upper critical values of Λ_3

		α		
N	0.1	0.05	0.025	0.01
20	6.46	7.18	7.86	8.75
30	7.75	8.61	9.41	10.45
40	8.67	9.62	10.52	11.66
50	9.39	10.40	11.37	12.60
60	9.97	11.05	12.07	13.37
70	10.47	11.59	12.67	14.02
80	10.89	12.06	13.18	14.58
90	11.27	12.47	13.63	15.07
100	11.61	12.84	14.04	15.52
125	12.32	13.63	14.89	16.46
150	12.90	14.27	15.59	17.22
175	13.40	14.81	16.18	17.87
200	13.83	15.28	16.70	18.43

Non-parametric test for testing three against two abrupt changes in the mean bout length. See subsection 3.3.1. Adapted from Meelis *et al.* (1990) with permission of Elsevier Science Publishers BV.

Table A13 Upper critical values of Λ_1^*

		α		
N	0.1	0.05	0.025	0.01
10	6.35	7.50	8.97	11.00
20	6.99	8.33	9.83	11.75
30	7.33	8.75	10.27	12.14
40	7.56	9.03	10.56	12.39
50	7.73	9.23	10.77	12.58
60	7.86	9.39	10.93	12.73
70	7.97	9.51	11.07	12.84
80	8.07	9.62	11.17	12.94
90	8.15	9.71	11.27	13.02
100	8.22	9.79	11.35	13.09
125	8.36	9.95	11.51	13.24
150	8.48	10.07	11.63	13.35
175	8.57	10.17	11.73	13.43
200	8.64	10.25	11.81	13.50
300	8.87	10.48	12.02	13.70
400	9.01	10.62	12.15	13.82
500	9.11	10.72	12.23	13.89
600	9.19	10.80	12.29	13.94
700	9.25	10.86	12.33	13.99
800	9.31	10.91	12.37	14.02
900	9.35	10.95	12.39	14.05
1000	9.39	10.98	12.41	14.07
2000	9.61	11.14	12.48	14.14

Likelihood ratio test for one against zero change points for exponentially distributed bout lengths. See subsection 3.3.1. Adapted from Haccou and Meelis (1988) and Meelis *et al.* (1990) with permission of Gordon and Breach Science Publishers S.A. and Elsevier Science Publishers BV, respectively.

Table A14 Upper critical values of Λ_2*

		α		
N	0.1	0.05	0.025	0.01
10	5.88	7.07	8.22	9.92
20	7.16	8.61	9.95	11.70
30	7.99	9.56	10.97	12.76
40	8.62	10.25	11.68	13.51
50	9.13	10.79	12.23	14.09
60	9.56	11.23	12.66	14.56
70	9.93	11.60	13.02	14.95
80	10.26	11.93	13.33	15.28
90	10.55	12.21	13.59	15.57
100	10.82	12.46	13.82	15.83
125	11.40	12.99	14.29	16.36
150	11.89	13.41	14.65	16.79
175	12.31	13.77	14.94	17.13
200	12.68	14.07	15.18	17.42

Likelihood ratio test for two against one change point for exponentially distributed bout lengths. See subsection 3.3.1. Adapted from Haccou and Meelis (1988) and Meelis *et al.* (1990) with permission of Gordon and Breach Science Publishers S.A. and Elsevier Science Publishers BV, respectively.

Table A15 Upper critical values of Λ_3*

		α		
N	0.1	0.05	0.025	0.01
10	7.86	9.37	10.76	11.92
20	20.17	22.96	25.21	27.61
30	33.69	37.45	40.26	43.74
40	47.64	52.15	55.38	59.77
50	61.71	66.83	70.36	75.56
60	75.75	81.36	85.15	91.03
70	89.68	95.70	99.70	106.17
80	103.46	109.81	114.01	120.98
90	117.05	123.68	128.07	135.46
100	130.45	137.31	141.89	149.64
125	163.05	170.36	175.41	183.80
150	194.40	202.00	207.56	216.30
175	224.55	232.34	238.47	247.32
200	253.56	261.49	268.24	277.02

Likelihood ratio test for three or more against two change points for exponentially distributed bout lengths. See subsection 3.3.1. Adapted from Haccou and Meelis (1988) and Meelis *et al.* (1990) with permission of Gordon and Breach Science Publishers S.A. and Elsevier Science Publishers BV, respectively.

Table A16 Upper critical values of *W*

	α			
n	0.00	0.90	0.95	0.99
3	0.58	12.71	25.45	127.32
4	0.52	5.34	7.65	17.28
5	0.47	4.18	5.39	9.46
6	0.44	3.73	4.60	7.17
7	0.41	3.48	4.20	6.14
8	0.39	3.32	3.95	5.56
9	0.37	3.21	3.78	5.19
10	0.36	3.14	3.66	4.93
15	0.30	2.97	3.36	4.32
20	0.26	2.90	3.28	4.13
25	0.24	2.89	3.23	3.94
30	0.22	2.86	3.19	3.86
35	0.20	2.88	3.21	3.87
40	0.19	2.88	3.17	3.77
45	0.18	2.86	3.18	3.79
50	0.17	2.87	3.16	3.79

Likelihood ratio test for a change point in a sequence of normally distributed bout lengths with common, but unknown, variance. See subsection 3.3.1. The critical values for *n* = 3 (1) 10 are exact, and for *n* > 10 obtained by simulation. Adapted from Worsley (1979) with permission of the American Statistical Association.

Table A17 Upper critical values of $\sqrt{n}\,D_n$

n	0.95	0.90	0.80	0.70	0.50	α 0.30	0.20	0.10	0.05	0.01	0.001
2	0.4422	0.4525	0.4719	0.5115	0.6133	0.7119	0.7788	0.8392	0.8673	0.8887	0.8934
3	0.3982	0.4407	0.5022	0.5483	0.6313	0.7140	0.7808	0.8852	0.9540	1.0398	1.0905
4	0.4144	0.4561	0.5090	0.5531	0.6326	0.7369	0.8014	0.8884	0.9687	1.1148	1.2429
5	0.4212	0.4589	0.5121	0.5553	0.6434	0.7418	0.8056	0.9044	0.9884	1.1464	1.3001
6	0.4227	0.4611	0.5148	0.5610	0.6479	0.7459	0.8133	0.9141	1.0007	1.1631	1.3466
7	0.4244	0.4630	0.5189	0.5651	0.6504	0.7508	0.8196	0.9211	1.0084	1.1797	1.3708
8	0.4260	0.4657	0.5219	0.5674	0.6530	0.7554	0.8241	0.9260	1.0153	1.1902	1.3897
9	0.4279	0.4682	0.5239	0.5691	0.6557	0.7587	0.8275	0.9303	1.0212	1.1986	1.4037
10	0.4299	0.4700	0.5254	0.5708	0.6584	0.7613	0.8303	0.9343	1.0258	1.2057	1.4146
12	0.4328	0.4726	0.5281	0.5741	0.6623	0.7653	0.8351	0.9402	1.0327	1.2164	1.4315
14	0.4348	0.4746	0.5306	0.5770	0.6652	0.7687	0.8390	0.9446	1.0381	1.2243	1.4435
16	0.4364	0.4764	0.5329	0.5793	0.6675	0.7716	0.8420	0.9482	1.0424	1.2304	1.4526
18	0.4378	0.4782	0.5347	0.5810	0.6695	0.7738	0.8444	0.9511	1.0458	1.2352	1.4597
20	0.4392	0.4797	0.5362	0.5826	0.6713	0.7757	0.8465	0.9536	1.0486	1.2392	1.4656
22	0.4405	0.4810	0.5374	0.5839	0.6728	0.7774	0.8484	0.9557	1.0511	1.2425	1.4704
24	0.4416	0.4821	0.5386	0.5852	0.6741	0.7788	0.8499	0.9575	1.0531	1.2454	1.4745
26	0.4425	0.4830	0.5396	0.5863	0.6753	0.7801	0.8513	0.9591	1.0549	1.2478	1.4780
28	0.4434	0.4839	0.5405	0.5873	0.6763	0.7813	0.8525	0.9605	1.0565	1.2500	1.4810
30	0.4441	0.4847	0.5414	0.5881	0.6772	0.7823	0.8537	0.9617	1.0580	1.2519	1.4837
35	0.4458	0.4864	0.5432	0.5900	0.6793	0.7845	0.8560	0.9644	1.0609	1.2558	1.4891
40	0.4472	0.4879	0.5447	0.5915	0.6809	0.7862	0.8579	0.9665	1.0633	1.2588	1.4933
45	0.4484	0.4891	0.5460	0.5928	0.6823	0.7877	0.8594	0.9682	1.0652	1.2613	1.4966
50	0.4494	0.4901	0.5470	0.5939	0.6834	0.7890	0.8608	0.9696	1.0668	1.2634	1.4996
60	0.4510	0.4918	0.5488	0.5957	0.6853	0.7910	0.8629	0.9720	1.0694	1.2665	1.5053
70	0.4523	0.4931	0.5502	0.5971	0.6868	0.7925	0.8645	0.9737	1.0714	1.2692	*
80	0.4534	0.4942	0.5513	0.5983	0.6880	0.7938	0.8658	0.9752	1.0729	1.2711	*
90	0.4543	0.4951	0.5522	0.5992	0.6890	0.7949	0.8669	0.9764	1.0742	1.2722	*
100	0.4550	0.4959	0.5530	0.6000	0.6898	0.7957	0.8678	0.9773	1.0753	1.2743	*

*Value not available.

Kolmogorov–Smirnov test for testing for exponentiality with unknown mean. See subsection 4.6.2.
Adapted from Durbin (1975) with permission of the Biometrika Trustees.

Table A18 Large-sample upper critical values of T_n

			α		
n	0.15	0.10	0.05	0.025	0.01
∞	0.926	0.990	1.094	1.190	1.308

Improved large-sample approximation for the Kolmogorov-Smirnov test for testing for exponentiality with unknown mean. See subsection 4.6.2. Adapted from Pearson and Hartley (1972) with permission of the Biometrika Trustees.

Table A19 Upper critical values of W^2

			α		
	0.15	0.10	0.05	0.025	0.01
W^2	0.149	0.177	0.224	0.273	0.337

The Cramér-von Mises test for testing for exponentiality with unknown mean. See subsection 4.6.2. Adapted from Stephens (1976) with permission of the Institute of Mathematical Statistics.

Table A20   Upper critical values of *W*

						α					
n	0.995	0.99	0.975	0.95	0.90	0.50	0.10	0.05	0.025	0.01	0.005
3	0.2519	0.2538	0.2596	0.2697	0.2915	0.5714	0.9709	0.9926	0.9981	0.9997	0.99993
4	0.1241	0.1302	0.1434	0.1604	0.1891	0.3768	0.7514	0.8581	0.9236	0.9680	0.9837
5	0.0845	0.0905	0.1048	0.1187	0.1442	0.2875	0.5547	0.6682	0.7590	0.8600	0.9192
6	0.0610	0.0665	0.0802	0.0956	0.1173	0.2276	0.4292	0.5089	0.5842	0.6775	0.7501
7	0.0514	0.0591	0.0700	0.0810	0.0986	0.1874	0.3474	0.4162	0.4852	0.5706	0.6426
8	0.0454	0.0512	0.0614	0.0710	0.0852	0.1625	0.2934	0.3497	0.4033	0.4848	0.5428
9	0.0404	0.0442	0.0537	0.0633	0.0751	0.1415	0.2553	0.3005	0.3454	0.4015	0.4433
10	0.0369	0.0404	0.0487	0.0568	0.0678	0.1225	0.2178	0.2525	0.2879	0.3391	0.3701
11	0.0339	0.0380	0.0447	0.0528	0.0616	0.1112	0.1934	0.2265	0.2619	0.3039	0.3314
12	0.0311	0.0358	0.0410	0.0494	0.0567	0.1009	0.1723	0.2019	0.2364	0.2716	0.2978
13	0.0287	0.0337	0.0382	0.0460	0.0528	0.0925	0.1563	0.1829	0.2113	0.2422	0.2642
14	0.0265	0.0317	0.0362	0.0428	0.0496	0.0847	0.1417	0.1647	0.1862	0.2131	0.2315
15	0.0247	0.0298	0.0344	0.0398	0.0466	0.0778	0.1285	0.1485	0.1669	0.1926	0.2123
16	0.0233	0.0280	0.0326	0.0374	0.0438	0.0728	0.1187	0.1355	0.1542	0.1770	0.1931
17	0.0222	0.0264	0.0310	0.0352	0.0412	0.0684	0.1099	0.1257	0.1423	0.1614	0.1794
18	0.0212	0.0250	0.0294	0.0332	0.0388	0.0640	0.1015	0.1164	0.1311	0.1483	0.1668
19	0.0203	0.0238	0.0278	0.0314	0.0368	0.0600	0.0935	0.1071	0.1199	0.1374	0.1452
20	0.0196	0.0227	0.0264	0.0302	0.0352	0.0570	0.0884	0.1002	0.1121	0.1286	0.1369
21	0.0190	0.0217	0.0250	0.0290	0.0337	0.0540	0.0839	0.0948	0.1054	0.1198	0.1288
22	0.0185	0.0208	0.0238	0.0278	0.0323	0.0516	0.0794	0.0894	0.0988	0.1118	0.1213
23	0.0181	0.0201	0.0230	0.0266	0.0310	0.0492	0.0749	0.0836	0.0933	0.1043	0.1142
24	0.0177	0.0194	0.0224	0.0256	0.0298	0.0468	0.0704	0.0788	0.0882	0.0984	0.1071
25	0.0173	0.0188	0.0218	0.0248	0.0286	0.0447	0.0668	0.0749	0.0836	0.0927	0.1000
26	0.0169	0.0182	0.0213	0.0240	0.0274	0.0426	0.0636	0.0712	0.0791	0.0885	0.0948
27	0.0165	0.0177	0.0208	0.0232	0.0264	0.0407	0.0606	0.0678	0.0747	0.0843	0.0896
28	0.0161	0.0172	0.0203	0.0225	0.0256	0.0391	0.0576	0.0649	0.0706	0.0801	0.0859
29	0.0157	0.0168	0.0198	0.0219	0.0249	0.0377	0.0555	0.0621	0.0671	0.0759	0.0822
30	0.0153	0.0164	0.0193	0.0213	0.0242	0.0364	0.0536	0.0593	0.0643	0.0719	0.0786
31	0.0149	0.0160	0.0188	0.0207	0.0235	0.0352	0.0518	0.0569	0.0615	0.0686	0.0753
32	0.0145	0.0156	0.0183	0.0201	0.0229	0.0340	0.0491	0.0547	0.0591	0.0661	0.0722
33	0.0141	0.0152	0.0178	0.0195	0.0223	0.0329	0.0475	0.0527	0.0573	0.0636	0.0691
34	0.0137	0.0148	0.0173	0.0190	0.0217	0.0319	0.0459	0.0507	0.0555	0.0611	0.0660
35	0.0133	0.0144	0.0168	0.0185	0.0211	0.0309	0.0444	0.0488	0.0537	0.0588	0.0639
36	0.0129	0.0141	0.0164	0.0180	0.0205	0.0300	0.0429	0.0470	0.0519	0.0567	0.0608
37	0.0125	0.0138	0.0160	0.0176	0.0200	0.0291	0.0414	0.0454	0.0501	0.0546	0.0578
38	0.0122	0.0135	0.0156	0.0172	0.0195	0.0283	0.0400	0.0440	0.0483	0.0525	0.0553
39	0.0120	0.0133	0.0152	0.0168	0.0190	0.0275	0.0386	0.0426	0.0465	0.0512	0.0531
40	0.0118	0.0131	0.0148	0.0164	0.0186	0.0267	0.0375	0.0414	0.0447	0.0499	0.0510
41	0.0116	0.0129	0.0144	0.0161	0.0182	0.0260	0.0364	0.0402	0.0430	0.0476	0.0493
42	0.0114	0.0127	0.0140	0.0158	0.0178	0.0253	0.0355	0.0389	0.0417	0.0464	0.0482
43	0.0112	0.0125	0.0137	0.0155	0.0174	0.0248	0.0346	0.0379	0.0405	0.0452	0.0471
44	0.0110	0.0123	0.0134	0.0152	0.0170	0.0243	0.0338	0.0369	0.0394	0.0440	0.0460
45	0.0108	0.0121	0.0131	0.0149	0.0166	0.0238	0.0329	0.0359	0.0385	0.0423	0.0449
46	0.0106	0.0119	0.0129	0.0146	0.0162	0.0233	0.0320	0.0349	0.0376	0.0416	0.0438
47	0.0104	0.0117	0.0127	0.0143	0.0158	0.0228	0.0311	0.0340	0.0367	0.0394	0.0427
48	0.0103	0.0115	0.0125	0.0141	0.0155	0.0223	0.0303	0.0332	0.0358	0.0382	0.0416
49	0.0102	0.0113	0.0123	0.0139	0.0152	0.0218	0.0295	0.0324	0.0349	0.0371	0.0405
50	0.0101	0.0111	0.0122	0.0137	0.0149	0.0213	0.0288	0.0317	0.0340	0.0360	0.0394
51	0.0100	0.0109	0.0120	0.0135	0.0147	0.0209	0.0282	0.0310	0.0331	0.0349	0.0383
52	0.0099	0.0107	0.0119	0.0133	0.0145	0.0205	0.0276	0.0303	0.0323	0.0341	0.0373
53	0.0097	0.0106	0.0118	0.0131	0.0143	0.0201	0.0270	0.0296	0.0315	0.0332	0.0363
54	0.0095	0.0104	0.0116	0.0129	0.0141	0.0197	0.0264	0.0289	0.0307	0.0329	0.0353
55	0.0094	0.0103	0.0115	0.0127	0.0139	0.0193	0.0258	0.0282	0.0299	0.0321	0.0343

Table A20 Continued

| | | | | | α | | | | | |
n	0.995	0.99	0.975	0.95	0.90	0.50	0.10	0.05	0.025	0.01	0.005
56	0.0093	0.0102	0.0113	0.0125	0.0137	0.0189	0.0252	0.0275	0.0292	0.0313	0.0333
57	0.0092	0.0101	0.0112	0.0123	0.0135	0.0185	0.0247	0.0268	0.0285	0.0306	0.0324
58	0.0091	0.0100	0.0110	0.0121	0.0133	0.0182	0.0242	0.0262	0.0279	0.0301	0.0318
59	0.0090	0.0098	0.0109	0.0119	0.0131	0.0179	0.0238	0.0257	0.0274	0.0296	0.0312
60	0.0089	0.0095	0.0108	0.0117	0.0129	0.0176	0.0234	0.0252	0.0270	0.0291	0.0306
61	0.0088	0.0093	0.0107	0.0115	0.0127	0.0173	0.0230	0.0247	0.0266	0.0286	0.0301
62	0.0087	0.0092	0.0105	0.0113	0.0125	0.0170	0.0226	0.0242	0.0262	0.0281	0.0296
63	0.0086	0.0091	0.0104	0.0112	0.0123	0.0167	0.0222	0.0238	0.0257	0.0276	0.0291
64	0.0085	0.0090	0.0102	0.0111	0.0121	0.0164	0.0218	0.0234	0.0252	0.0271	0.0286
65	0.0084	0.0089	0.0101	0.0109	0.0119	0.0161	0.0215	0.0230	0.0247	0.0266	0.0281
66	0.0082	0.0088	0.0099	0.0108	0.0117	0.0159	0.0211	0.0225	0.0242	0.0261	0.0276
67	0.0081	0.0087	0.0098	0.0107	0.0115	0.0157	0.0207	0.0221	0.0237	0.0256	0.0271
68	0.0080	0.0086	0.0096	0.0105	0.0114	0.0155	0.0204	0.0217	0.0232	0.0251	0.0266
69	0.0079	0.0085	0.0095	0.0104	0.0113	0.0152	0.0198	0.0213	0.0227	0.0246	0.0261
70	0.0078	0.0084	0.0094	0.0103	0.0111	0.0150	0.0194	0.0209	0.0222	0.0241	0.0256
71	0.0077	0.0083	0.0093	0.0102	0.0109	0.0147	0.0191	0.0205	0.0218	0.0237	0.0251
72	0.0076	0.0082	0.0092	0.0101	0.0108	0.0145	0.0188	0.0201	0.0214	0.0232	0.0246
73	0.0075	0.0081	0.0091	0.0100	0.0107	0.0143	0.0185	0.0198	0.0211	0.0228	0.0241
74	0.0074	0.0080	0.0090	0.0098	0.0106	0.0141	0.0182	0.0195	0.0208	0.0224	0.0236
75	0.0073	0.0079	0.0089	0.0097	0.0105	0.0139	0.0179	0.0192	0.0205	0.0220	0.0231
76	0.0073	0.0078	0.0088	0.0096	0.0104	0.0137	0.0176	0.0189	0.0202	0.0217	0.0227
77	0.0072	0.0077	0.0087	0.0095	0.0103	0.0135	0.0173	0.0186	0.0199	0.0214	0.0223
78	0.0071	0.0077	0.0086	0.0093	0.0101	0.0134	0.0170	0.0183	0.0196	0.0211	0.0219
79	0.0070	0.0076	0.0085	0.0092	0.0100	0.0132	0.0168	0.0180	0.0193	0.0208	0.0215
80	0.0070	0.0075	0.0084	0.0091	0.0099	0.0131	0.0166	0.0177	0.0190	0.0205	0.0211
81	0.0069	0.0074	0.0083	0.0090	0.0098	0.0129	0.0164	0.0175	0.0187	0.0202	0.0207
82	0.0068	0.0074	0.0082	0.0088	0.0097	0.0128	0.0162	0.0173	0.0184	0.0199	0.0203
83	0.0067	0.0073	0.0081	0.0087	0.0096	0.0126	0.0160	0.0170	0.0181	0.0196	0.0199
84	0.0067	0.0073	0.0080	0.0086	0.0095	0.0125	0.0158	0.0168	0.0178	0.0193	0.0196
85	0.0066	0.0072	0.0079	0.0085	0.0094	0.0123	0.0156	0.0166	0.0174	0.0190	0.0193
86	0.0066	0.0071	0.0078	0.0085	0.0093	0.0122	0.0154	0.0164	0.0172	0.0187	0.0190
87	0.0065	0.0071	0.0077	0.0084	0.0092	0.0120	0.0152	0.0162	0.0170	0.0184	0.0187
88	0.0065	0.0070	0.0077	0.0084	0.0091	0.0119	0.0150	0.0160	0.0168	0.0181	0.0185
89	0.0064	0.0070	0.0076	0.0083	0.0090	0.0117	0.0148	0.0158	0.0166	0.0179	0.0183
90	0.0064	0.0069	0.0075	0.0082	0.0089	0.0116	0.0147	0.0156	0.0164	0.0176	0.0181
91	0.0063	0.0068	0.0075	0.0082	0.0088	0.0114	0.0145	0.0154	0.0162	0.0173	0.0179
92	0.0063	0.0068	0.0074	0.0081	0.0087	0.0113	0.0143	0.0153	0.0160	0.0171	0.0177
93	0.0062	0.0067	0.0073	0.0081	0.0086	0.0112	0.0141	0.0151	0.0158	0.0168	0.0175
94	0.0062	0.0067	0.0073	0.0080	0.0085	0.0110	0.0139	0.0149	0.0156	0.0165	0.0173
95	0.0061	0.0066	0.0072	0.0079	0.0084	0.0109	0.0138	0.0147	0.0154	0.0163	0.0171
96	0.0061	0.0065	0.0072	0.0078	0.0083	0.0108	0.0136	0.0145	0.0153	0.0161	0.0169
97	0.0060	0.0065	0.0071	0.0077	0.0082	0.0107	0.0134	0.0143	0.0152	0.0159	0.0167
98	0.0060	0.0064	0.0070	0.0076	0.0081	0.0105	0.0133	0.0142	0.0151	0.0157	0.0165
99	0.0059	0.0064	0.0070	0.0075	0.0080	0.0104	0.0132	0.0140	0.0150	0.0155	0.0163
100	0.0059	0.0063	0.0069	0.0074	0.0079	0.0103	0.0131	0.0139	0.0149	0.0153	0.0161

The Shapiro–Wilk W test for goodness of fit for a sample of two-parameter exponentially distributed random variables. See subsection 4.6.2. Adapted from Shapiro and Wilk (1972).

Table A21 Critical values of Barlow's test

			α		
$n-1$	0.10	0.05	0.025	0.01	0.005
2	1.553	1.684	1.776	1.859	1.900
3	2.157	2.331	2.469	2.609	2.689
4	2.753	2.953	3.120	3.300	3.411
5	3.339	3.565	3.754	3.963	4.097
6	3.917	4.166	4.376	4.610	4.762
7	4.489	4.759	4.988	5.244	5.413
8	5.056	5.346	5.592	5.869	6.053
9	5.619	5.927	6.189	6.487	6.683
10	6.178	6.504	6.781	7.097	7.307
11	6.735	7.077	7.369	7.702	7.924
12	7.289	7.647	7.953	8.302	8.535
13	7.841	8.214	8.533	8.898	9.142
14	8.391	8.779	9.111	9.491	9.745
15	8.939	9.341	9.685	10.079	10.344
16	9.486	9.901	10.257	10.666	10.940
17	10.032	10.460	10.827	11.249	11.532
18	10.576	11.016	11.395	11.830	12.122
19	11.118	11.571	11.961	12.408	12.709
20	11.660	12.125	12.525	12.985	13.295
21	12.201	12.678	13.088	13.559	13.877
22	12.741	13.229	13.649	14.132	14.458
23	13.280	13.779	14.208	14.704	15.037
24	13.818	14.328	14.767	15.273	15.615
25	14.355	14.876	15.324	15.841	16.190

Barlow's 'cumulative total time on test' for testing against increasing or decreasing termination rate. See subsection 4.6.3. Partially adapted from Barlow *et al.* (1972) with permission of John Wiley & Sons Limited.

Table A22 Upper critical values of LR_n

	α		
n	0.10	0.05	0.01
10	7.967	9.458	12.833
20	8.404	9.914	13.320
30	8.633	10.151	13.573
40	8.785	10.308	13.738
50	8.897	10.424	13.860
60	8.985	10.515	13.956
70	9.057	10.589	14.034
80	9.118	10.652	14.100 *
90	9.170	10.706	14.156
100	9.217	10.753	14.206

* Corrected value.

Test of Matthews and Farewell for an abrupt change in the termination rate. See subsection 4.6.4. The upper critical values are due to Worsley (1988). Adapted from Worsley (1988) with permission of The Biometric Society.

Table A23 Upper critical values of $n(1-R_n^2)$

	ε										
α	0.00	0.05	0.10	0.15	0.20	0.25	0.30	0.35	0.40	0.45	0.50
0.10	1.06	1.10	1.17	1.25	1.33	1.45	1.59	1.78	2.06	2.43	2.97
0.05	1.39	1.43	1.56	1.66	1.77	1.94	2.12	2.39	2.77	3.25	4.05
0.01	2.27	2.34	2.53	2.66	2.88	3.15	3.44	3.87	4.49	5.38	6.64

Chen's correlation goodness-of-fit test for censored data. See subsection 4.7.2 for further explanation. ε denotes the expected proportion of censors.
The table is adapted from Chen (1984) with permission of the Biometrika Trustees.

Table A24   Upper critical values of K_3

n_1	n_2	n_3	0.25	0.20	α 0.15	0.10	0.05
1	1	4	3.571	3.571	–	–	–
1	1	5	3.857	3.857	3.857	–	–
1	2	2	3.600	3.600	–	–	–
1	2	3	3.524	3.524	3.857	4.286	–
1	2	4	3.161	3.161	3.750	4.500	–
1	2	5	2.867	3.333	3.783	4.200	5.000
1	3	3	3.143	3.286	4.000	4.571	5.143
1	3	4	2.764	3.208	3.764	4.056	5.208
1	3	5	2.951	3.218	3.378	4.018	4.960
1	4	4	2.967	3.267	3.867	4.167	4.967
1	4	5	2.913	3.087	3.524	3.987	4.986
1	5	5	2.909	3.236	3.527	4.109	5.127
2	2	2	3.714	3.714	4.571	4.571	–
2	2	3	3.429	3.929	4.464	4.500	4.714
2	2	4	3.125	3.667	4.167	4.458	5.333
2	2	5	3.240	3.360	4.093	4.373	5.160
2	3	3	3.139	3.778	4.111	4.556	5.361
2	3	4	3.111	3.444	4.000	4.511	5.444
2	3	5	3.022	3.414	3.942	4.651	5.251
2	4	4	3.054	3.464	4.009	4.554	5.454
2	4	5	2.914	3.364	3.818	4.541	5.273
2	5	5	3.023	3.392	3.862	4.508	5.338
3	3	3	3.289	3.467	4.267	4.622	5.600
3	3	4	3.027	3.391	3.836	4.709	5.727
3	3	5	2.970	3.442	3.927	4.533	5.648
3	4	4	2.932	3.417	3.848	4.546	5.598
3	4	5	2.953	3.318	3.831	4.549	5.631
3	5	5	2.936	3.429	3.807	4.545	5.706
4	4	4	3.038	3.500	3.962	4.654	5.692
4	4	5	2.918	3.330	3.910	4.619	5.618
4	5	5	2.886	3.311	3.883	4.523	5.643
5	5	5	2.960	3.420	3.860	4.560	5.780

The Kruskal-Wallis non-parametric k-sample test. Table of the small-sample critical values for $k = 3$. See subsection 5.3.1. Adapted from Hollander and Wolfe (1973) with permission of John Wiley & Sons Limited.

Table A25 Upper critical values of r_s

N	0.25	0.10	0.05	0.025	α 0.01	0.005	0.0025	0.001	0.0005
4	0.600	1.000	1.000						
5	0.500	0.800	0.900	1.000	1.000				
6	0.371	0.657	0.829	0.886	0.943	1.000	1.000		
7	0.321	0.571	0.714	0.786	0.893	0.929	0.964	1.000	1.000
8	0.310	0.524	0.643	0.738	0.833	0.881	0.905	0.952	0.976
9	0.267	0.483	0.600	0.700	0.783	0.833	0.867	0.917	0.933
10	0.248	0.455	0.564	0.648	0.745	0.794	0.830	0.879	0.903
11	0.236	0.427	0.536	0.618	0.709	0.755	0.800	0.845	0.873
12	0.224	0.406	0.503	0.587	0.671	0.727	0.776	0.825	0.860
13	0.209	0.385	0.484	0.560	0.648	0.703	0.747	0.802	0.835
14	0.200	0.367	0.464	0.538	0.622	0.675	0.723	0.776	0.811
15	0.189	0.354	0.443	0.521	0.604	0.654	0.700	0.754	0.786
16	0.182	0.341	0.429	0.503	0.582	0.635	0.679	0.732	0.765
17	0.176	0.328	0.414	0.485	0.566	0.615	0.662	0.713	0.748
18	0.170	0.317	0.401	0.472	0.550	0.600	0.643	0.695	0.728
19	0.165	0.309	0.391	0.460	0.535	0.584	0.628	0.677	0.712
20	0.161	0.299	0.380	0.447	0.520	0.570	0.612	0.662	0.696
21	0.156	0.292	0.370	0.435	0.508	0.556	0.599	0.648	0.681
22	0.152	0.284	0.361	0.425	0.496	0.544	0.586	0.634	0.667
23	0.148	0.278	0.353	0.415	0.486	0.532	0.573	0.622	0.654
24	0.144	0.271	0.344	0.406	0.476	0.521	0.562	0.610	0.642
25	0.142	0.265	0.337	0.398	0.466	0.511	0.551	0.598	0.630
26	0.138	0.259	0.331	0.390	0.457	0.501	0.541	0.587	0.619
27	0.136	0.255	0.324	0.382	0.448	0.491	0.531	0.577	0.608
28	0.133	0.250	0.317	0.375	0.440	0.483	0.522	0.567	0.598
29	0.130	0.245	0.312	0.368	0.433	0.475	0.513	0.558	0.589
30	0.128	0.240	0.306	0.362	0.425	0.467	0.504	0.549	0.580
31	0.126	0.236	0.301	0.356	0.418	0.459	0.496	0.541	0.571
32	0.124	0.232	0.296	0.350	0.412	0.452	0.489	0.533	0.563
33	0.121	0.229	0.291	0.345	0.405	0.446	0.482	0.525	0.554
34	0.120	0.255	0.287	0.340	0.399	0.439	0.475	0.517	0.547
35	0.118	0.222	0.283	0.335	0.394	0.433	0.468	0.510	0.539
36	0.116	0.219	0.279	0.330	0.388	0.427	0.462	0.504	0.533
37	0.114	0.216	0.275	0.325	0.383	0.421	0.456	0.497	0.526
38	0.113	0.212	0.271	0.321	0.378	0.415	0.450	0.491	0.519
39	0.111	0.210	0.267	0.317	0.373	0.410	0.444	0.485	0.513
40	0.110	0.207	0.264	0.313	0.368	0.405	0.439	0.479	0.507
41	0.108	0.204	0.261	0.309	0.364	0.400	0.433	0.473	0.501
42	0.107	0.202	0.257	0.305	0.359	0.395	0.428	0.468	0.495

Table A25 Continued

N	0.25	0.10	0.05	0.025	α 0.01	0.005	0.0025	0.001	0.0005
43	0.105	0.199	0.254	0.301	0.355	0.391	0.423	0.463	0.490
44	0.104	0.197	0.251	0.298	0.351	0.386	0.419	0.458	0.484
45	0.103	0.194	0.248	0.294	0.347	0.382	0.414	0.453	0.479
46	0.102	0.192	0.246	0.291	0.343	0.378	0.410	0.448	0.474
47	0.101	0.190	0.243	0.288	0.340	0.374	0.405	0.443	0.469
48	0.100	0.188	0.240	0.285	0.336	0.370	0.401	0.439	0.465
49	0.098	0.186	0.238	0.282	0.333	0.366	0.397	0.434	0.460
50	0.097	0.184	0.235	0.279	0.329	0.363	0.393	0.430	0.456

Spearman's rank correlation. See subsection 5.4.2. Adapted from Zar (1972).

Table A26 Critical values of T_g at $\alpha = 0.05$

				n				
s	3	4	5	6	7	8	9	10
10	–	42	32	24	–	–	–	–
11	–	55	43	33	25	–	–	–
12	–	70	56	44	34	26	–	–
13	105	87	71	49	45	35	27	21
14	126	106	76	62	50	40	36	28
15	149	111	91	75	61	49	41	33
16	174	132	110	84	66	58	50	40
17	201	153	119	97	81	67	55	49
18	230	178	136	114	94	80	66	56
19	261	187	159	123	103	85	73	63
20	294	214	172	140	120	100	86	72
21	299	239	191	161	129	109	93	81
22	334	270	214	172	146	124	106	90
23	371	301	235	193	159	137	115	101
24	410	316	252	212	176	150	130	112
25	415	345	281	227	195	163	141	123
26	492	382	308	252	210	180	156	136
27	537	417	325	269	231	197	167	147
28	584	438	356	294	*	212	184	162
29	593	471	389	319		*	199	173
30	642	510	408	340			*	188
31	693	553	439	367				*
32	746	580	476	*				
33	797	617	501					
34	854	660	534					
35	913	709	571					
36	974	758	*					
37	987	783						
38	1050	830						
39	1115	883						
40	1176	*						
41	1245							
42	1316							
43	1387							
44	1406						* Not tabulated.
45	1481							
46	1558							
47	1629							
48	1710							

Homogeneity test for the geometric distribution. See subsection 5.4.3. Adapted from Meelis (1974) with permission of the American Statistical Association.

Table A27 The coefficients $a_{i,n}$

i	2	3	4	5	6	7	8	9	10
1	0.7071	0.7071	0.6872	0.6646	0.6431	0.6233	0.6052	0.5888	0.5739
2	–	0.0000	0.1667	0.2413	0.2806	0.3031	0.3164	0.3244	0.3291
3	–	–	–	0.0000	0.0875	0.1401	0.1743	0.1976	0.2141
4	–	–	–	–	–	0.0000	0.0561	0.0947	0.1224
5	–	–	–	–	–	–	–	0.0000	0.0399

i	11	12	13	14	15	16	17	18	19	20
1	0.5601	0.5475	0.5359	0.5251	0.5150	0.5056	0.4968	0.4886	0.4808	0.4734
2	0.3315	0.3325	0.3325	0.3318	0.3306	0.3290	0.3273	0.3253	0.3232	0.3211
3	0.2260	0.2347	0.2412	0.2460	0.2495	0.2521	0.2540	0.2553	0.2561	0.2565
4	0.1429	0.1586	0.1707	0.1802	0.1878	0.1939	0.1988	0.2027	0.2059	0.2085
5	0.0695	0.0922	0.1099	0.1240	0.1353	0.1447	0.1524	0.1587	0.1641	0.1686
6	0.0000	0.0303	0.0539	0.0727	0.0880	0.1005	0.1109	0.1197	0.1271	0.1334
7	–	–	0.0000	0.0240	0.0433	0.0593	0.0725	0.0837	0.0932	0.1013
8	–	–	–	–	0.0000	0.0196	0.0359	0.0496	0.0612	0.0711
9	–	–	–	–	–	–	0.0000	0.0163	0.0303	0.0422
10	–	–	–	–	–	–	–	–	0.0000	0.0140

i	21	22	23	24	25	26	27	28	29	30
1	0.4643	0.4590	0.4542	0.4493	0.4450	0.4407	0.4366	0.4328	0.4291	0.4254
2	0.3185	0.3156	0.3126	0.3098	0.3069	0.3043	0.3018	0.2992	0.2968	0.2944
3	0.2578	0.2571	0.2563	0.2554	0.2543	0.2533	0.2552	0.2510	0.2499	0.2487
4	0.2119	0.2131	0.2139	0.2145	0.2148	0.2151	0.2152	0.2151	0.2150	0.2148
5	0.1736	0.1764	0.1787	0.1807	0.1822	0.1836	0.1848	0.1857	0.1864	0.1870
6	0.1399	0.1443	0.1480	0.1512	0.1539	0.1563	0.1584	0.1601	0.1616	0.1630
7	0.1092	0.1150	0.1201	0.1245	0.1283	0.1316	0.1346	0.1372	0.1395	0.1415
8	0.0804	0.0878	0.0941	0.0997	0.1046	0.1089	0.1128	0.1162	0.1192	0.1219
9	0.0530	0.0618	0.0696	0.0764	0.0823	0.0876	0.0923	0.0965	0.1002	0.1036
10	0.0263	0.0368	0.0459	0.0539	0.0610	0.0672	0.0728	0.0778	0.0822	0.0862
11	0.0000	0.0122	0.0228	0.0321	0.0403	0.0476	0.0540	0.0598	0.0650	0.0697
12	–	–	0.0000	0.0107	0.0200	0.0284	0.0358	0.0424	0.0483	0.0537
13	–	–	–	–	0.0000	0.0094	0.0178	0.0253	0.0320	0.0381
14	–	–	–	–	–	–	0.0000	0.0084	0.0159	0.0227
15	–	–	–	–	–	–	–	–	0.0000	0.0076

Table A27 Continued

i	31	32	33	34	35	36	37	38	39	40
1	0.4220	0.4188	0.4156	0.4127	0.4096	0.4068	0.4040	0.4015	0.3989	0.3964
2	0.2921	0.2898	0.2876	0.2854	0.2834	0.2813	0.2794	0.2774	0.2755	0.2737
3	0.2475	0.2463	0.2451	0.2439	0.2427	0.2415	0.2403	0.2391	0.2380	0.2368
4	0.2145	0.2141	0.2137	0.2132	0.2127	0.2121	0.2116	0.2110	0.2104	0.2098
5	0.1874	0.1878	0.1880	0.1882	0.1883	0.1883	0.1883	0.1881	0.1880	0.1878
6	0.1641	0.1651	0.1660	0.1667	0.1673	0.1678	0.1683	0.1686	0.1689	0.1691
7	0.1433	0.1449	0.1463	0.1475	0.1487	0.1496	0.1505	0.1513	0.1520	0.1526
8	0.1243	0.1265	0.1284	0.1301	0.1317	0.1331	0.1344	0.1356	0.1366	0.1376
9	0.1066	0.1093	0.1118	0.1140	0.1160	0.1179	0.1196	0.1211	0.1225	0.1237
10	0.0899	0.0931	0.0961	0.0988	0.1013	0.1036	0.1056	0.1075	0.1092	0.1108
11	0.0739	0.0777	0.0812	0.0844	0.0873	0.0900	0.0924	0.0947	0.0967	0.0986
12	0.0585	0.0629	0.0669	0.0706	0.0739	0.0770	0.0798	0.0824	0.0848	0.0870
13	0.0435	0.0485	0.0530	0.0572	0.0610	0.0645	0.0677	0.0706	0.0733	0.0759
14	0.0289	0.0344	0.0395	0.0441	0.0484	0.0523	0.0559	0.0592	0.0622	0.0651
15	0.0144	0.0206	0.0262	0.0314	0.0361	0.0404	0.0444	0.0481	0.0515	0.0546
16	0.0000	0.0068	0.0131	0.0187	0.0239	0.0287	0.0331	0.0372	0.0409	0.0444
17	–	–	0.0000	0.0062	0.0119	0.0172	0.0220	0.0264	0.0305	0.0343
18	–	–	–	–	0.0000	0.0057	0.0110	0.0158	0.0203	0.0244
19	–	–	–	–	–	–	0.0000	0.0053	0.0101	0.0146
20	–	–	–	–	–	–	–	–	0.0000	0.0049

Table A27 Continued

					n					
i	41	42	43	44	45	46	47	48	49	50
1	0.3940	0.3917	0.3894	0.3872	0.3850	0.3830	0.3808	0.3789	0.3770	0.3751
2	0.2719	0.2701	0.2684	0.2667	0.2651	0.2635	0.2620	0.2604	0.2589	0.2574
3	0.2357	0.2345	0.2334	0.2323	0.2313	0.2302	0.2291	0.2281	0.2271	0.2260
4	0.2091	0.2085	0.2078	0.2072	0.2065	0.2058	0.2052	0.2045	0.2038	0.2032
5	0.1876	0.1874	0.1871	0.1868	0.1865	0.1862	0.1859	0.1855	0.1851	0.1847
6	0.1693	0.1694	0.1695	0.1695	0.1695	0.1695	0.1695	0.1693	0.1692	0.1691
7	0.1531	0.1535	0.1539	0.1542	0.1545	0.1548	0.1550	0.1551	0.1553	0.1554
8	0.1384	0.1392	0.1398	0.1405	0.1410	0.1415	0.1420	0.1423	0.1427	0.1430
9	0.1249	0.1259	0.1269	0.1278	0.1286	0.1293	0.1300	0.1306	0.1312	0.1317
10	0.1123	0.1136	0.1149	0.1160	0.1170	0.1180	0.1189	0.1197	0.1205	0.1212
11	0.1004	0.1020	0.1035	0.1049	0.1062	0.1073	0.1085	0.1095	0.1105	0.1113
12	0.0891	0.0909	0.0927	0.0943	0.0959	0.0972	0.0986	0.0998	0.1010	0.1020
13	0.0782	0.0804	0.0824	0.0842	0.0860	0.0876	0.0892	0.0906	0.0919	0.0932
14	0.0677	0.0701	0.0724	0.0745	0.0765	0.0783	0.0801	0.0817	0.0832	0.0846
15	0.0575	0.0602	0.0628	0.0651	0.0673	0.0694	0.0713	0.0731	0.0748	0.0764
16	0.0476	0.0506	0.0534	0.0560	0.0584	0.0607	0.0628	0.0648	0.0667	0.0685
17	0.0379	0.0411	0.0442	0.0471	0.0497	0.0522	0.0546	0.0568	0.0588	0.0608
18	0.0283	0.0318	0.0352	0.0383	0.0412	0.0439	0.0465	0.0489	0.0511	0.0532
19	0.0188	0.0227	0.0263	0.0296	0.0328	0.0357	0.0385	0.0411	0.0436	0.0459
20	0.0094	0.0136	0.0175	0.0211	0.0245	0.0277	0.0307	0.0335	0.0361	0.0386
21	0.0000	0.0045	0.0087	0.0126	0.0163	0.0197	0.0229	0.0259	0.0288	0.0314
22	-	-	0.0000	0.0042	0.0081	0.0118	0.0153	0.0185	0.0215	0.0244
23	-	-	-	-	0.0000	0.0039	0.0076	0.0111	0.0143	0.0174
24	-	-	-	-	-	-	0.0000	0.0037	0.0071	0.0104
25	-	-	-	-	-	-	-	-	0.0000	0.0035

The coefficients $a_{i,n}$ for the Shapiro-Wilk test for departure from normality. See subsection 6.2.1. Adapted from Pearson and Hartley (1972) with permission of the Biometrika Trustees.

Table A28 Upper critical values of *W*

n	0.99	0.98	0.95	0.90	α 0.50	0.10	0.05	0.02	0.01
3	0.753	0.756	0.767	0.789	0.959	0.998	0.999	1.000	1.000
4	0.687	0.707	0.748	0.792	0.935	0.987	0.992	0.996	0.997
5	0.686	0.715	0.762	0.806	0.927	0.979	0.986	0.991	0.993
6	0.713	0.743	0.788	0.826	0.927	0.974	0.981	0.986	0.989
7	0.730	0.760	0.803	0.838	0.928	0.972	0.979	0.985	0.988
8	0.749	0.778	0.818	0.851	0.932	0.972	0.978	0.984	0.987
9	0.764	0.791	0.829	0.859	0.935	0.972	0.978	0.984	0.986
10	0.781	0.806	0.842	0.869	0.938	0.972	0.978	0.983	0.986
11	0.792	0.817	0.850	0.876	0.940	0.973	0.979	0.984	0.986
12	0.805	0.828	0.859	0.883	0.943	0.973	0.979	0.984	0.986
13	0.814	0.837	0.866	0.889	0.945	0.974	0.979	0.984	0.986
14	0.825	0.846	0.874	0.895	0.947	0.975	0.980	0.984	0.986
15	0.835	0.855	0.881	0.901	0.950	0.975	0.980	0.984	0.987
16	0.844	0.863	0.887	0.906	0.952	0.976	0.981	0.985	0.987
17	0.851	0.869	0.892	0.910	0.954	0.977	0.981	0.985	0.987
18	0.858	0.874	0.897	0.914	0.956	0.978	0.982	0.986	0.988
19	0.863	0.879	0.901	0.917	0.957	0.978	0.982	0.986	0.988
20	0.868	0.884	0.905	0.920	0.959	0.979	0.983	0.986	0.988
21	0.873	0.888	0.908	0.923	0.960	0.980	0.983	0.987	0.989
22	0.878	0.892	0.911	0.926	0.961	0.980	0.984	0.987	0.989
23	0.881	0.895	0.914	0.928	0.962	0.981	0.984	0.987	0.989
24	0.884	0.898	0.916	0.930	0.963	0.981	0.984	0.987	0.989
25	0.888	0.901	0.918	0.931	0.964	0.981	0.985	0.988	0.989
26	0.891	0.904	0.920	0.933	0.965	0.982	0.985	0.988	0.989
27	0.894	0.906	0.923	0.935	0.965	0.982	0.985	0.988	0.990
28	0.896	0.908	0.924	0.936	0.966	0.982	0.985	0.988	0.990
29	0.898	0.910	0.926	0.937	0.966	0.982	0.985	0.988	0.990
30	0.900	0.912	0.927	0.939	0.967	0.983	0.985	0.988	0.990
31	0.902	0.914	0.929	0.940	0.967	0.983	0.986	0.988	0.990
32	0.904	0.915	0.930	0.941	0.968	0.983	0.986	0.988	0.990
33	0.906	0.917	0.931	0.942	0.968	0.983	0.986	0.989	0.990
34	0.908	0.919	0.933	0.943	0.969	0.983	0.986	0.989	0.990
35	0.910	0.920	0.934	0.944	0.969	0.984	0.986	0.989	0.990

Table A28 Continued

n	0.99	0.98	0.95	0.90	α 0.50	0.10	0.05	0.02	0.01
36	0.912	0.922	0.935	0.945	0.970	0.984	0.986	0.989	0.990
37	0.914	0.924	0.936	0.946	0.970	0.984	0.987	0.989	0.990
38	0.916	0.925	0.938	0.947	0.971	0.984	0.987	0.989	0.990
39	0.917	0.927	0.939	0.948	0.971	0.984	0.987	0.989	0.991
40	0.919	0.928	0.940	0.949	0.972	0.985	0.987	0.989	0.991
41	0.920	0.929	0.941	0.950	0.972	0.985	0.987	0.989	0.991
42	0.922	0.930	0.942	0.951	0.972	0.985	0.987	0.989	0.991
43	0.923	0.932	0.943	0.951	0.973	0.985	0.987	0.990	0.991
44	0.924	0.933	0.944	0.952	0.973	0.985	0.987	0.990	0.991
45	0.926	0.934	0.945	0.953	0.973	0.985	0.988	0.990	0.991
46	0.927	0.935	0.945	0.953	0.974	0.985	0.988	0.990	0.991
47	0.928	0.936	0.946	0.954	0.974	0.985	0.988	0.990	0.991
48	0.929	0.937	0.947	0.954	0.974	0.985	0.988	0.990	0.991
49	0.929	0.937	0.947	0.955	0.974	0.985	0.988	0.990	0.991
50	0.930	0.938	0.947	0.955	0.974	0.985	0.988	0.990	0.991

The Shapiro-Wilk test for departure from normality. See subsection 6.2.1. Adapted from Pearson and Hartley (1972) with permission of the Biometrika Trustees.

APPENDIX II

HARD- AND SOFTWARE

Hard- and software systems for recording and analysing behaviour are rapidly being developed. We here mention three systems: Camera, The Observer and The Analyst. For detailed information see the addresses below.

Camera is a hard- and software package for recording behaviour and preliminary analysis with extended video facilities.

The Observer is a software package for observing, coding and analysis with optional video interface. The Observer and Camera systems can be coupled.

The Analyst is a software package for advanced statistical analysis of behavioural records. It can import Camera as well as Observer files.

Camera: IEC ProGAMMA, P.O. Box 841, 9700 AV Groningen, The Netherlands
Phone ..3150636900; Fax ..3150636687
or Dr. M.R. Kruk, Ethopharmacology Group, Wassenaarseweg 72, 2333 AL Leiden, The Netherlands
Phone ..3171276239; Fax ..3171276292.

The Observer: Costerweg 5, 6702 AA Wageningen, The Netherlands
Phone ..31837097677; Fax ..31837024496.

The Analyst: Dr. F.H.D. van Batenburg, Institute of Theoretical Biology
P.O. Box 9516, 2300 RA Leiden, The Netherlands
Phone ..3171274972; Fax ..3171274900.

REFERENCES AND BIBLIOGRAPHY

Aalen, O.O., and Hoem, J.M. (1978). Random time changes for multivariate counting processes. *Scandinavian Actuarial Journal*, 81–101.

Abramowitz, M., and Stegun, A. (eds) (1972). *Handbook of mathematical functions*, (9th edn). Dover Publications Inc., New York.

Adke, S.R., and Manjunath, S.M. (1984). *Introduction to finite Markov processes*. Wiley Eastern Ltd, New Delhi.

Aitchison, J. (1986). *The statistical analysis of compositional data*. Chapman and Hall, London.

Aitkin, M., Anderson, D., Francis, B., and Hinde, J. (1989). *Statistical modelling in Glim*. Oxford University Press, Oxford.

Aldous, D.J. (1978). Weak convergence of randomly indexed sequences of random variables. *Mathematical Proceedings of the Cambridge Philosophical Society*, **83**, 117–26.

Altmann, S.A. (1965). Sociobiology of rhesus monkeys. II: Stochastics of social communication. *Journal of Theoretical Biology*, **8**, 490–522.

Anderson, T.W., and Goodman, L.A. (1963). Statistical inference about Markov chains. In *Readings in mathematical psychology* (ed. R.D. Luce, R.R. Bush, and E. Galanter), pp. 241–62. Wiley & Sons, New York.

Baker, R.J., Clarke, M.R.B., and Lane, P.W. (1985). Zero entries in contingency tables. *Computational Statistics & Data Analysis*, **3**, 33–45.

Barlow, R.E., Bartholomew, D.J., Bremner, J.M., and Brunk, H.D. (1972). *Statistical inference under order restrictions: the theory and applications of isotonic regression*. Wiley & Sons, New York.

Barnett, V., and Lewis, T. (1980). *Outliers in statistical data*. Wiley & Sons, New York.

Basawa, I.V., and Prakasa Rao, B.L.S. (1980). *Statistical inference for stochastic processes*. Academic Press, London.

Bendre, S.M., and Kale, B.K. (1985). Masking effect on tests for outliers in exponential models. *Journal of the American Statistical Association*, **80**, 1020–5.

Billingsley, P. (1974). *Statistical inference for Markov processes*. University of Chicago Press, Chicago.

Bouza, C.N. (1987). A ratio estimator of the kappa index of agreement between two observations. *Biometrical Journal*, **8**, 1011–15.

Bowman, K.O., and Shenton, L.R. (1988). *Properties of estimators for the gamma distribution*. Marcel Dekker, New York.

Bressers, W.M.A., Meelis, E., Haccou, P., and Kruk, M.R. (1991). When did it really start or stop: the impact of censored observations on the analysis of durations. *Behavioural Processes*, **23**, 1–20.

Burke, M.D. (1982). Tests for exponentiality based on randomly censored data. In *Nonparametric statistical inference* (ed. B.V. Gnedenko, M.L. Puri, and I. Vincze), pp. 89–102. North-Holland, Amsterdam.

Carpenter, G.A. (ed.) (1987). *Some mathematical questions in biology: circadian rhythms.* The American Mathematical Society, Providence, Rhode Island.

Chatfield, C. (1980). *The analysis of time series: an introduction*, (2nd edn). Chapman and Hall, London.

Chen, C.-H. (1984). A correlation goodness-of-fit test for randomly censored data. *Biometrika*, **71**, 315–22.

Chen, Y.Y., Hollander, M., and Langberg, N.A. (1982). Small-sample results for the Kaplan-Meier estimator. *Journal of the American Statistical Association*, **77**, 141–4.

Chung, K.L. (1967). *Markov chains with stationary transition probabilities*, (2nd edn). Springer-Verlag, New York.

Cochran, W.G. (1954). Some methods for strengthening the common χ^2 tests. *Biometrics*, **10**, 417–51.

Cohen, J. (1960). A coefficient of agreement for nominal scales. *Educational and Psychological Measurement*, **20**, 37–46.

Colgan, P.W. (ed.) (1978). *Quantitative ethology.* Wiley & Sons, New York.

Colgan, P.W. and Zayan, R. (eds) (1986). *Quantitative models in ethology.* Privat, IEC, Toulouse.

Cox, D.R. (1975). Partial likelihood. *Biometrika*, **62**, 269–76.

Cox, D.R., and Hinkley, D.V. (1974). *Theoretical statistics.* Chapman and Hall, London.

Cox, D.R., and Lewis, P.A.W. (1978). *The statistical analysis of series of events*, (3rd edn). Chapman and Hall, London.

Cox, D.R., and Oakes, D. (1984). *Analysis of survival data.* Chapman and Hall, London.

Darling, D.A. (1953). On a class of problems related to the random division of an interval. *The Annals of Mathematical Statistics*, **24**, 239–53.

David, H.A., and Moeschberger, M.L. (1978). *The theory of competing risks.* Griffin, London.

De Jonge, G., and Ketel, N.A.J. (1981). An analysis of copulatory behaviour of *Microtus agrestis* and *M. arvalis* in relation to reproductive isolation. *Behaviour*, **78**, 227–59.

Deshpande, J.V. (1983). A class of tests for exponentiality against increasing failure rate average alternatives. *Biometrika*, **70**, 514–18.

Dienske, H., Luxemburg, P.J.C.M. van, De Jonge, G., Metz, J.A.J., and Ribbens, L.R. (1981). Studying effects of maternal care in rhesus monkeys on different levels of resolution. In *Primate behaviour and sociobiology* (ed. A.B. Chiarelli and R.S. Corruccini), pp. 75–80. Springer-Verlag, New York.

Dienske, H., and Metz, J.A.J. (1977). Mother-infant body contact in macaques, a time interval analysis. *Biology of Behaviour*, **2**, 3–30.

Dienske, H., Metz, J.A.J., Luxemburg, P.J.C.M. van, and De Jonge, G. (1980). Mother-infant body contact in macaques, II: Further steps towards a representation as a continuous time Markov chain. *Biology of Behaviour*, **5**, 61–94.

Dixon, W.J., and Massey Jr, F.J. (1969). *Introduction to statistical analysis*, (3rd edn). McGraw-Hill Book Company, New York.

Dunn, O.J. (1964). Multiple comparisons using rank sums. *Technometrics*, **6**, 241–52.

Durbin, J. (1975). Kolmogorov-Smirnov tests when parameters are estimated with applications to tests of exponentiality and tests on spacings. *Biometrika*, **62**, 5–22.

Eisenhart, C., Hastay, M.W., and Wallis, W.A. (eds) (1947). *Selected techniques of statistical analysis*. McGraw-Hill Book Company, New York.

Everitt, B.S. (1968). Moments of the statistic kappa and weighted kappa. *British Journal of Mathematical and Statistical Psychology*, **21**, 97–103.

Feinstein, A.R. (1975). Clinical biostatistics XXXI. On the sensitivity, specificity, and discrimination of diagnostic tests. *Clinical Pharmacology and Therapeutics*, **17**, 104–16.

Feller, W. (1968). *An introduction to probability theory and its applications*, Vol. 1, (3rd edn). Wiley & Sons, New York.

Feller, W. (1971). *An introduction to probability theory and its applications*. Vol. 2, (2nd edn). Wiley & Sons, New York.

Ferguson, T.S. (1967). *Mathematical statistics: a decision theoretic approach.* Academic Press, New York.

Fieller, N.R.J. (1976). *Some problems related to the rejection of outlying observations.* Ph.D. Thesis, University of Sheffield.

Fienberg, S.E. (1980). *The analysis of cross-classified categorial data*, (2nd edn). MIT Press, Cambridge, Massachusetts.

Fisher, R.A. (1932). *Statistical methods for research workers*, (4th edn). Oliver and Boyd, Edinburgh.

Fleiss, J.L. (1971). Measuring nominal scale agreement among many raters. *Psychological Bulletin*, **76**, 378–82.

Fleiss, J.L., Cohen, J., and Everitt, B.S. (1969). Large sample standard errors of kappa and weighted kappa. *Psychological Bulletin*, **72**, 323–7.

Fleming, T.R., and Harrington, D.P. (1991). *Counting processes and survival analysis.* Wiley & Sons, New York.

Gibson, R.N. (1980). A quantitative description of the behaviour of wild juvenile plaice (*Pleuronectes platessa* L.). *Animal Behaviour*, **28**, 1202–16.

Gill, R.D. (1980). Nonparametric estimation based on censored observations of a Markov renewal process. *Zeitschrift für Wahrscheinlichkeitstheorie und verwandte Gebiete*, **53**, 97–116.

Gill, R.D. (1986). The total time on test plot and the cumulative total time on test statistic for a counting process. *The Annals of Statistics*, **14**, 1234–9.

Gombay, E., and Horvath, L. (1990). Asymptotic distributions of maximum likelihood tests for change in the mean. *Biometrika*, **77**, 411–14.

Goosen, C., and Metz, J.A.J. (1980). Dissecting behaviour: relations between autoaggression, grooming and walking in a macaque. *Behaviour*, **75**, 97–132.

Guiasu, S. (1971). On the asymptotic distribution of the sequences of random variables with random indices. *The Annals of Mathematical Statistics*, **42**, 2018–28.

Habib, M.G., and Thomas, D.R. (1986). Chi-square goodness-of-fit tests for randomly censored data. *The Annals of Statistics*, **14**, 759–65.

Haccou, P. (1987). *Statistical methods for ethological data.* Ph.D. Thesis, Leiden University.

Haccou, P., Bavel, E.T. van, and Kruk, M.R. (1985). Markov chain description and analysis of changes induced by hypothalamic stimulation in a male CPBWEzob rat at intensities below attack threshold. In *Mathematical methods and representations in ethological aggression research* (ed. M.R. Kruk and P.F. Brain), pp. 31–56. Ethopharmacology, University of Leiden.

Haccou, P., De Vlas, S.J., Alphen, J.J.M., and Visser, M.E. (1991). Information processing by foragers: effects of intra-patch experience on the leaving tendency of *Leptopilina heterotoma*. *Journal of Animal Ecology*, **60**, 93–106.

Haccou, P., Dienske, H., and Meelis, E. (1983). Analysis of time-inhomogeneity in Markov chains applied to mother-infant interactions of rhesus monkeys. *Animal Behaviour*, **31**, 927–45.

Haccou, P., Kruk, M.R., Meelis, E., Bavel, E.T. van, Wouterse, K.M., and Meelis, W. (1988*a*). Markov models for social interactions: analysis of electrical stimulation in the hypothalamic aggression area of rats. *Animal Behaviour*, **36**, 1145–63.

Haccou, P., and Meelis, E. (1988). Testing for the number of change points in a sequence of exponential random variables. *Journal of Statistical Computation and Simulation*, **30**, 285–98.

Haccou, P., Meelis, E., and Geer, S. van de (1988*b*). The likelihood ratio test for the change point problem for exponentially distributed random variables. *Stochastic Processes and their Applications*, **27**, 121–39.

Haccou, P., Meelis, E., Langeler, E.G., and Dienske, H. (1988*c*). Detection of low dose effects of psychopharmaca: application of a semi-Markov model to rhesus monkey behaviour. *Behavioural Processes*, **17**, 145–66.

Hájek, J., and Šidák, Z. (1967). *Theory of ranks tests*. Academic Press, New York.

Hallin, M., and Mélard, G. (1988). Rank-based tests of randomness against first-order serial dependence. *Journal of the American Statistical Association*, **83**, 1117-28.

Hauske, G. (1967). Stochastische und rhytmische Eigenschaften spontan auftretender Verhaltensweisen eines Fisches. *Kybernetik*, **4**, 26–36.

Hazlett, B.A. (ed.) (1977). *Quantitative methods in the study of animal behavior*. Academic Press, New York.

Hemelrijk, C.K., and Ek, A. (1991). Reciprocity and interchange of grooming and 'support' in captive chimpanzees. *Animal Behaviour*, **41**, 923–35.

Henderson, R. (1990). A problem with the likelihood ratio test for a change-point hazard rate model. *Biometrika*, **77**, 835–43.

Hochberg, Y. (1988). A sharper Bonferroni procedure for multiple tests of significance. *Biometrika*, **75**, 800–2.

Hollander, M., and Wolfe, D.A. (1973). *Nonparametric statistical methods*. Wiley & Sons, New York.

Hommel, G. (1989). A comparison of two modified Bonferroni procedures. *Biometrika*, **76**, 624–5.

Hooff, J.A.R.A.M. van (1982). Categories and sequences of behavior: methods of description and analysis. In *Handbook of methods in nonverbal behavior research* (ed. K.R. Scherer and P. Ekman), pp. 362–439. Cambridge University Press, Cambridge.

Horn, S.D. (1977). Goodness-of-fit tests for discrete data: a review and an application to health impairment scale. *Biometrics*, **33**, 237–48.

Huntingford, F.A. (1984). *The study of animal behaviour*. Chapman and Hall, London.

James, B., James, K.L., and Siegmund, D. (1987). Tests for a change-point. *Biometrika*, **74**, 71–83.

Johnson, N.L., and Kotz, S. (1969). *Distributions in statistics. Discrete distributions*. Houghton Mifflin Company, Boston.

Johnson, N.L., and Kotz, S. (1970). *Distributions in statistics. Continuous univariate distributions* - 1. Houghton Mifflin Company, Boston.

Kalbfleisch, J.D., and Prentice, R.L. (1980). *The statistical analysis of failure time data*. Wiley & Sons, New York.

Kalbfleisch, J.G. (1979). *Probability and statistical inference II*. Springer-Verlag, New York.

Kaplan, E.L., and Meier, P. (1958). Nonparametric estimation from incomplete observations. *Journal of the American Statistical Association*, **53**, 457–81.

Karlin, S., and Taylor, H.M. (1975). *A first course in stochastic processes*, (2nd edn). Academic Press, London.

Kemeny, J.G., and Snell, J.L. (1976). *Finite Markov chains*. Springer-Verlag, New York.

Kendall, M., and Ord, J.K. (1990). *Time series*, (3rd edn). Edward Arnold, Hodder & Stoughton, London.

Kimber, A.C. (1985). Tests for the exponential, Weibull and Gumbel distributions based on the stabilized probability plot. *Biometrika*, **72**, 661–3.

Kincaid, W.M. (1962). The combination of $2 \times m$ contingency tables. *Biometrics*, **18**, 224–8.

Krauth, J. (1984). A modification of kappa for interobserver bias. *Biometrical Journal*, **26**, 435–45.

Kruk, M.R., Laan, C.E. van der, Meelis, W., Phillips, R.E., Mos, J., and Poel, A.M. van der (1984). Brain-stimulation induced agonistic behaviour: a novel paradigm in ethopharmacological aggression research. In *Ethopharmacological aggression research* (ed. K.A. Miczek, M.R. Kruk, and B. Olivier), pp. 157–77. Alan R. Liss, New York.

Kruk, M.R., Poel, A.M. van der, and De Vos-Frerichs, T.P. (1979). The induction of aggressive behaviour by electrical stimulation in the hypothalamus of male rats. *Behaviour*, **70**, 292–322.

Kruk, M.R., Poel, A.M. van der, Meelis, W., Hermans, J., Mostert, P.G., Mos, J., and Lohman, A.H.M. (1983). Discriminant analysis of the localization of aggression-inducing electrode placements in the hypothalamus of male rats. *Brain Research*, **260**, 61–79.

Kullback, S., Kupperman, M., and Ku, H.H. (1962). Tests for contingency tables and Markov chains. *Technometrics*, **4**, 573–608.

LaMotte, L.R. (1983). Fixed-, random- and mixed-effects models. In *Encyclopedia of statistical sciences*, Vol. 3 (ed. S. Kotz and N.L. Johnson), pp. 137–41. Wiley & Sons, New York.

Lancaster, H.O. (1949). The combination of probabilities arising from data in discrete distributions. *Biometrika*, **36**, 370–82.

Lancaster, H.O. (1961). The combination of probabilities: an application of orthonormal functions. *Australian Journal of Statistics*, **3**, 20–33.

Landis, J.R., and Koch, G.G. (1975*a*). A review of statistical methods in the analysis of data arising from observer reliability studies (Part I). *Statistica Neerlandica*, **29**, 101–23.

Landis, J.R., and Koch, G.G. (1975*b*). A review of statistical methods in the analysis of data arising from observer reliability studies (Part II). *Statistica Neerlandica*, **29**, 151–61.

Lewis, T., and Fieller, N.R.J. (1979). A recursive algorithm for null distributions for outliers: I. Gamma samples. *Technometrics*, **21**, 371–6.

Light, R.J. (1971). Measures of response agreement for qualitative data: some generalizations and alternatives. *Psychological Bulletin*, **76**, 365–77.

Likeš, J. (1966). Distribution of Dixon's statistics in the case of an exponential population. *Metrika*, **11**, 46–54.

Liptak, T. (1958). On the combination of independent tests. *Magyar Tudomany Akad. Mat. Kutato Int. Kozl.*, **3**, 171–97.

Marden, J.I. (1985). Combining independent one-sided noncentral t or normal mean tests. *The Annals of Statistics*, **13**, 1535–53.

Mardia, K.V., Kent, J.T., and Bibby, J.M. (1979). *Multivariate analysis*. Academic Press, London.

Martin, P., and Bateson, P. (1986). *Measuring behaviour. An introductory guide.* Cambridge University Press, Cambridge.

Matthews, D.E., and Farewell, V.T. (1982). On testing for a constant hazard against a change-point alternative. *Biometrics*, **38**, 463–8.

McCullagh, P., and Nelder, J.A. (1983). *Generalized linear models.* Chapman and Hall, London.

Meelis, E. (1974). Testing for homogeneity of k independent negative binomial distributed random variables. *Journal of the American Statistical Association*, **69**, 181–6.

Meelis, E., Bressers, W.M.A., and Haccou, P. (1991). Non-parametric testing for the number of change points in a sequence of independent random variables. *Journal of Statistical Computation and Simulation*, **39**, 129–37.

Meelis, E., Haccou, P., and Bressers, W.M.A. (1990). Detection of time-inhomogeneity in behavioural processes: tests for multiple abrupt changes in boutlengths. *Behavioural Processes*, **22**, 121–32.

Metz, J.A.J. (1974). Stochastic models for the temporal fine structure of behaviour sequences. In *Motivational control systems analysis* (ed. D.J. McFarland), pp. 5–86. Academic Press, New York.

Metz, J.A.J. (1981). *Mathematical representations of the dynamics of animal behaviour. An expository survey.* Ph.D. Thesis, Leiden University.

Metz, J.A.J., Dienske, H., De Jonge, G., and Putters, F.A. (1983). Continuous-time Markov chains as models for animal behaviour. *Bulletin of Mathematical Biology*, **45**, 643–58.

Miller, R.G. (1981*a*). *Simultaneous statistical inference*, (2nd edn). Springer-Verlag, New York.

Miller, R.G. (1981*b*). *Survival analysis.* Wiley & Sons, New York.

Moeschberger, M.L., and Klein, J.P. (1985). A comparison of several methods of estimating the survival function when there is extreme right censoring. *Biometrics*, **41**, 253–9.

382   *References and bibliography*

Moore, E.H., and Pyke, R. (1968). Estimation of the transition distributions of a Markov renewal process. *Annals of the Institute of Statistical Mathematics*, **20**, 411–24.

Morrison, D.F. (1967). *Multivariate statistical methods*. McGraw-Hill Book Company, New York.

Nelson, K. (1965). The temporal patterning of courtship behaviour in the glandulocaudine fishes (*Ostariophysi, Characidae*). *Behaviour*, **24**, 90–146.

Oosterhoff, J. (1969). *Combination of one-sided statistical tests*. Mathematisch Centrum, Amsterdam.

O'Reilly, F.J., and Stephens, M.A. (1982). Characterizations and goodness of fit tests. *Journal of the Royal Statistical Society*, B **44**, 353–60.

Patel, J.K., Kapadia, C.H., and Owen, D.B. (1976). *Handbook of statistical distributions*. Marcel Dekker, New York.

Pearson, E.S., D'Agostino, R.B., and Bowman, K.O. (1977). Tests for departure from normality: comparison of powers. *Biometrika*, **64**, 231–46.

Pearson, E.S., and Hartley, H.O. (1954). *Biometrika tables for statisticians*, Vol. 1, Cambridge University Press, Cambridge.

Pearson, E.S., and Hartley, H.O. (1972). *Biometrika tables for statisticians*, Vol. 2, Cambridge University Press, Cambridge.

Peterson, E.L. (1980). The temporal pattern of mosquito flight activity. *Behaviour*, **72**, 1–25.

Pettitt, A.N. (1979). A non-parametric approach to the change-point problem. *Applied Statistics*, **28**, 126–35.

Priestley, M.B. (1981). *Spectral analysis and time series*. Academic Press, London.

Putters, F.A., Metz, J.A.J., and Kooijman, S.A.L.M. (1984). The identification of a simple function of a Markov chain in a behavioural context: barbs do it (almost) randomly. *Nieuw Archief voor Wiskunde*, **4**, 110–23.

Rao, C.R. (1973). *Linear statistical inference and its applications*, (2nd edn). Wiley & Sons, New York.

Ridder, G. (1987). *Life cycle patterns in labor market experience: a statistical analysis of labor market histories of adult man*. Ph.D. Thesis, University of Amsterdam.

Royston, J.P. (1982). An extension of Shapiro and Wilk's *W* test for normality to large samples. *Applied Statistics*, **31**, 115–24.

Samanta, M., and Schwarz, C.J. (1988). The Shapiro-Wilk test for exponentiality based on censored data. *Journal of the American Statistical Association*, **83**, 528–31.

Savage, I.R. (1956). Contributions to the theory of rank order statistics - the two-sample case. *The Annals of Mathematical Statistics*, **27**, 590–615.

Scheffé, H. (1959). *The analysis of variance*. Wiley & Sons, New York.

Sen, A., and Srivastava, M.S. (1990). *Regression analysis, theory methods, and applications*. Springer-Verlag, New York.

Shapiro, S.S., and Wilk, M.B. (1965). An analysis of variance test for normality (complete samples). *Biometrika*, **52**, 591–611.

Shapiro, S.S., and Wilk, M.B. (1972). An analysis of variance test for the exponential distribution (complete samples). *Technometrics*, **14**, 355–70.

Shorack, G.R. (1972). The best test of exponentiality against gamma alternatives. *Journal of the American Statistical Association*, **67**, 213–14.

Siegel, S., and Castellan Jr, N.J. (1988). *Nonparametric statistics for the behavioral sciences*, (2nd edn). McGraw-Hill Book Company, New York.

Simes, R.J. (1986). An improved Bonferroni procedure for multiple tests of significance. *Biometrika*, **73**, 751–4.

Sokal, R.R., and Rohlf, F.J. (1981). *Biometry: the principles and practice of statistics in biological research*, (2nd edn). W.H. Freeman and company, San Francisco.

Spurrier, J.D. (1984). An overview of tests for exponentiality. *Communications in Statistics*, A **13**, 1635–54.

Srivastava, M.S., and Khatri, C.G. (1979). *An introduction to multivariate statistics*. North-Holland, New York.

Stephens, M.A. (1974). EDF statistics for goodness of fit and some comparisons. *Journal of the American Statistical Association*, **69**, 730–7.

Stephens, M.A. (1976). Asymptotic results for goodness-of-fit statistics with unknown parameters. *The Annals of Statistics*, **4**, 357–69.

Stoer, J., and Bulirsch, R. (1980). *Introduction to numerical analysis*. Springer-Verlag, New York.

Strogatz, S.H. (1986). *The mathematical structure of the human sleep-wake cycle*. Springer-Verlag, New York.

Tippett, L.H.C. (1931). *The method of statistics*. Williams and Norgate, London.

Tukey, J.W. (1952). Allowances for various types of error rates. Unpublished IMS address. Virginia Polytechnic Institute, Blacksburg.

Wald, A., and Wolfowitz, J. (1943). An exact test of randomness in the non-parametric case based on serial correlation. *The Annals of Mathematical Statistics*, **14**, 378–88.

Wellner, J.A. (1985). A heavy censoring limit theorem for the product limit estimator. *The Annals of Statistics*, **13**, 150–62.

Westberg, M. (1985). Combining independent statistical tests. *The Statistician*, **34**, 287–96.

Wetherill, G.B. (1986). *Regression analysis with applications*. Chapman and Hall, London.

Wiley, R.H., and Hartnett, S.A. (1980). Mechanisms of spacing in groups of juncos: measurement of behavioural tendencies in social situations. *Animal Behaviour*, **28**, 1005–16.

Wilkinson, B. (1951). A statistical consideration in psychological research. *Psychological Bulletin*, **48**, 156–8.

Wilks, S.S. (1962). *Mathematical statistics*. Wiley & Sons, New York.

Winfree, A.T. (1980). *The geometry of biological time*. Springer-Verlag, New York.

Wong, W.H. (1986). Theory of partial likelihood. *The Annals of Statistics*, **14**, 88–123.

Worsley, K.J. (1979). On the likelihood ratio test for a shift in location of normal populations. *Journal of the American Statistical Association*, **74**, 365–7.

Worsley, K.J. (1988). Exact percentage points of the likelihood-ratio test for a change-point hazard-rate model. *Biometrics*, **44**, 259–63.

Zar, J.H. (1972). Significance testing of the Spearman rank correlation coefficient. *Journal of the American Statistical Association*, **67**, 578–80.

Zelen, M., and Severo, N.C. (1972). Probability functions. In *Handbook of mathematical statistics* (ed. M. Abramowitz and I.A. Stegun), pp. 925–97. Dover, New York.

AUTHOR INDEX

SUBJECT INDEX

Numbers in bold print indicate the pages where subjects are discussed more extensively